Memoirs of The New York Botanical Garden

Volume 124

Launched in 1900 by The New York Botanical Garden (NYBG), and with 120 volumes published as of 2019, Memoirs has traditionally focused on taxonomy and floristics, with an emphasis on plants of the Western Hemisphere. Most titles comprised a single volume, but quite a few were multi-volume titles and constitute some of the most iconic and substantial monographic publications of NYBG. In recent years, the Memoirs series has expanded to encompass field guides and botanical history as well. All in-print volumes of Memoirs through volume 120 are available exclusively through NYBG Shop: https://nybgshop.org/nybg-press/. Volumes 121 and going forward are available exclusively through Springer.

Ronald H. Petersen

William Alphonso Murrill: The Naturalist

Springer

Ronald H. Petersen
Ecology & Evolutionary Biology
University of Tennessee
Knoxville, TN, USA

ISSN 0077-8931 ISSN 2662-2858 (electronic)
Memoirs of The New York Botanical Garden
ISBN 978-3-031-93342-4 ISBN 978-3-031-93343-1 (eBook)
https://doi.org/10.1007/978-3-031-93343-1

This Springer imprint is published by the registered company Springer Nature Switzerland AG
The registered company address is: Gewerbestrasse 11, 6330 Cham, Switzerland

Preface

For the casual naturalist, William Alphonso Murrill's name is associated with a plethora of new species of fungi, his mysterious resignation from the New York Botanical Garden, his disappearance from New York, and rediscovery as a free spirit in Florida. Pivotal were his years in New York City, of course, but the process through which he arrived there, the people who acted to shape him, and his activities while there, leading to his life-changing trauma, have not been previously adequately summarized. Three academic biologists were especially important to his development: George Francis Atkinson at Cornell, Lucien Marcus Underwood at Columbia, and Nathaniel Lord Britton at the New York Botanical Garden. Other individuals appeared omnipresent to the story: John Torrey, John Newberry, and Seth Low, the latter President of Columbia University and Mayor of New York City. The Torrey Botanical Club played an important role in Murrill's integration in the city and the spread of his name to the wider biology community. Edna, the love of his life and his wife, remains a mysterious personage. The articulation of forces, individuals, and the times combined to produce an interesting man attempting to forge his own path as the Naturalist.

Knoxville, TN, USA | Ronald H. Petersen

Acknowledgments

A manuscript of this size and detail cannot be compiled without considerable input by other authorities on various subjects. Two men have been particularly generous with their materials: William "Wallace" Martin, Professor Emeritus at Randolph-Macon University, long-time collector of Murrilliana, and David Rose, former archivist at the New York Botanical Garden, who weighed in substantially (and continue to do so).

Don Pfister at Harvard, Greg Mueller from the Chicago Botanical Garden, and Egon Horak in Austria, all provided archival material on Rolf Singer. Stephen Simon, Archivist at the New York Botanical Garden, always freely responded to numerous requests for materials and answers to historical questions. Other folks who contributed substantive materials: Brad Bolman, Princeton Program in Advanced Studies; Christine Jankowski at the Lloyd Library, Cincinnati; Roy Halling, reviewer on behalf of the Garden; and Diana Murphy at the New York State Museum in Albany for information on C. H. Peck.

I owe a deep debt of gratitude to Bill Buck, Editor of the Memoirs series for the Garden. Apparently, he saw some worth in this project, although the amount of secretarial editing turned out to be mammoth. Many of his hours were spent in formatting and matching the text to the bibliography, and my thanks are wholehearted.

The mass of Murrill's larger publications in the database of the Hathi Trust is immense, and made possible much of the research on these strange (and rare) resources. Equally, Mycoportal (and its mycological director, Andrew Miller) provided much detailed and revelatory information. In mycological nomenclature, *Index Fungorum* and its Director, Paul Kirk, furnished accurate information otherwise arcane at best.

Slavomir Adamcik and Bart Buyck generously shared their bibliography on Murrill *Russula* types.

Two men, now deceased, were most generous: James W. Kimbrough, at the University of Florida, wrote and lectured on Murrill, and provided the script and illustrations of his lecture. Emory Simons, elder statesman of mycology, was more than willing to chat over mycological history, including William Alphonso Murrill.

Over the past 7 years, Karen Hughes, my wife and love, has listened to tale after tale of mycological history, always with patience and pertinent questions. I offer my unmitigated gratitude.

To all these, and others, sincere thanks.

Contents

The Early Years

1 The Making of the Man 3

The Big Apple

2 Tilling the Soil 27

3 Planting the Seed 45

Murrill of the Garden

4 Home and Away 81

5 "All That Glitters..." 111

6 The Final Straw 135

The Florida Years

7 Florida Freedom 165

8 Poking and Proding 185

9 The Twilight Years 209

10 A Postlude 233

Appendices 237

References 275

Index 295

The Early Years

Chapter 1
The Making of the Man

In the Appalachian Mountains of Virginia, mid-October is unpredictable. The warm, cloudless remnants of summer could extend warmth. Gray-bottomed occlusion from the northwest could drop wet snow on high elevation; the Peaks of Otter and Apple Orchard Mountain north of Bedford can take on white summits, largely melted by the afternoon. Harvest is a daily gamble. When tobacco was a money crop, it was hung to cure for the November–December auctions, making possible a calculation of the year's hard work and therefore the oracles of the next. For rural families, October brought the birth of winter conceptions, William Alphonso Murrill being one.

William was a second son, born October 13, 1869, to Samuel L. and Virginia D. Woodruff Murrill (1945a; Weber 1961). "William" was commonly sprinkled in the larger family, but "Alphonso" was Virginia's cousin. Mid-October was the season for tobacco harvest and curing, and the birth, with help from "Aunt Minnie," could not disrupt the "all hands present" ritual. Big brother Ashby was hardly a year old and still a suckling. But after all, tobacco was the chief money crop for the farm—nothing to be trifled with (Fig. 1.1).

Like so many rural post-Civil War families, the Murrills were multiplying, seven live births in 12 years: four boys, three girls. The parents opened a girls' school in Lynchburg, but Samuel later took a job as manager of a large farm outside town. Eventually, the Murrills came to own their own small farm in the vicinity of Blacksburg (more below). A formal roster of Samuel and Virginia's offspring follows: Ashby (1868), William Alphonso (1869), Minnie Douglas (1872), Howard Agassiz (1873), Anna Elizabeth (1876), Virginia Wood (1878), and Pitt Samuel (1881) (William Martin pers. comm.).

At the time of William's birth, the Murrills ran a farm near Lynch's Station, a crossroads which still exists a few miles south of Lynchburg on the railroad. The James River winds and twists its way from northwest to southeast, so 15 "river miles" north of Lynchburg is difficult to judge, and two miles from it does not reveal

R. H. Petersen, *William Alphonso Murrill: The Naturalist*, Memoirs of The New York Botanical Garden 124, https://doi.org/10.1007/978-3-031-93343-1_1

Fig. 1.1 Curing tobacco. October curing tobacco in Virginia hill country. (Getty image)

whether east or west (or north or south, depending on its direction when used as the starting point; Fig. 1.2).

In his autobiography (Murrill 1945a), some personal characteristics of William are mentioned in passing, but give small glimpses of the man. Although the slant of the mature handwriting would indicate right-handedness, he may have been left-handed, if his description of operating a cross-cut saw with Ashby is any indication. He was probably mesomorphic when young, but with a significant spurt of limbs (and feet) in early teenage. He almost surely spoke with a Southern mountain drawl, for whenever he wrote about the speech of family or friends, he juggled letters and apostrophes just as did Horace Kephart (1913) in his characterization in "Our Southern Highlanders," written only in Murrill's mid-life, and just down the Appalachians. Murrill was not only well-versed on nature by living in its midst, but must have been inquisitive of parents and teachers, and remained quite proud of the

Fig. 1.2 James River The James River snaking its way near Lynchburg, Virginia. (Getty image)

extensive lore he could relate. By teenage, he possessed most of the "horse sense" for rural, temperate living. He attended Sunday school and church and seemed not to resent it. He took piano lessons and must have become quite adept at it. His mother was more than the average educated farmer's wife, a college-graduate in fact, and baptized William with poetry from which he could recite—and write—verses at length. By mid-teenage, he had eaten squirrel, groundhog, raccoon, 'possum, rabbit, weasel, snake, turtle, several avians, domestic and hunted, eels and several varieties of fish and scads of personally cultivated vegetables, led by potatoes. He had captured numerous insects and watched their metamorphoses. He fought off all the diseases of youth—measles, mumps, whooping cough, each treated with remedies from nature. He knew the basics of rock and soil types and the agriculture they could support, and he could hold forth on the heavens—galaxies, stars, planets, meteorites, constellations. Perhaps most pervasive of all, he was born and raised a Confederate, referring to his father's role in the Civil War and some

Fig. 1.3 Meadow and fence. Pastoral boundaries in Virginia hill country. (Getty image)

memorabilia in his house when a child (Murrill 1918a, 1945a; Fig. 1.3). "The North" was an unsettling apparition, from its lingering, strange political world view to its packed neighborhoods, museums and galleries.

Billy made friends easily, it seems. He had his own siblings, of course, but boys his own age were always in the neighborhood, which could mean on a farm a mile away, but all converging on school each day, clear or inclement. But Billy was also friends with girls, although they had habits of their own, not the rough and tumble of the boys. In fact, Billy had at least three "sweethearts" over his young years, from whom one, Maggie, he had to part when his family decamped for far-distant Blacksburg, at William's tender age of twelve (Murrill 1918a).

By the time he was prepared for his first (but not his last) departure from the Virginia hill country (Fig. 1.3), Billy had attended three different schools, all one-room affairs, the last of which bore the name "Possum-trot." Three types of competition were left behind, spelling, recitation, and arithmetic. Billy excelled in all of them, for a while with the fierce competition of his sweetheart of the moment, Maggie (Murrill 1918a). His less intellectual world also played a part of those days. He could daydream a world of living beings with whom he could converse or learn more about their worlds. One aged, wise, and kind being on whom he could call for comfort and advice was "mother oak." With only a few days left to nestle on her leaves between her spreading roots, he could reveal his love of nature and receive her blessings for his continuing growth. For her part, she pledged nature's support in his intentions to learn more of her world. This conversation, when he was twelve, was revealed to us from New York City in 1918, when he was 49.

His more rational timeline, though, was recouped in the closing days of "big school." His decision for a life occupation as a teacher had been reached. He would also be a "Naturalist" (with a capital N), but had not yet crystalized this impulse.

In Billy's recollection of those years, there is almost no mention of race. This was not plantation country, where Blacks often outnumbered whites (Pierce and Little 2020). Murrill referred to them as negroes (without a capital n) and described them only as workers. In one instance, he opined: "We like the negroes much more and treat them a lot better than your [Northern] people do, when it comes to facts, because we understand and appreciate them" (Murrill 1918a).

The Emancipation Proclamation had little effect on the hill country folk, except that it seemed to usurp the rights of the state. Billy's father had fought in several battles during the War, only a few to victory, but Confederate sentiment didn't seem of paramount importance for the hill-country farm folk.

The Murrill family might easily be branded as nomadic. In a decade, it moved from the big farm north of Lynchburg to Bedford County, a somewhat more northerly locale visited repeatedly by William in adulthood, to the tiny crossroads of Bonsack, just outside of Roanoke. "At the age of twelve the Naturalist became a student at the Virginia Agricultural and Mechanical College, located in Blacksburg. His father was now manager of the Miller Farm, south of the town, and the family occupied a small house surrounded by broad cultivated fields and stretches of virgin timberland" (Murrill 1945a).

Arrangements must have been made aforehand for Billy to enroll at the prep school, considering his age. The school was associated with Virginia's land-grant school, founded in 1872. As a Confederate state, Virginia might not have been ready for such an idea both for political and financial reasons, neither of which affected William. In fact, the first six presidents of the College had previously served as officers in the Army of the Confederacy (VPI history 2023), and the College was all-male. William was a commuter, walking to and from school daily, summer and winter, and in 1887, age 18, he graduated with highest honors and a Bachelor of Science degree. Had he followed the normal routine of the day, summer holidays (and perhaps a week at harvest time) would have been spent at home, adding a pair of cheap hands to the farm chores.

A socio-economic move from farm manager to farm owner was not common, but Sam Murrill succeeded. "A few years later the family moved to a roomy house north of town, which was a real home because they owned it. There were six acres of land; enough for a cow, a garden, a good orchard and a stable with two horses" (Murrill 1945a). This was to become the destination of frequent pilgrimages for many years.

In his autobiography, written at the age of 75, Murrill (1945a) looked back at those early years in regard to music. "The Naturalist inherited his love for music mostly from his father, who had helped Stuart lead the singing at the head of his cavalry troop and whose strong voice [William] had heard since childhood in the songs at Sunday-school and Church. Before William could talk he would sing.

We know that he continued his piano lessons and performed at a recital. When he played the piano, organ or guitar he always sang with the instrument if the music allowed it. In Sunday-school, Church, Glee Clubs, at parties, in camps, on the road—everywhere he sang. Before ten he attended a singing school in the country…"

> The removal of tonsils was formerly a matter of much discussion. The Naturalist would probably have been a baritone if he had happened upon the right doctor while he was still young. Although offered voice training free he could not take full advantage of it because of his tonsils…. (Murrill 1945a).

The male Murrills were coached by their father about their life choices. William himself wrote of the bargain: "Now," Billy's Father had said to each of his four boys, 'If you don't drink whiskey, use tobacco, or say bad words before you are twenty-one, I will give you a fine riding-horse and a gold watch.' The boys thought a good deal of this promise, but they thought a lot more of the good example set them by their Father. If he had done any one of these things, a whole team of horses couldn't have held them back from doing just the same thing.

"Many years after, the boys decided they would rather go to college than have riding-horses and gold watches, so that was agreed upon, and it was also agreed that the oldest boy should wear the family watch when he got to be twenty-one and pass it on to the other boys in rotation—just as is sometimes done with trousers. So the old family watch kept time for each one in turn, getting them out of bed in the morning and bringing them home at night, like a faithful old servant; and they were very proud to wear it" (Murrill 1919a). Whether the story played out in reality is no longer available, but upon graduation from VAMC, William moved on to Randolph-Macon College, many miles away on the Piedmont Plateau, just north of Richmond, the ex-Confederate capital. If nothing more, the unique "Northern Neck" enunciation of the natives would have brought a grin, but the state of Black citizens would have erased it.

Unlike VAMC, all male with a corps of cadets, and emphasis on agriculture and mechanical subjects, Randolph-Macon was a small, liberal arts, Methodist-related, albeit male school. Founded in Boydton, Virginia, close to the North Carolina border, the institution, minus physical plant, was moved to escape the anticipated ravages of the Civil War (Randolph-Macon history 2023). Perhaps a location near the Confederate capital, Richmond, was thought safe. The curriculum was heavily structured around Latin and Greek, with other philosophical seasoning of religious study and science. Murrill took to it, and degrees followed: in 1889 a Bachelor of Science, in 1890 a Bachelor of Arts, and in 1891 the Master of Arts (Fig. 1.4). George Weber (1961), surely based on conversations with Murrill in the last decades of Murrill's life, wrote: "During these academic years he always helped his father on the farm during the summer, closely observing nature, and studied Latin, Greek, music and philosophy."

William Martin (see acknowledgments) added the following. At Randolph-Macon, Murrill lived in a boarding house "near the railroad tracks," run by a Mrs. Hill, also a friend of the College President, W.W. Smith, whose brother, Richard was a

Fig. 1.4 Randolph-Macon Class. Class photo at Randolph-Macon College, 1891. William Murrill is no. 6. Note the fashionable flat-crown straw hats for summer. (Image courtesy of William Martin)

professor, and probably a teacher to Murrill. Murrill's roommate (he wrote in his autobiography) was a divinity student, now thought to have been young Andrew W. Sledd, son of a Methodist clergyman influential in college affairs. Young Andrew developed into an early and controversial leader in the civil rights movement, and eventually the first and youngest president of the University of Florida at Lake City, one of four campuses merged to form the University of the State of Florida in Gainesville, a tenure lasting only 4 years. A peripatetic scholar of classical languages and The New Testament, Sledd continued his academic career, finally succumbing to a heart attack in 1939 (Sledd Wiki 2024). Sledd Hall residence, at the University of Florida, named only after his death, is part of his legacy. As a roommate, Sledd was no "shrinking violet" for Murrill.

Murrill related that the Principal of the Bowling-Green (Virginia) Female Seminary paid a visit to Randolph-Macon College in the spring of his senior year, and among social calls, must have expressed his idea of hiring an unmarried male teacher for his girls. A bit chancy for obvious reasons, he inquired whether any of the soon-to-graduate young men (no longer boys) might qualify. Why he was directed to Murrill is missing data, but William later pictured the "interview" as "a little general conversation and a few questions," resulting in an offer of $500 plus board for the upcoming academic year. These terms being apparently acceptable, the deal was sealed. Any equivocations by William also remain untold (Murrill 1919a).

If a line between Richmond and Fredericksburg were divided into thirds, Ashland (and Randolph-Macon) was at the southern third mile marker and Bowling Green was at the northern. The Seminary stood atop a knoll—in essence an island on the outskirts of town. Murrill was introduced to the rest of the staff, either female or married, but cross-gender interaction was reserved for the classroom. "The Naturalist had completed courses at two colleges containing a great variety of subjects, so he was given the senior classes in English, Latin, French, German, and Mathematics as well as all the work in Physics, Chemistry, Zoology and Botany. There were a few other subjects in his teaching curriculum, but those mentioned are sufficient to prove his audacity" or minimize any self-doubts. The girls included William's younger sister, Minnie, so any untoward rumors could abe pipe-lined home.

Murrill took special enthusiasm in excursions, always Natural History, regardless of the title. Very much later, he wrote: "The fields and woods were used as both laboratory and museum, with the advantage of actual life and a natural setting. Nature was studied and loved both as a whole and as made up of infinite details, all of which were beautiful and wonderful." Although he might not have been exposed to the pedagogical controversy of the day, his teaching method was close to the growingly popular, imported "new Botany" (see below).

A major event of his time at Bowling Green was a trip to Washington (with "his" girls) to hear the music of the visiting New York Symphony Orchestra under the baton of the young Frank Damrosch, in whose father's name Carnegie Hall was being constructed in New York. Murrill would have been dismissive had he been told that by the end of the decade, he would live in that very city, free to enjoy the music in person.

In the spring of 1893, the Naturalist received a letter from the President of the Wesleyan Female Institute in Staunton, Virginia, offering him a position there at a marginally higher salary. He accepted and left Bowling Green, hopefully leaving a good impression. He carried with him a much-prized, handsome set of Shakespeare's works, a present from his Sunday School Class. His collections of rocks, flowers, bugs, skeletons and the like were apparently left to the next teacher (Murrill 1919a).

Staunton, a good distance from his family home, was at least in the same general geophysical part of Virginia (Murrill 1919a). The town seemed a Mecca for institutions, from the School for the Deaf, Dumb and Blind and the Insane Asylum, to Capt. Kable's Military Academy for boys and the Baptist, Lutheran and Methodist seminaries for girls, the latter named for the Methodist preacher, John Wesley, and the Naturalist's home-base (Murrill 1919a, 1945a). All this on top of the local public school system. While his domicile was small, his teaching facilities were more than adequate.

There is no indication that the Wesleyan headmaster and Murrill had personal contact before the letter of invitation, but what better impression could there be than a man of 24 with degrees including a Master of Arts, who seemed able to teach nearly everything and with not so much as a blemish on his personal behavior (Fig. 1.5). The headmaster took a chance and asked the Naturalist to spend the summer in Staunton, helping with planning the upcoming year's schedules, announcements and student recruitment—as far away as Arkansas and Missouri. His expenses were paid, of course, by the school, and may have been his first experience with the American West. Part of the deal was permission to visit Chicago and the World's Fair, in full swing. As Murrill (1919a) later wrote: "This was a wonderful

Fig. 1.5 Murrill at Staunton. Mustached William Murrill with research paraphernalia during his teaching time in Staunton, Virginia. (Image courtesy of Jim Kimbrough 2003)

opportunity and very much appreciated. Nothing could have given him a better insight into the best that the world offered than those busy days in the famous White City on the shore of Lake Michigan." Probably unknown to him, a later friend and icon, William Ashbrook Kellerman, editor of the JOURNAL OF MYCOLOGY, recently added to the faculty of Ohio State University, had won praise for his exhibit of the timber trees of Ohio.

Also at the World's Fair, an electric car was introduced by William Morrison from Iowa. Some crude battery vehicles had preceded, but Morrison's contraption featured rechargeable batteries, rack-and-pinion steering and high-spoked wheels for traversing rutted roads. Murrill remembered that such vehicles came to Staunton early in his residency, equaled in curiosity only by Mrs. Henry St. George Tucker's bicycle, which was rumored to have cost $150.

During the academic year, some peculiarity of class scheduling left William with free Mondays which he frequently used to travel eastward to Charlottesville and the campus of the University of Virginia. There, as often as possible, he listened to lectures in science, particularly those by a Professor John William Mallet, the Irish theoretical chemist, on the emerging Atomic Theory: nuclei seemed to be composed of electrons of negative charge, circling a larger, positively charged particle. "The zoology museum, the medical school, the chemistry laboratories, and the observatory were highly influential in generating the decision that he would be a naturalist" (Weber 1961).

Among the several books he read during this period was "The life of Agassiz," written by the wife of Harvard zoologist Louis Agassiz (about whom more significantly below). Murrill came to believe that "the true scientist is the noblest of the race, without greed or hatred, an altruist and a lover of beauty, justice, and truth"

(Murrill 1919a). In some ways, science was theism in different clothing and religion must be comfortable with observable data. Agassiz was attractive enough to provide William's younger brother Howard, with Agassiz as a middle name in 1873.

Another use for free time was the newly established Young Men's Christian Association—the YMCA—which Murrill described as "well housed in a handsome building, and … one of the most efficient institutions in the city. The gymnasium, library, and reading-room was [sic] much used by the Naturalist."

The Naturalist remembered that he taught a Sunday School class "of several splendid girls," each of whom he described. Three stood out. "Virginia and Anna, the nieces of a great artist, were pretty, dainty, clever, good, and demure, about as near perfection as healthy young mortals ever get." A rare opportunity to probe Murrill's ideal.

"Nelle had been a pupil of Dr. Underwood in Indiana." Surely this was Lucien Marcus Underwood, at that moment on faculty of the Alabama Polytechnic College in Auburn (Petersen 2019), but who would soon be appointed to the faculty of Columbia University, where he would meet Murrill in New York. Even sooner, ferns would act as an interlocutor. So Murrill and Underwood may have known of each other since perhaps 1893 or so. Neither could have predicted Underwood's influential role in the botanical community, both domestic and international, or the role he would play as mentor in Murrill's time in New York City.

Another period away from the classroom was summer vacation, during one of which Murrill attended Summer Normal School at Bedford City, about halfway between Roanoke and Lynchburg. As its title indicated, classes were on various subdisciplines of pedagogy, and one of these might have been the teaching of natural history. The instructor was a Prof. Apgar from Trenton, New Jersey, who made a mighty impression on the Naturalist. Apgar not only lectured ("In drawing on the blackboard, he used both hands at once and used them quickly and skillfully."), but led excursions west to the Peaks of Otter (Fig. 1.6) and north to Apple Orchard Mountain. Both are part of the Appalachian chain, not far from the present-day Blue Ridge Parkway and Bedford, Murrill's home just a few years previous. Apgar showed the Naturalist a Carolina hemlock (*Tsuga caroliniana*)—the first seen by Murrill, and fortuitous considering Murrill's later Ph.D. dissertation on fertilization in *Tsuga*. These exposures to Apgar surely bolstered Murrill's growing conviction to make natural history his life's work (Murrill 1919a; Fig. 1.7).

Murrill's (1919a) "The Naturalist in a Boarding School" proceeds in chronological order as condensed above, but at a particular point, Murrill (writing in 1919 at age 50) abruptly changed styles, stating a particular date followed by a few lines of poetry brought to memory by it. For example, the first entry (p. 75) is March 10, 1895, followed by a short paragraph on a partial eclipse of the sun and an elegiac image therefrom. With even more spartan words: "April 14, 1895. I took a party of girls to Washington and spent Easter week showing them the sights. They had a wonderful time." No mention of his own experience.

Notable: "April 29, 1895: I agreed to return next [academic] year [starting fall, 1895] for the following courses: Zoology, followed by Botany; Physics, followed by Astronomy; Chemistry, followed by Geology; three classes of German." And with

Fig. 1.6 Peaks of Otter. An autumnal view of the Peaks of Otter, a favorite Murrill destination. (Image courtesy Virginia Tourist Bureau)

Fig. 1.7 Shenandoah Park. High-altitude spring rhododendrons in Shenandoah National Park, Virginia. (Image courtesy, Shenandoah National Park)

that, he spent the summer of 1895 in Blacksburg with family, day after day shooting and stuffing birds, hiking, studying Strasburger's book and preparing for the new school year in Staunton.

Just as abruptly, Murrill (1919a) interwove some paragraphs without dates. The first of these might be deduced to be tied to his birthday (October 13). "How I feel at 26. Young as ever, with work hardly yet begun. The disappointment as to our salary made me a little reckless for a time, but I can wait two more years, I guess, for Johns Hopkins. I will learn all the Biology I can before that time. Our school may come out, but, if she does not, money is not life. Shall I be rich? Never. As to religious beliefs and duties, I hope I am true, if negligent. As to love, I am wedded to Science. My life work shall be biology." This passage, pointing in several directions not previously introduced, is important, especially as written in 1919, in his 21st year of marriage, not 1895 as his memory disinters.

Immediately, though, Murrill expanded on his remark about Johns Hopkins. "The Naturalist visited Dr. Remsen, whose textbook he was using in chemistry; Dr. Gildersleeve, the great Latin scholar; Dr. Brooks, the animal embryologist; and Dr. Humphrey, the botanist, to arrange work at The Hopkins for a doctor's degree, and Dr. Humphrey contributed a number of plant types to be thoroughly studied and drawn in connection with work outlined in Strasburger's 'Practical Botany.' On a subsequent visit, Dr. Lotsy suggested trying the experiments given in Oels' 'Physiology' …" (Murrill 1919a). It was Gildersleeve, a family friend, who arranged for a fellowship aimed at Murrill. Faithful to his promise, Murrill spent hours over late winter and spring of 1896 reading Strasburger's book and working out whatever experiments he could with limited equipment (Murrill 1919a).

At two locations in the Virginia mountains, Murrill collected *Asplenium ebenoides*, a spleenwort fern assumed to be quite rare and a new report for Virginia. When his discovery became known, he was asked by Willard N. Clute, Editor, to contribute a paper to FERN BULLETIN, which he did (Murrill 1897), but reproducing most again in the "Staunton Book" (Murrill 1919b). It was his virginal contribution to scientific writing, in which he held that *A. ebenoides* was a discrete species, not a hybrid as had been conjectured by some. In the paper, he acknowledged help from Underwood, who had just arrived at Columbia in New York, and who proceeded to request herbarium material of all Virginia ferns. Murrill also sent a couple living plants to Cornell, where they were successfully maintained in the greenhouse.

In the "Boarding School" book (Murrill 1919a), there follow several adages and short stories making various, unrelated points. Chapters XIV (The English language), XV (How to live wisely); XVI (Character); XVII (A love story in outline); XVIII (Condensed paragraphs); XIX (Quotations relating to man) are all a collection of quotes, aphorisms, maxims, short rhymes from poets, Shakespeare, Bible, etc., but with no information about the author except deductions about his state of mind in 1919. The last dates are in spring, 1896. If there is a rational theme to this 136-page soliloquy, it eludes this reader, except perhaps as a self-effacing demonstration of otherwise negligible tidbits—as young men might display baseball cards, political lapel buttons, marbles or penknives.

Murrill's last-minute switch from zoology at Johns Hopkins to botany at Cornell—made on the night of March 10, 1896—was expanded only slightly in Murrill's autobiography. "When he was about ready to enter Hopkins, however, he was met by two Cornell men, Coville and Howard, who argued that new positions in biological teaching would be open to botanists rather than zoologists and he should therefore go to Cornell and take graduate work under [George Francis] Atkinson. To make their argument more convincing they actually secured a fellowship for him at Cornell to match the one offered by Hopkins. The result was a victory for Cornell which was never regrated." William entered Cornell University in the fall of 1897, where he had, indeed, been awarded a fellowship in botany.

Even during the days of decision for Murrill—Johns Hopkins and zoology versus Cornell and botany—there was a muffled backdrop of war in the headlines. The wonderous magic of telegraphy carried the news from the Atlantic coast where news was written, to the interior—say, Staunton, Virginia. Workers in the Spanish Caribbean were rioting, especially in Cuba. But the United States had deep economic roots in Cuba, from the sugar cane industry to the furniture factory of Elizabeth Britton's family near Havana. More below.

THE AWAKENING. As remembered in maturity from his boyhood days, William had sought to bring nature under his mental control, whether tracing the burrows of moles, shooting birds for their skins, pinning and spreading butterflies and illustrating their wing patterns to exercise who was master of their beauty, to recognizing all the ferns of Virginia. He had crowned himself as The Naturalist; third person singular substituted for his given name (or even "I"). In Bowling Green and Staunton (Murrill 1919a, b), he had been the king of his classroom, passing on his experience to his vassals. Now, suddenly, he was to be baptized into a world of science—often as opposed to his brand of Natural History—as a serf rather than king, gathering materials for his professor in a milieu of academics. He related that he arrived "with a ton of fungi," almost all of parasitic types gathered during the previous summer but previously almost unmentioned in his descriptions of life to that point. MycoPortal records from the Cornell Plant Pathology Herbarium vouch for the feast of Virginia fungi shipped north.

It would be easy to judge his expertise with a cruise through the lengthy lists of plants (not animals) collected during his Staunton days (Murrill 1919b), only to recognize that in the fall of 1897 he would meet those who might have spent years studying a single small genus in the list, but who were at a loss to recite even a single Victorian quote describing the emotions of a cloudburst.

EDNA LEE. Perhaps most poignant, we know nothing substantial of Edna Lee Luttrell from William; the observer had to wait for William's obituary and Gleason's (1961) "thumbnail sketches." From the latter we read: "His wife, whom he had married while she was in her teens…" (he was 28) indicate a difference of some years between the two, perhaps mirrored in differences in life view and aspirations for the future. The summer preceding their September 1, 1897, wedding went this way:

"After the school closed at Staunton, I went to Blacksburg and devoted myself especially to collecting, pressing and studying all the parasitic fungi to be found within a radius of about twenty miles of the place, often walking thirty miles a day over the mountains and finishing off with a game of tennis." Moreover, his two-line description of leaving Virginia reads: "Early in the autumn I departed for Cornell University with a ton or so of fungi and a fund of health and experience sufficient to last for many years. My record for the summer was over a thousand miles." MycoPortal records from Cornell divulge a single specimen collected in Cascadilla by Murrill on August 29. Edna's name is not attached.

Murrill (1918b) often wrote in allegory. In "Three young Crusoes," the thinly veiled Edna was described: "Edna was thirteen [only one year younger than William, another of the Crusoes], with golden hair and eyes of the deepest blue. She was slightly under size for her age, so that the two boys [William and Henry] seemed older than they really were by comparison. All her life previous to this fateful voyage had been spent in a Virginia town [Falls Church] where she had learned something about books and a great deal about housekeeping and the care of babies" (Fig. 1.8).

Later in the same book: "When their college courses were finished … William [became] a Professor of Natural History in one of the Virginia colleges. A few months after his appointment, William (Fig. 1.9) married Edna and took her to New York to spend the Christmas holidays." If accepted, these passages run counter to those of Gleason (1961) and Weber (1961).

By far, the most revealing narration is Chap. XVII of "The Naturalist in a boarding school" (Murrill 1919a). Although written in melodic style in Murrill's 50th year, the facts ring true (i.e. "The Girl was almost twenty, and he was just twenty-seven.")

Fig. 1.8 Edna in Virginia. Edna, female lead of "Three Young Crusoes" suspected to be Edna Lutrell, William Murrill's wife

Fig. 82. Edna at her home in Virginia.

Fig. 1.9 Henry and William. Henry and William, male leads of "Three Young Crusoes," the latter suspected to be William Murrill. Note the fashionable Bowler hat

Edna was born in Tennessee; her father moved to accept a job in Washington, DC, and the family moved to a small town, Falls Church, quite near the District. Her middle name might indicate an "aristocratic" Southern association. Her name (as "Mrs. Murrill") is mentioned occasionally in William's reports on Caribbean, European, Mexican and Pacific coast travel and, thanks to MycoPortal, recorded on herbarium labels of specimens collected. It is quite possible that her withdrawal from the marriage was so traumatic that Murrill (1945a) could not mention her name even in advanced age—his death certificate noted "never married."

Photos of the hypothetical characters, Henry, William and Edna, appear toward the end of "Three young Crusoes" with no parenthetical cautions. It may be the only such picture of Edna.

CORNELL. Telltale evidence indicates that Murrill visited Cornell twice in 1896, a year before enrollment (Fig. 1.10). In June, a single specimen of *Russula emetica* was gathered in Ithaca Cemetery, and in August, two boletes were found in Cascadilla (MycoPortal CUP), both by Murrill. No data include a co-collector, and both trips were probably made during vacation periods.

William Murrill and George Atkinson were both engaged in introducing young people to natural history, including botany, but Atkinson's tuning was at least an octave higher than Murrill's. Exemplifying the role of naturalist, Atkinson had made a tour of several Southern colleges, and published very widely, especially on insects

Fig. 1.10 Cornell Cascadilla. Image of early Cornell College, with bridge across Cascadilla Gorge near Sage College Hall, probably circa 1885. (Image courtesy Cornell University Archive)

and phytopathology (Atkinson 1886a, b, 1887a, b, c, d, e, f). For his part, Murrill had had more than one in-person interview in Baltimore with biology professors at Johns Hopkins, so an equal familiarity might be suspected at Cornell. The postponement of the Johns Hopkins matriculation had granted time to be introduced to the Cornell choice, and such a life-changing prospect deserved extra scrutiny.

INBRED FACULTY. When Murrill arrived in Ithaca in 1897, Cornell botany was already an inbred department. An article in a popular magazine reported that when Albert Prentiss signed up to teach botany at the new school in 1868, Botany was a Department of one. By Prentiss' death in 1896, the Department had grown to several men, as follows: George Francis Atkinson (1854–1918), Professor, student of Dudley and Prentiss; Willard Winfield Rowlee (1861–1923), Instructor, student of Prentiss; Elias Judah Durand (1879–1922), B.A. Botany, Cornell, 1893, D.Sci. under Atkinson, 1895; Benjamin Minge Duggar (1872–1956), plant physiologist, born in Alabama, educated at Univ. Alabama 1887–1889, Mississippi

A&M, B.S. 1891, M.S. Alabama Polytech. 1892, Ph.D. student at Cornell in 1897, Ph.D. 1898. Murrill was listed merely as "scholar," the only one so named (Petersen 2021). Perhaps the most interesting was Karl McKay Wiegand (1874–1942), B.S. Cornell 1894, Ph.D. Cornell, 1898. Wiegand remained at Cornell until 1907, when he moved to Wellesley College, but returned to Cornell to head the second botany department (this one in the School of Agriculture) which eventually absorbed the College of Arts and Sciences' Botany Department from which he had received his Ph.D.

To find one's self in the midst of men who already know each other's territories, weaknesses, strengths, irritants and mannerisms surely requires patience and fortitude. Murrill was tied to Atkinson, who needed fresh material for his classes and who enjoyed collecting excursions with his students, both traits well within Murrill's scope.

ATKINSON. Sources seem to agree that George Atkinson [Appendix 1] was not an ebullient soul; quiet, perhaps withdrawn, self-sensitive. Murrill (1945a) wrote: "As an investigator and a teacher of research students Professor Atkinson had few equals. His interest, encouragement, help and forceful driving energy never relaxed. The teaching of undergraduates was to him a chore; working with graduates was a pleasure. Some called him hard and disagreeable; the Naturalist never found him so, but one of the best friends and teachers he ever had." Moreover (Murrill 1945a). Although jealous and disagreeable in disposition, he was well-nigh perfect in directing investigation, possessing great energy and enthusiasm both for his own work and that of his students. I collected scores of mushrooms which he photographed for his mushroom book, and was amazed at the untiring patience he always exhibited." Charles Thom, Atkinson's biographer, added that his lectures were dull and boring (Thom 1956) but that frequent fieldtrips were exciting and fun. Students worked into the night arranging and studying the fungi collected during the day. Later, Curtis Gates Lloyd heaped criticism in his direction, and Atkinson suffered psychologically (Petersen 2023). On field trips, Murrill must have been selective in his comments on interesting but non-germane objects (Petersen 2021).

At Cornell, the death of Albert Prentiss in 1896 (Atkinson 1896a, b) catapulted Professor Atkinson into the Botany Department Head's position. His teaching duties continued unabated, but other responsibilities accrued, especially the relationship with the Experiment Station (Atkinson 1892a). Atkinson's publication output diminished somewhat, but this was the result of more time-consuming research in the month and year before. In 1897 came commentaries on pedagogy (Atkinson 1892b, 1897a, b) and in the following year (Atkinson 1898) more in this vein. These were not his last significant contributions on this subject.

On Murrill's part, he remembered being rather self-confidant. Many years later he (Murrill 1945a) wrote: "The Naturalist could have been fairly successful in several different fields, such as Mathematics, Chemistry, Literature, Music or Philosophy, but Nature had a special appeal for him and he felt certain that his choice would finally be made in some department of Biology.."

To sum up, the Cornell Botany Department was involved in change on several fronts. There was pressure to modernize teaching to reflect the "New Botany." The relationship with the Geneva Experiment Station had Liberty Hyde Bailey bringing agricultural research into the department, and Benjamin Duggar acting as the Department's ambassador to this wide field. These notions would all be exacerbated over the subsequent years but only with painful stresses throughout (Petersen 2021).

WAR. As the situation in the Caribbean grew increasingly ominous through 1897, the U.S. stood by, both coached and hampered by the Monroe Doctrine, Spain being the latest target. At least on the surface, the U.S. had to sympathize with the revolutionaries and democratic government. To exercise its interest, power, and presence, the battleship USS Maine was sent to berth in Havana, only to have a mysterious explosion sink the ship and bring the pot to a rolling boil, causing Congress to approve U.S. intervention "to end the fighting in Cuba between the rebels and Spanish forces, and to establish a 'stable government' that would 'maintain order' and ensure the 'peace and tranquility and the security' of Cuban and U.S. citizens on the island.... the U.S. Congress passed a joint resolution that acknowledged Cuban independence, demanded that the Spanish government give up control of the island, foreswore any intention on the part of the United States to annex Cuba, and authorized [the President] to use whatever military measures he deemed necessary to guarantee Cuba's independence" (Spanish American War 2023). This was April, 1898, and Murrill was finishing his first academic year at Cornell—his own critical decision also having been made.

Murrill did not see fit to mention the war in his autobiography, nor did it seem to perturb Cornell's botany curriculum. The war's process, Teddy Roosevelt, the "Rough Riders," the Black soldiers and their heroism all take a back seat to this narrative, but the terms under which the war ended had importance to the plot. The U.S. government took over the government of Cuba, Puerto Rico, the Philippines and minor Pacific Islands. Cuba would attain independence as soon as the U.S. government gave it permission, but Puerto Rico was a U.S. territory and would have a hand in deciding its own governmental fate. The Philippines would be granted independence appropriately. All three were now open to U.S. jurisdiction, which Nathaniel Britton would construe included extensive botanical exploration in his mission to uncover the botany of the American tropics.

Only two papers (Atkinson 1893a; Atkinson and Schrenk 1893) had dealt with fungi, and it could have raised an eyebrow for Murrill, for it emanated from the immediate region of Edna's birth and early life. More important, Atkinson's first consuming botanical group was the ferns (Atkinson 1893b, c, d, 1894a, b), surely part of the attraction of Cornell to Murrill. Prentiss's death had passed the mycology course to Atkinson, and just as Murrill arrived at Cornell, Atkinson was shifting his personal efforts from parasitic fungal relationships to "fleshy fungi"—shelf and bracket fungi, pileate forms of hedgehog fungi, gilled forms and leather fungi. This rendered Murrill's "ton of fungi" from Virginia into fodder for class examination, but Murrill understood that Atkinson was gathering collections and photographs of candidates for a "mushroom book." Except for Prentiss's course, Atkinson was largely self-educated in these forms, but he obtained help from Albert P. Morgan in Ohio, himself self-taught, and both men sought help from Charles Horton Peck, also self-taught and author of numerous new species in his annual reports as New York State Botanist. Over the next decade several illustrated guidebooks would appear (Petersen 2020a, b) and a rush was on to produce photographic illustrations. Atkinson's pictures were among the better, even though contrast and size were less than perfect. His first contribution was concurrent with the arrival of Murrill (Atkinson 1897d), so Murrill's bounty was intended for a much-enlarged anticipated sequel (Atkinson 1900a, 1903). Encouragement must have come from unexpected sources—C. G. Lloyd, who introduced photographs in his writings as early as 1898, and Charles McIlvaine, who was to include photographs in his magnum opus in 1900 (McIlvaine and Macadam 1973).

The rivalry between watercolor ("aquarelle") and photographic illustrations was heated but unequal as long as photographs were black and white. In his early days at the Garden, wife Edna illustrated fresh collections, and in his later years, Murrill himself used a camera for illustrative purposes. But the whole subject is well beyond this narrative (but see Petersen 2020b; Petersen and Knudsen 2015).

Another small caveat: Weber (1961) categorically reported William Murrill's marriage to Edna as Sept. 1, 1897, but a MycoPortal record from the Cornell Plant Pathology Herbarium shows a single specimen of *Gyrocephalus rufus* collected by Murrill at Cascadilla on August 29, 1897.

From his first-year assistantship for Atkinson, Murrill became the Graduate Assistant for the undergraduate botany class in his second year. He wrote (Murrill 1945a)—with attention to his use of third person: "It was the duty of the Naturalist to assist students with their microscope work and to prepare the specimens used in class. Tree diseases were also studied intensively, together with the large group of woody fungi causing decay in timber. This was the beginning of the Naturalist's study of the polypores. Photography was essential, so he learned it; using the Professor's excellent work as a model." On more personal activities: "During the first year the Naturalist lived with Mrs. Dean, a short walk from the university but surrounded by farms and woodlands. Fortunately, there was a short cut through Professor [Liberty Hyde] Bailey's pear orchard. An invitation to join the 'Horticultural Lazy Club' was promptly accepted, which carried the privilege of testing samples of fruit. This club met weekly and was thoroughly enjoyable as well as educational. Professor Bailey was a remarkable man and a valuable friend. He wore a long rough overcoat but never a dress suit. His books have been best sellers for many, many years. When the flour in his barrel runs low he writes another one" (Murrill 1945a).

Although Murrill was headquartered in Botany, "The chairman of his committee was Professor [John Henry] Comstock, [the entomologist] and his major was Plant Cytology with minors in Mycology and Plant Histology." This alignment surely explains the content of his Ph.D. thesis, "The development of the archegonium and fertilization in the hemlock spruce (*Tsuga canadensis* Carr.)" (Murrill 1900).

"The next two years were spent in Cascadilla Building [quarters for the Botany Department], overlooking the deep gorge by that name which separated the campus from the town. ... Once a week the summer students in Botany spent the day in some delightful spot and came back loaded with various kinds of specimens ... [Elias] Durand could name most of the mosses and many of the discomycetes; Wiegand had begun to specialize on certain groups of flowering plants; and the Naturalist took care of the ferns, trees and fungi in general. Professor Atkinson was not a trained taxonomist, so he wisely avoided the subject—as much as possible."

Unlike 1897, represented by only a single Murrill specimen accessioned into the Cornell herbarium, 1898 was a banner year. Collecting began in May with some representative ascomycetes, but August and September were rich in agarics, polypores, hydnums and resupinates. Fewer collections were made in October and November, but included a few agarics usually considered winter or early spring taxa. Several locales were visited, some more than once. What help in identification was needed or sought cannot be known from specimen label data, but surely Atkinson and fellow students were included on some of the excursions.

Remembering that Atkinson's position included responsibility for overseeing the research at the Experiment Station—a duty usually relegated to his senior graduate student—Murrill wrote: "During his third year at Cornell the Naturalist was given the position vacated by [Benjamin] Duggar, requiring the investigation of plant diseases. First he accumulated a large number of bulletins bearing on the subject and then began some special studies of peach diseases and fungi affecting shade trees. The latter field was developed quite extensively, resulting in a bulletin which was widely distributed and had considerable influence" (Murrill 1902a). It is amusing to consider that Murrill passed through Dr. Liberty Hyde Bailey's pear orchard in his initial Cornell year, and sampled the fruit together with other students.

"Little need be said here about major work in Plant Cytology. Indeed, it is often stated that a man gets so tired of his major subject that he is apt to turn to something else when he finished his thesis. ...The long hours spent by the Naturalist in collecting and fixing material, sectioning, staining, studying mounts under high powers of the microscope, drawing the stages in infinite detail, coming to conclusions and writing the whole thing up in orderly fashion was an education in itself. The value of original research cannot be overestimated." His preoccupation is reflected by a paucity of Murrill's collections in Cornell herbarium for 1899.

One anecdote seems especially revealing. "There was little opportunity at Cornell for campus activities or social life. The Naturalist mingled with his immediate associates and went to a few house parties but had no time for much else. On Sundays he attended religious services in Sage Chapel, where great preachers of different denominations were heard and an opportunity given to meet them during the following week. The Naturalist called on [one of them] during one of his visits and discussed certain matters related to Christian conduct. [The Naturalist] was advised to be liberal and charitable; to indulge in social activities such as dancing and card-playing with good people under the proper circumstances and at proper times; and to avoid assuming the pose of a critic or puritan, since such individuals exerted little or no influence for good among people in general. The advice of the good bishop was accepted without question although it differed radically from some ideas instilled in the mind of the Naturalist from childhood. His first step was to take dancing lessons. Later he composed many original dances and wrote a book on certain phases of the subject which was distributed to ten thousand girl scouts." This was retrospectively written in 1945 at age 75 with no reference to a wife. His pamphlet for the Girl Scouts takes precedence over memories of Edna—did she also meet the preacher and take dancing lessons?

Equally succinct, Weber (1961) reported: "Their only child, a son born in 1899, died in infancy." Again, an event not mentioned by the Naturalist in his autobiography. Edna's pregnancy must have complicated life for the young couple; for William, the limitations of time for needed attention; for Edna, inordinate time alone, perhaps with medical difficulties. That a male child brought delight and love cannot be questioned, but his health might have been fragile very early. His death—in infancy, not at birth—must have been tragic for the couple, but brought no temporal relief for William, and perhaps reinforced postpartum depression for Edna.

What Murrill (1945a) DID write: "In addition to his diploma the Doctor took away with him a Sigma Xi key; a thesis on 'Fertilization in the Hemlock Spruce," later published in the 'Annals of Botany;' a manuscript copy of his bulletin on 'Shade Trees;' [but see Atkinson 1901b] and the good wishes of many helpful friends." Not a comment on melancholia or stressed partnership. In his memory (Murrill 1945a) the Naturalist merely closed a door.

THE BIG APPLE. Following immediately after the above, Murrill (1945a) wrote: "Out in the wide, wide world he went, hunting for a job. New York City seemed to offer the best prospects; moreover, he wanted to go there anyway and, after a few rude jolts, he contacted a Virginian who happened to be a school principal. That settled it. He was given a chance to teach Biology in a high school at a bigger salary than he had ever dreamed of, and there would be an automatic raise every year he made good." Dissecting these cheerful words, some details may be teased out.

According to Weber (1961), the high school in question was Dewitt Clinton High School, at that time all male and located in downtown Manhattan (later it moved to the Bronx, at the other end of Mosholu Parkway from The Garden). He found living quarters at 439 Manhattan Ave., just below Morningside Heights and the recently moved Columbia University. Nearby lived Lucien Underwood, Professor of Botany at Columbia and a correspondent of Murrill on ferns.

Given Murrill's repeated pilgrimages back to western Virginia over the next decades, his words "moreover, he wanted to go there anyway" (written about New York City) seem a bit discordant. Perhaps he wished to visit the city as a tourist, but obviously he was job-hunting. From his job at Bowling Green to his appointment at Wesleyan Seminary, to his choice of botany versus zoology, and Cornell versus Johns Hopkins, his move to New York City and employment now appear like a last-minute or impulsive decision—as usual, with no mention of Edna. In this case, the lynchpin person was "a Virginian."

So William Alphonso Murrill, offspring of a conservative Christian family in conservative Christian western Virginia, but seeking just how conservative his

Christian conduct should be shaped in this strange, Northern, urban, suspicious society, found himself surrounded by new acquaintances and students, but few friends (save, perhaps, Underwood). Whether he even had a wife is in doubt; that a son was lost almost simultaneously seems without doubt. We find him writing "moreover, he waned to go there anyway" about New York City. If there ever was a den of iniquity, this was the chaotic city of the Union, and he faced classrooms of young men after some years of educating young women. He sought to make his way as the Naturalist. "Mother oak" might be expected to fade with his childhood.

The Big Apple

Chapter 2
Tilling the Soil

Doctor William A. Murrill, Ph.D., left Cornell University in 1900, ready to sample the metropolis on its own turf. Perhaps wife, Edna, was with him, but Murrill wrote exclusively in the third person, singular, so only occasional circumstantial evidence testifies to his marital status. One way or another, it was not as though "Zeus threw a lightning bolt and the city appeared in the full glory of its existence." Instead, its sociological and political status was complicated, and the situation for "naturalists" was not readymade. The total scene had developed over time, and Murrill was to be injected at a specific time and role. It was as though a field had been tilled before the seed could be sowed.

AN OLDER SIBLING. The New York Botanical Garden was not the first such establishment in New York. Both the American Museum of Natural History and the Metropolitan Museum of Art preceded the Garden, and consumed aliquots of the City's largesse of money and attitude. Some history is worth investigating.

Commercial enterprises, like clans, can often be traced genealogically. There was a "museum" in New York town a few years after the Constitution was approved late in the eighteenth century. John Pintard founded the American Museum on King Street (street name probably changed post-war). (Scudder's Museum Wiki 2023), with the patronage of the Tammany Society (the progenitor of the political organization of the nineteenth century) of which Pintard was a member (Tammany Hall Wiki 2023). The Museum changed its name when Pintard's partner, Gardner Baker, took charge, and in short order, the Museum was moved to Pearl and Broad Streets in 1794, was called "The Exchange," and upon Baker's death fell into the hands of William I. Walden. The next owner, Edward Savage, called it the "Columbian Gallery of Painting and City Museum" in 1802, and hired John Scudder to curate the collections. Scudder bought the collection, moved it to Chatham Street in 1810, and named it "Scudder's American Museum." The next move was to space in the

R. H. Petersen, *William Alphonso Murrill: The Naturalist*, Memoirs of The New York Botanical Garden 124, https://doi.org/10.1007/978-3-031-93343-1_2

city's poor house in 1817. Scudder died in 1821, and the museum passed to his heirs, who moved it to the triangle of Broadway, Park Row and Ann Street, where it remained for an extended period. Family squabbles finally resulted in another sale, this to P. T. Barnum, the noted showman, in 1841. The establishment's genealogy ceased in a catastrophic fire in 1865.

Surely the most spectacular exhibition hall was the Crystal Palace (Fig. 2.1), a steel and glass conservatory modelled after its namesake in London. Built in 1853, and owned by Barnum, it burned to the ground in 1858. Juxtaposed was a tall observatory which would serve as the inspiration of the Eiffel Tower in Paris.

Aptly written: "Foreshadowing trends in American commercial amusement, the Museum gathered exhibitions and amusements that previously had been offered in separate milieus. In an urban culture characterized by increasing difference—in taste, in subject, and in audience—it was the first to combine sensational entertainment and gaudy display with instruction and moral uplift. For a twenty- five cent admission, visitors viewed an ever-revolving series of 'attractions,' from the patchwork Feejee Mermaid to the diminutive and articulate Tom Thumb, but the museum also promoted educational ends, including natural history in its menageries, aquaria, and taxidermy exhibits; history in its paintings, wax figures, and memorabilia; and temperance reform and Shakespearean dramas in its 'Lecture Room' or theater (Fig. 2.2). Thus, the Museum drew its audience from a wide range of social classes and strove to assure that the lecture room and salons would be one of the few respectable public spaces for middle-class women. However, until the Civil War, African Americans were, except for certain days, barred from the Museum, as they were from most antebellum New York commercial amusements. In an unprecedented fashion, the Museum encompassed in one place immigrants and native-born

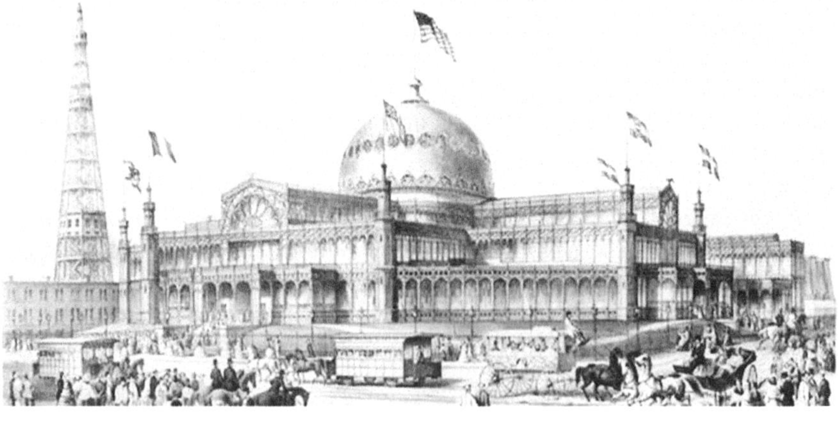

Fig. 2.1 Crystal Palace Exterior. Crystal Palace and observation tower (left), mid-nineteenth century. (Google Images)

Fig. 2.2 (**a**) Crystal Palace interior. Inside view of Crystal Palace, mid-nineteenth century. Note two floor, ornate steel construction and high windows for interior ventilation. (**b**) Façade of a Coney Island museum, early nineteenth century. (Google Images)

Fig. 2.3 "Little Italy," a photo of street with Market Carts. Classic image of "Little Italy," depicting the crowded conditions of ghetto life in lower Manhattan, early twentieth century. (Google Images)

working-class and middle-class, men and women, city residents and rural visitors—and yet, repeatedly and increasingly, foundered on the social tensions arising from gender, ethnic, class, and racial differences." (Lost Museum Archive 2023; Fig. 2.3).

While the fire of 1865 deprived New Yorkers of a popular pastime, it also liberated some philanthropic minds to consider how New York City, the population and commercial center of the states but a poor reflection of the capitals of Europe, could be coached toward an uplifted and refined cultural milieu (Fig. 2.4). Sloan (1980) summed it up this way: "During the 1860s a relatively small number of New York cultivated elite launched a comprehensive program of cultural uplift and civic improvement; among them were intellectuals like Frederic Law Olmsted and Henry W. Bellows, journalists like E. L. Godkin, Horace Greeley, and Parke Godwin, professional men like the lawyers Joseph Choate, Charles Daly, and Addison Brown (1908), a number of public-spirited businessmen-philanthropists like William E. Dodge and Morris K. Jessup, and a few city officials like Andrew H. Green, head of the New York City Park Commission." The large omission from the list was William Magar Tweed (1823–1879), Boss of Manhattan (Fig. 2.5), New York State Senator, who held court from Tammany Hall, by then much more influential than at the end of the eighteenth century (Tweed Wiki 2023).

Fig. 2.4 Early twentieth century Lawn Party. A weekend lawn party on Governor's Island in New York Harbor, circa. 1900. Note the womens' long, full skirts. (Google Images)

Fig. 2.5 William "Boss" tweed in 1870. Formal portrait of William Magear Tweed, politician of lower Manhattan, late nineteenth century. (Wikipedia)

That a museum of natural history was not the only such idea beginning to take root in the City, Sloan (1980) furnished the following summary. "Within just a few years of each other came the founding of the Columbia School of Mines—1864—and the American Museum of Natural History—1867—the chartering of the Torrey Botanical Club—1873—and the rejuvenation of the New Lyceum of Natural

Fig. 2.6 Louis Agassiz. Formal portrait of Louis Agassiz. Note the inquisitive eyes. (Wikipedia)

History under its new name, the New York Academy of Sciences—1876. Around these major institutions there began to form local scientific societies, about a score of which existed by the end of the 1880s."

Up the coast in Cambridge, Massachusetts, Louis Agassiz (Fig. 2.6), the Swiss naturalist by now famous on two continents, had a laboratory, graduate students, and public acclaim. Agassiz displayed at least three attributes which set him apart from other faculty at Harvard: a charismatic personality expressed publicly and privately, a careful research ethic admired by his colleagues, and ability to raise money to support his laboratory and collecting expeditions to add to the immense collections already in hand. He had his mind set on a "comparative zoology" museum for Boston.

Agassiz had come to America in 1848, a year of European political disturbances, to deliver a series of Lowell lectures, but a warm welcome persuaded him to remain. His already large specimen collection joined him, and in 1853, Harvard bought the collection, thus anchoring Agassiz to the College. In 1856, "an admirer" willed $50,000 to Harvard for creation of a Museum of Comparative Zoology. Additional funding and a newly erected building followed. In November, 1860, $240,000 had been raised for the Museum. For Agassiz, life was good.

Among Agassiz's carefully chosen graduate students was Albert S. Bickmore, graduate of Ivy League Dartmouth University, and well-connected among New England wealthy families (Kennedy 1968). His academics were acceptable, but what proved at least as important, he became enraptured of the museum concept. Perhaps unanticipated, though, Bickmore reckoned that the center of money, population and "culture" in the states was New York City, and he conjured, like the apostle Paul, to take the gospel to "foreign shores." But first, the Civil War intervened, and Bickmore served in the Union Army.

One antebellum proponent of a natural history museum in New York had been Andrew Haskell Green, Treasurer of the newly created Central Park, and a pivotal advocate for such a museum within the Park. The fight for the Park itself had not been easy, but the State Legislature was won over and the Park was approved in 1853. Even in 1861, Green was public in his approval of the park and even cultural institutions within it. The War put an end to the movement.

Bickmore "dipped his toe in the water" of New York City just after the War, but not only was the reception tepid, but he was on the verge of a 3-year expedition to Java, Japan, China and Siberia. Upon his return in 1867, he tried again, and after a stuttering start, was put in touch with Theodore Roosevelt, Senior, who became an enthusiast and drew other elites into the circle (Osborn 1911). The list of founders, nineteen in all, included a mixture of philanthropists, and often congruent, collectors of various natural historical objects. Simultaneously, two large and valuable European collections came on market and were purchased (Osborn 1911). The proposed charter was introduced to the state legislature in Albany and surprisingly, was approved by William "Boss" Tweed without objection. On April 6, 1869, the governor signed the bill and the American Museum of Natural History began its legal existence, more than 20 years before The New York Botanical Garden would undergo the same process. Neither was to have the sideshow of curiosities featured by Barnum, but each was pledged to public education and scientific research, each had its evangelist, each its band of philanthropic apostles, each found a site on public lands, and each grew to take its place among the world's greatest scientific establishments. It must be noted that in those years, Columbia College and its associates were 30 years away from becoming a single unit in a single location. In fact, John Torrey (Fig. 2.7) was still alive and the Lyceum botanists were mulling over the possibility of forming a club around him.

"Natural history," by its very expansive name, included botanical subjects, but even the founders of the American Museum would have explicitly agreed that their establishment was "zoocentric," and even tighter, topics to attract the "common

Fig. 2.7 John Torrey. John Torrey, leading New York botanist of the second half of the nineteenth century, "patron saint" of the Torrey Botanical Club. (Wikipedia)

man." Plant collections and botanical research were not high priority—hardly mentioned—and would remain so as long as the New York entity competed with the Boston Museum of "comparative zoology."

One new wrinkle was stitched into the charter. The grounds, previously called Manhattan Square, and any buildings erected thereon, and their upkeep, were the responsibility of the Parks Department, while the collections, their storage furniture, their curators and their salaries were the responsibility of the Museum Corporation. A partnership was thus established between the political entity and the academy. Over the impending years, the embryonic Botanical Garden and several other cultural/educational institutions were founded under the same general financial structures, for which two sources of revenue were tapped: in part by direct city appropriations, and in part by gifts from philanthropically minded citizens and by the income from these endowments, each under the control of their own governing boards (Merrill 1938).

COMMON GROUND: THE TORREY BOTANICAL CLUB. The botanists of the City might have pouted through the adulation for what "they"—the press and the public—called "natural history:" money flowed to the Museum and its research and its new exhibits called "habitat groups," all but ignoring botany. Curators were hired, collections were gathered and purchased, but if plants were involved, they were given "short shrift." Luckily, the Lyceum of Natural History still held plant collections, John Torrey was willing to open his home to informal gatherings, and John Newberry (see APPENDIX 1) was ready to give of his wide experience.

The Torrey Botanical Club became a diverse group of citizens—even this label may be limiting— brought together by their common interest in botany, and from them came the document which set in motion The New York Botanical Garden. How to describe them? In the late 1880s, almost half of the several hundred members were women, who spoke at meetings, read papers and sometimes exhibited horticultural attributes not supplied by their menfolk (Mickulas 2007). Even 30 years later, when the Botanical Garden was running low on funds, it was a separate "Ladies' Organization" from the Club that was conceived to boost subscriptions.

Certainly, across the Club's male membership, occupations were scattered. Pharmacists, teachers, gardeners, judges, accountants, preachers, florists could all speak the same language at the Club's meetings. Here, some members active over the course of this narrative have been included in APPENDIX 1. They ranged over a wide age difference, and at least three or four times as much could be written if the Torrey Botanical Club were the major actor in this narrative.

A significant homogeneity was a "given" in botanical communication in the last quarter of the nineteenth century; botany still equaled taxonomy. Derivative disciplines—comparative morphology, genetics, physiology, histology, ecology—would entice students in growing numbers at the end of the century (more below), but for the time being, field-based identification, naming and amassing collections kept the professionals and amateurs busy and content to interact.

ANN IDEA. Although described by active participants during the years of gestation of a botanical garden for New York City, their accounts do not quite match. Rusby (1906) related that in the years after the neglect of Hosack's Elgin Garden, lay botanists began cultivating their own small patches, but like the Elgin, even these were subject to the piecemeal accretion by the City. These circumstances, the memories of "physic gardens" of Europe, and the inattention of botanical science at the new American Museum, prompted renewed interest in such a project. As early as 1874 [not in bibliography], Rusby wrote, the Club appointed a committee to partner with the New York Pharmaceutical Association in a request to the City to establish such a garden in Central Park. Something similar had been established at Harvard under the curatorship of Asa Gray, and talk "around the old pot-stove in Professor Newberry's room, during his presidency of the [Torrey] Club" took on a sense of urgency. So, in the Torrey Botanical Club there was sentiment that New York City lacked a botanical garden of any size and sophistication (Moore 2000) rivaling those in the capitals of Europe.

The Torrey Botanical Club had already provided the milieu for the courtship of Nathaniel Britton and Elizabeth Knight (see Appendix 1 for both). In September of 1888, the Brittons paid a visit to the Royal Botanic Gardens at Kew, a suburb of London, England. The spacious grounds of the Gardens were landscaped with plants from over the worldwide British empire, and the Garden's "crown jewel" was Temperate House, a huge glass conservatory (see Fig. 4.9) then still being enlarged (Temperate House 2022; Merrill 1938). It reminded the visitor of The Crystal Palace of New York City. As they strolled, Elizabeth asked Nathaniel, "Why can't we have such a place in New York?" By the time they arrived home, the idea had been liberally nourished by the couple and had gained some weight. At the next (October) meeting of the Torrey Botanical Club, Elizabeth described the Kew Gardens and then repeated her question. The talk and the idea met with enthusiasm and at the November meeting a committee was appointed to investigate possibilities (Britton 1915, 1916). Among the eight committee members (all male) were Arthur Hollick (by then Ph.D., old friend of Nathaniel Britton and fellow graduate of Columbia Mines), H. H. Rusby (ethnobotanist), N.L. Britton (taxonomic botanist), Addison Brown (both botanist and federal judge; see Appendix 1), and John Strong Newberry (Britton's mentor, naturalist, and supervisor at Columbia School of Mines; see Appendix 1). By January, 1889, the Committee had written an appeal document, presented it to the Club, and was then charged with securing an agreement from the Department of Public Parks. The Parks Department consented, with the provision that the Committee ensure that funds for the garden's upkeep and development be raised privately, separate from public coffers. The New York State Legislature passed a bill of agreement, drawn up, in part, by Judge Brown, in April, 1891. The dream had become reality in less than 3 years.

The list of incorporators was long, 48 males in all, reflective of the time and the evangelistic prowess of the Committee: such luminaries as Seth Low (President of Columbia College; see Appendix 1), Charles A. Dana (Editor of the New York Tribune and publisher of the New York Sun), Cornelius Vanderbilt, J. Pierpont

Morgan, Andrew Carnegie, David Lydig and William C. Schermehorn (philanthropists), lawyer Addison Brown and botanist Nathaniel L. Britton.

EARLY RESISTANCE. To be sure, the original statutory structure of governance of the Garden was neither easily understood nor public. From the ample ranks of incorporators, a Board of Managers "responsible for conducting all business affairs and for the financial management of the organization" was elected to terms of 1–3 years, so the Managers could be completely replaced over a 3-year period. The Board of Managers, though, was quite rigidly structured: nine from the corporation at large, plus two ex-officio members (The Mayor of the City and the President of the Board of Parks Commission) plus six Scientific Directors. The Scientific Directors "would have the management and control of the scientific and educational departments of said corporation and the appointment of the Director-in-chief of said institution…" The Director-in-Chief's only limited statutory duty was to appoint a First Assistant and a Chief Gardener (Mickulas 2003).

The officers of the first Board of Managers were Cornelius Vanderbilt, President; Andrew Carnegie, Vice-president; J. Pierpont Morgan, Treasurer; and N. L. Britton, Secretary. The Mayor of New York was Seth Low (Appendix 1), President of Columbia University, one of the city's wealthy elite, and the first subscriber of support for the Garden. The relation between Garden governance and wealthy philanthropy was clear: equally clear was a close relationship between the Garden and Columbia University, solidified by Nathaniel Britton as Professor of Botany as well as Director-in-Chief. All this was codified by the state legislature in 1891.

Less blatant, but also supportive of the arrangement, the Torrey Botanical Club made Columbia its headquarters, and the College's herbarium (compiled and organized by Britton) and botanical library were to become the nucleus of the Garden collections. Finally, graduate courses of instruction were to be offered at the Garden, with credit toward advanced degrees to be granted by Columbia, all agreed to by 1895 (Mickulas 2003). More on this below.

But such a cozy comradery stuck in the throat of another New York City educator, Reverend Henry M. MacCracken, Chancellor of the University of the City of New York, these days better known as NYU. Since occupying the Chancellorship in 1891, he had striven to enlarge and enhance his university's stature in the city, always in the shadow of Columbia. Here was the newest affront in which Columbia obtained philanthropic attention, received public support and land, and automatically was bestowed a Botany Department, a faculty and attention.

It seemed impossible to challenge the Garden/Columbia partnership head on, but MacCracken took two sidesteps, actuated through an amendment to the Garden's charter quietly introduced in the state legislature. The amendment would alter the Garden's Board of Scientific Directors to include, in addition to the four representatives from Columbia, two members of the NYU faculty and one faculty member each from St. John's College (Fordham University), Manhattan College, the Normal College, and Barnard College. In fact, this 1892 bill would probably have passed smoothly except for the keen eye of a city newspaper man who informed Addison

Fig. 2.8 Frederick Law Olmstead. Formal portrait of Frederick Law Olmstead, city planner, father of Central Park and "Emerald Necklace" of Bronx parks. (Wikipedia)

Brown, jurist, Torrey Botanical Club member and chief architect of the Garden's charter document, who sounded the alarm to Low, Britton and others. Columbia responded by implying that under the new plan, current subscriptions might be declared null and void, upsetting philanthropists who had supported the Garden. For the time being, the fight subsided. Britton met with MacCracken in spring 1893, and found him no less serious about his previous intent. Only the tactics were changed, and as a result, William Gilman Thompson (1856–1927), NYU's professor of biology, was added to the Scientific Directors of the Garden (Mickulas 2003). Although not known professionally as a botanist (his medical degree was from Columbia), in the years after this narrative, Thompson rose to become President of the Garden from 1913 to 1926 (Thompson Wiki 2023; Britton 1928). Otherwise, the charter remained as written.

THE EARTHEN PLOT. Among the Bronx's "emerald necklace" of Frederick Law Olmsted (Fig. 2.8) parks, Bronx Park is centrally located between Van Cortlandt and Pelham Bay Parks, and has all the characteristics of having been assembled by a committee. The large land holdings west of the Bronx Parkway, now home to The New York Botanical Garden and the Bronx Zoo, were annexed by the City in 1874 (Mickulas 2003: another source refers to the 1884 New Parks Act; Bronx Park Wiki 2023). The eastern side of the present-day Bronx Park was tacked on later, and with a very different mission than the western portion (Fig. 2.9).

In present maps of the Bronx, it would be tempting to see the Bronx Parkway (running north-south) as dividing the tract, but it is really the Bronx River, running more or less parallel to the Parkway to its west that bisected the area long before such things as the Parkway. Because the eastern section does not play a major role in the Botanical Garden story, the narrative will refer to only the western territory.

The City-owned tract had come through the Lydig family, familiar for its mention on the title page of most Garden publications as "The David Lydig Fund,

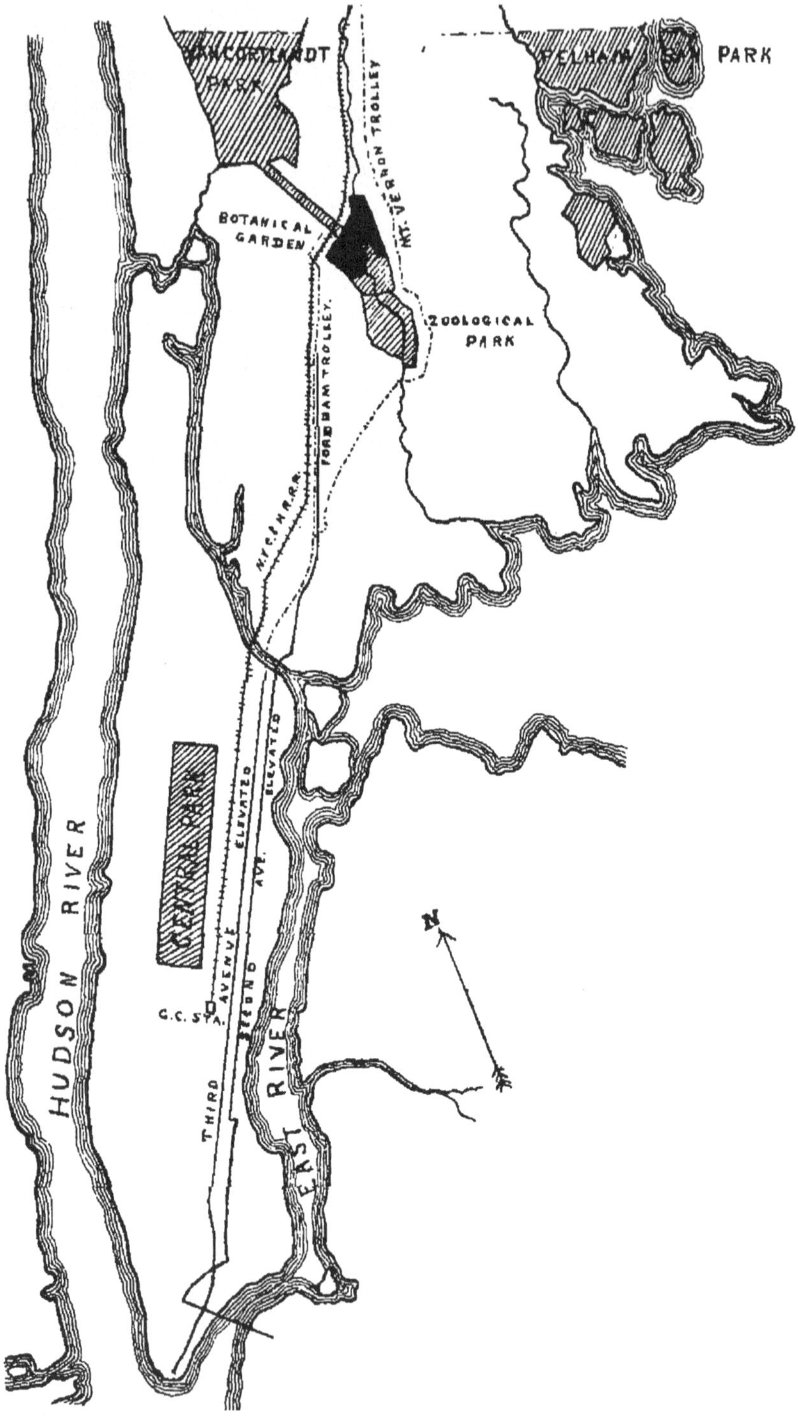

Fig. 2.9 Transit map, Manhattan and Bronx. 1900 transit map of access to the Garden from Manhattan. Depicted are Manhattan's Second and Third Avenue Elevated Tracks, both east of Central Park. Both were used by Murrill during his two jobs. In the Bronx, transit routes switched to the New York Central Railroad and Mt. Vernon Trolley. (Journal of the New York Botanical Garden)

bequeathed by Charles P. Daly." Maria Lydig was married to Judge Daly, and David Lydig was part of her father's lineage. Maria had an interest in the founding of the Garden, and Judge Daly acknowledged his wife's interest in his (Daly's) will.

According to another source (Bronx Park Wiki 2023), much of the land was acquired from St. John's College (now Fordham University; Fordham University 2023) which gave away the land on the promise that it would be used as zoological and botanical gardens. Another section was obtained from the family of Pierre Lorillard, the tobacco mogul.

The holdings of the Lorillard family and the Lorillard story are worth a few words, for they add a new knot in the weaving linking the Garden with other familiar museums, music halls and galleries created with the largesse of wealthy philanthropists (see Appendix 1).

A FINANCIAL TREMOR. Anchoring the Botanical Garden to a panel of wealthy subscribers was intentional and initially proved admirable, but within a short time an unpredictable financial slump broached some questions. As with most dated events, the "Financial crisis of 1893" showed its first tremors before, and aftershocks rumbled afterward. Moreover, while the national bank failures to modest fortunes made headlines locally, the true dimensions were international. In fact, the Columbian Exposition in Chicago, with all its splendor and glitz, took place in the midst of the crisis (Hoffman 1970).

The "Gilded Age" had seen the amassing of enormous fortunes of tycoons who sought control not only of corporations, but of scores of firms involved with similar commodities, often forming "trusts" during the process. Careful manipulation was necessary, lest a trust over-develop its segment of the economy, and thereby become inflated to the point of deflation (Panic of 1893 Wiki 2023). One such trust was the Philadelphia and Reading Railroad (P&R Railroad 2023), which aimed at controlling the transportation of anthracite coal from the Pennsylvania and Ohio deposits to the metroplex of Philadelphia, New Jersey and New York. A second trust was the National Cordage Company (NatCordCo 2023; Ohio History Central 2023), seeking a monopoly on hemp and fiber importation and production of rope and twine. Both trusts (and many others) over-built, over-invested, and in early 1893, found their profits tied up in operations and insufficient to pay their investors and their workers. The result was stockholders demoralized, selling their shares with predictable stock price plunge; citizens suddenly unsure of their savings, demanding their money, often as gold instead of paper from banks already stretched thin; banks calling in their loans to produce more capital and then borrowing from larger banks to provide the cash; bank failures leading to stock market stress—most of this domestic but soon infested by a largely unrelated international banking crisis (Steeples and Whitten 1998). In such a web of insecurity, it would be predictable that the Garden's Board of Managers might grow cautious about both giving and spending.

But if a hopefully eternal institution like the Garden was threatened, the temporary Chicago Exposition was even more seriously threatened. The exposition itself had gone through a fierce intercity fight between New York and Chicago, each with its roster of super-rich philanthropists vying to see which city could pledge more money to underwrite the event. That Chicago won at the last moment—with the last million raised less than a day before the deadline—must soon have soured the stomachs of the underwriters as the financial "panic" grew, and understated smiles from the New York cadre. The Exposition was dedicated October 1, 1892, before the "crisis" descended, but opened to the public May 1, 1893, just in time for dire headlines. During its 6-month life, over 21 million people went through the turnstiles. It was a headline-stealer and a living symbol of the resilience of Chicago and securing its label as the country's "second city."

HAND AND GLOVE. To complete the real estate part of the Garden's birth, financial contributions were to come from the City—$500,000 for land and structures to be built, with the proviso that by the end of 7 years, at least $250,000 was to be raised privately for staff and building contents. The first private subscription of $25,000 was from Judge Addison Brown (1908), and the second, of equal amount, came from the Columbia trustees via President Seth Low. The inaugural officers of the Board of Managers read like a page from "The Four Hundred" directory. Seth Low was appointed first Chairman of the Scientific Directors.

Not only was Columbia College entwined with the Garden through Seth Low and Nathaniel Britton (both, Appendix 1), but there soon followed an arrangement through which Columbia moved its herbarium and botanical library to the Garden, in return for unfettered use of the Garden facilities, albeit meager at first, by Columbia botanical faculty and advanced students (Underwood 1903). This arrangement exists to this day (Merrill 1938). Mickulas (2003, 2007) summarized the arrangement: "This crucial link between Columbia and the Garden was not entirely Low's work; aware of the benefits of such a collaboration, Britton himself proposed the outline of the arrangement to Low in the spring of 1895. Britton argued that the Columbia herbarium (the nucleus of which had been John Torrey's personal collection, which Britton himself had organized in taxonomic sequence) was of little use to basic undergraduate education. Rather, he proposed, the collection should be moved to the Bronx, to be housed in the new museum and available for graduate-level research to be conducted under the auspices of the Garden and the University. Ultimately, the arrangement between the two New York institutions proved so beneficial that Nicholas Murray Butler, Low's successor as Columbia president, further formalized their relationship, recalling Britton from his emeritus status to hold a seat on the University's Faculty of Pure Science."

The Garden was more than a sheaf of papers: its Director was expected to gather a staff of botanists, plan facilities, set a menu of research, and develop (or preserve) the grounds. For the immediate future, endowment money allowed (and required, if more was envisioned) visible progress, and in 1896, Britton left his faculty duties at Columbia (but not his title), immediately filled by Lucien Underwood, called up from Alabama Polytechnic Institute (Appendix 1).

By Murrill's arrival in the City in 1900, the botanical garden concept had come to life, its first director-in-chief had been appointed, its financial books balanced, and construction on the first of its buildings, the Museum Building, commenced. But

Fig. 2.10 David Hosack, nineteenth century. David Hosack, physician and botanist of New York, early nineteenth century. Portrait by Rembrandt Peale. (Wikipedia)

The Naturalist was only a high school teacher, professionally untethered except for Underwood at Columbia. His marital state, then reportedly 3 years old, remains uncommunicated to us. He was 31 years old and spoke with a lilt distinctly not New Yorkese.

THE NEW BOTANY. Rusby (1906), in describing the Torrey Botanical Club in its early days, wrote: "There came a time when New York experienced an invasion of botanists with concepts, knowledge, interests, and methods which were largely foreign to us. Their importations were of incalculable value to New York, and at the same time most urgently needed, and resulted in giving us a new, modern and broad botany. The event was not, however, free from unfortunate incidents. Laboratory work was given undue prominence. Ecological and other field work came to be largely neglected, and what might not inappropriately be called the disjointed period of the [Torrey] Club's history ensued."

Botany in New York had never been static, and the time span considered here was no exception. The "begats" of David Hosack to John Torrey to Nathaniel Britton chiefly dealt with plant taxonomy, for plant taxonomy was botany by simply another name. Being a student of medicine, Hosack related plants with medications, but he could recognize most of the local plants collected and described by Britton. But Hosack (Fig. 2.10) would not have taught plant physiology, ecology or mutation theory. The lines below attempt to sketch an important development in the discipline during the turn of the twentieth century.

Nineteenth century science and scientific methodology inherited a teaching protocol millennia old. The teacher lectured, the students took notes and recited as challenged by the teacher: examinations were based on memorization of lectures. Botany, as a discrete subject, was a handmaiden of medicine, with "physic gardens" displaying plants used in medication. Linnaeus's mathematical system based on sexual organs coupled with binomial nomenclature interrupted the otherwise placid continuum, but the audience was still limited to medically oriented workers. The questions to be answered were "what is this plant, and how is it used in medication?" It was at this point that New Yorker Amos Eaton (Fig. 2.11) began his program of courses in about 1810. Eaton was neither a physician nor a professor, but a teacher seeking students.

As an aid, he published a guide for botany students, the first such manual (Eaton 1820), and led field trips to investigate natural history *in situ*. David Hosack, an M.D. at New York's College of Physicians and Surgeons, planted his Elgin Garden (Fig. 2.12) and used it to instruct younger students, including John Torrey. The most detailed narrative of American "new botany" pedagogy has been by Rudolph (1996).

Fig. 2.11 Amos Eaton, nineteenth century. Formal sketch of Amos Eaton, teacher and botanist, first half, nineteenth century. Note the inquisitive eyes. (Wikipedia)

Fig. 2.12 Elgin Botanic Garden. David Hosack's Elgin Botanic Garden, mid-town Manhattan. Note the glass conservatory at center, flanked by hothouses for propagation. (Torrey Botanical Club Archive)

As the nineteenth century dawned and domestic agriculture saw a steady increase, botanical courses became included in educational curricula, but lapsed into the traditional lecture-recitation regimen. Plant health was important and plant pathology became subject to research, one revelation of which was that fungi and bacteria could cause plant diseases. New questions sprouted. How do these diseases occur? What were the effects? Why were occasional plants able to thrive when most died? Once the Civil War ended, Land Grant colleges were established, charged with teaching agricultural as well as mechanical and military arts.

Torrey-esque taxonomy found a new botany based on more than the Linnaeus system, but floristics continued with American western expansion. Harvard's Asa Gray inserted some laboratory work for selected graduate students (Rudolph 1996) but microscopes were scarce and exposure brief.

Charles Edwin Bessey took his degree from Michigan Agricultural College in 1869 (Murrill's birth year), migrated to Iowa Agricultural College, and with little money, tried to develop a laboratory-based course. It modelled two three-month visits to Harvard where Bessey had studied with Gray and Farlow, and was introduced to "well-equipped" teaching laboratories. Farlow had just returned from Europe, where he had observed botany courses where the students were all equipped with microscopes and apparatus for chemical and physical experimentation with plants. Farlow found the atmosphere exhilarating and returned to Harvard with a new enthusiasm. It was the introduction of "The New Botany" to the U.S. (Rudolph 1996), with Bessey as an evangelist. The first edition of his textbook appeared in 1880 (Bessey 1880), with at least two later editions.

What was it that Farlow found exciting about continental teaching of botany? At least one major influence can be identified (Rudolph 1996). First, Julius Sachs (1870; Sachs Wiki 2023) had produced a textbook which required students to master knowledge of basic methodology in microscopy, plant anatomy, and plant physiology, chiefly the role of water. German-speaking teaching laboratories were crowded with graduate students comparing results, sharing data, and asking questions (Rudolph 1996)—in short, acting as their own teachers

While Farlow's observations were of graduate students, Bessey's time at Harvard imbued in him the idea that undergraduates could profit by the same treatment. His textbook (Bessey 1880) was slanted somewhat toward agricultural applications, understandable in his midwestern locations over those years. Another group of professors exposed to Farlow's experiences included John Merle Coulter, Charles Reid Barnes and Joseph C. Arthur, who were publishing a journal, Botanical Gazette, which could feature "new botany" reports. Coulter and Barnes would go on to establish botany strong in botano-experimental subjects at the University of Chicago (Petersen 2025), bound to form a counterweight to the Columbia University/New York Botanical Garden combination.

In the closing years of the nineteenth century, it was becoming evident that plant biology, botany, could start with an individual plant—as does taxonomy and botanical nomenclature—and reach both inward to experimental study of metabolic function and tropisms as well as outward to plant associations (with similar considerations in zoology) to continental-sized biomes. That both extensions from the individual were possible in botany teaching, reflecting the pedagogical situation already exhibited in Europe. A rapid scan of BOTANICAL GAZETTE, especially book reviews, revealed the growth of experimental methodology in plant research at the end of the nineteenth century. A somewhat more detailed search shows the dominance of European, mostly German, literature in these disciplines.

A typical pathway is distilled in the following: Charles Barnes, Co-Editor of BOTANICAL GAZETTE, wrote a review of William Francis Ganong's volume on botanical pedagogy (1899a, THE TEACHING BOTANIST, 1st edition), which extolled "The New Botany." (1) Native of maritime Canada, Ganong received an A.B; from Harvard in 1887, and was exposed to the Harvard philosophy of pedagogy. (2) Ganong interrupted his U.S. education to earn a Ph.D. from Munich in 1894, and begin his career at Smith College. (3) Barnes would become an apostle of this new teaching philosophy, and foster its introduction to the Chicago laboratory. Barnes (1899a) wrote: "The [Ganong] book is divided into two parts: the first part is a collection of eight essays on botanical pedagogies full of excellent ideas, useful suggestions, and earnest admonitions. Many of these reforms botanists have long been advocating. Now the ideas are pithily put and in permanent form, they will become more widely efficient than heretofore. Those who have proclaimed the new gospel will be delighted to have this book to which they may refer teachers seeking some help. It will save reams of letters and hours of talk."

Almost back to back, Barnes (1899b) reviewed John Merle Coulter's PLANT RELATIONS. True to his beliefs, Barnes wrote: "Granting the ecological aspect to be the ideal botanical course, the question is whether we are yet far enough away from the floristic or pseudo-floristic teaching to justify an attempt to reach so remote an ideal. The writer [CRB] has already committed himself to the view that the simpler morphology and physiological topics should be first presented in an elementary course and therefore only state the pedagogical problem. To its solution Dr. W.F. Ganong contributes interesting arguments in a book elsewhere reviewed."

Between Barnes' reviews was one by Chamberlain (1899), another of the Chicago group, of Fischer's book on plant histology. In just a few pages, the century-ending status of botany was summarized.

Rusby (1906), of course, was describing "the new botany" coming from Europe via American scientists pursuing just such exposure after their American Ph.D.s. In fact, Rusby described Dr. Emily L. Gregory. "Thoroughly educated in the best modern schools of Germany, and especially a disciple of Schwendener, she became a missionary of advanced work and methods. She founded the botanical department of Barnard College and established there a botanical center which has since steadily grown in strength and influence, and is now one of our most important botanical possessions." It would seem that the ladies at Barnard had exposure to "The New Botany" before the boys across the street at Columbia.

So this was the New York botany scene into which William Murrill injected himself (and perhaps Edna). The American Museum of Natural History, a mature and populated Torrey Botanical Club, a prestigious Columbia University, and a public Botanical Garden being delivered into existence by Nathaniel Britton.

Chapter 3
Planting the Seed

Bearing in mind Murrill's repeated visits to Johns Hopkins and Cornell before his enrollment at the latter, it would have been expected that once he was offered the job teaching in New York City, he would pay the "Virginian" principal an introductory visit, but no such report can be found.

Murrill's autobiography would have us assume that the period between his graduation and arrival in New York City was seamless, but a seamless report also cannot be made. To begin, although Murrill graduated from Cornell presumably in spring of 1900, it is possible that he did not attend the graduation ceremony, for two mycological specimens testify that he was in Virginia collecting plant pathogens accessioned in the Cornell Herbarium (MycoPortal, CUP). Even those data are suspect, however, for one of the specimens is labelled as collected in "Blacksburg, Fairfax County," but these two locations are well-separated.

Much more secure, however, was time spent in Blacksburg, Virginia, from mid-August to mid-September where he logged numerous collections of higher fungi, mostly agarics (MycoPortal).

And there is the problem of his New York address. When he was elected to active membership in the Torrey Botanical Club on Thursday, May 14, 1901, his address was reported as 232 W 114th Street (Manhattan), near the southeastern end of Morningside Park. Soon thereafter, though, he started a correspondence with Charles Horton Peck (Fig. 3.1), the State Botanist in Albany, and the return address on his penny post card was 439 Manhattan Ave, some blocks north of the previous location: date, December 5, 1901 (Fig. 3.2). Perhaps he moved precipitously, but what of Edna? The Manhattan Avenue address, Murrill wrote, was near that of Dr. Lucien Underwood (Appendix 1), with whom he was already on friendly terms over fern topics (Murrill 1897), and with whom a friendly in-person relationship and collaboration was soon established. The most efficient solution to all this might be that Murrill did not arrive in New York until late in 1900, perhaps in time for the winter semester.

R. H. Petersen, *William Alphonso Murrill: The Naturalist*, Memoirs of The New York Botanical Garden 124, https://doi.org/10.1007/978-3-031-93343-1_3

Fig. 3.1 Charles Horton Peck. Charles Horton Peck, State Botanist of New York, acted as a mentor to the botanical community, including Murrill, who visited Peck regularly during their years of shared interests. (Wikipedia)

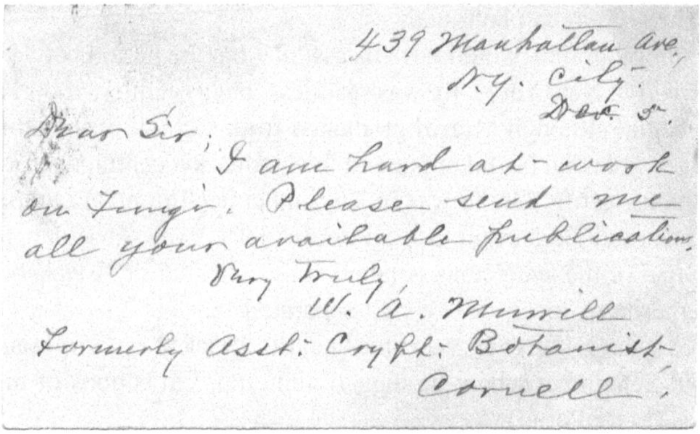

439 Manhattan Ave,
N.Y. City.
Dec. 5.
Dear Sir;
I am hard at work on Fungi. Please send me all your available publications.
Very truly,
W. A. Murrill
Formerly Asst. Crypt. Botanist
Cornell.

Fig. 3.2 Early Peck-Murrill correspondence. Scan of Murrill's postcard to Peck dated Dec. 5, 1901, showing Murrill's address, request, handwriting and signature. (Courtesy New York State Museum)

Parenthetically, Underwood was well-established at Columbia by 1901, and used his position as editor of the Bulletin of the Torrey Botanical Club (Underwood 1898a, b) to lambast some European authors, a lesson not lost on Murrill. Underwood was also mentoring his first post-graduate student, Violetta S. White (1901) on a group of fungi hardly considered in his sphere of interest.

Murrill must have thought a request to a luminary such as Peck required more than his name, for he ended his very brief message—"I am hard at work on fungi. Please send me all your available publications"—"Formerly Asst. Crypt. Botanist Cornell." Peck, as was his way, acquiesced with numerous papers and reports.

In his role as assistant to George Atkinson (Appendix 1) at Cornell, Murrill had collected bracket fungi, which were common and greatly varied in form. He later wrote that this experience was the start of his own interest in the group. Now in

New York and soon with access to Underwood's herbarium, microscope and library—Columbia's herbarium was of little use in fungi—he took up the group in earnest, and his first mycological paper (Murrill 1902b) paired his activity in the Torrey Botanical Club and his polypore research.

Soon after the 1900–1901 school term opened, Murrill began to ferret out the possibilities of a European trip for the summer of 1901; "This realization of a dream came much sooner than he had expected. …[he] attended the Botanical Congress at Geneva and the Zoological Congress in Berlin." Now, the only problem with this is that there was no recorded botanical congress in Geneva in 1901 through 1903 (Nicolson 1991; Briggs 1991). In fact, an important 1900 Congress in Paris had postponed nomenclatural questions until the next Congress. The question lingers; what meeting did Murrill (1945a) remember for his autobiography?

His trip to Europe in summer, 1901, is vouchered by two collections at the Garden, both made in early August, but neither well-annotated. The first, from Sweden, was noted as *Collybia* on inside notes, but *Clitocybe* on the specimen label. The second, gathered at the Tower of London, was annotated as *P*[*olyporus*] *squamosus*, a common early summer fungus. Less than a month later, three additional specimens (MycoPortal) indicate his presence in Falls Church, Virginia, probably at the home of Edna's family. For his first collecting season in New York, it was either sparse or unrecorded.

Franklin Sumner Earle (Appendix 1), Professor of Biology at Alabama Polytechnic Institute in Auburn, was about to be invited to join The New York Botanical Garden to curate fungi and plant pathology, but perhaps he was not aware of this turn of affairs as he went in the field on New Year's Day, 1901. It was a good day, and he returned with numerous specimens, some of which were gilled fungi, theretofore not his specialty. He continued to spend time in the field through the "winter" and spring, but by September was in New York getting settled. On September 20 (Friday), he took a stroll around the Garden's grounds (Figs. 3.3 and 3.4), gathering several specimens, mostly higher fungi. Sensing it was a good season, he went out again the following Wednesday, making an equally good haul (MycoPortal NY). He would soon cross paths with Murrill.

An adventure into fleshy fungi by Underwood (1898c) had really been an exercise in nomenclature, taking issue with Berlin mycologists. Now a mature 48 (Underwood 1900), well-settled at Columbia, a member of the Board of Scientific Directors of the NYBG (Britton 1908a), and a reputable nomenclaturalist, he felt free to pen a diatribe on several American and European cryptogamic botanists who doubted his conclusions.

Murrill (1902b): "It is evident, even to the beginner in mycology, that the present classification of the higher fungi is not satisfactory." With this, he reported what he had found in a year of research and challenged the mycological community to do better. "Before this large group can be reduced to order, full descriptions must be had of all its species, the type specimens in Sweden, England, Germany, and elsewhere must be examined, and sufficient knowledge gained of the life and habits of each plant to determine its natural position in the group." Without a particular taxonomic reference, he accepted and promoted the "type method" put forward in the

Fig. 3.3 Garden's Museum Building under construction. Construction underway on the Garden's Museum Building, circa 1900. (Journal of the New York Botanical Garden)

several American nomenclatural codes, not yet accepted by the Europeans, but in which Underwood played an important role. Perhaps not widely recognized, Murrill's career as a mycologist had begun. Simultaneously, the paper was reprinted as a contribution by The New York Botanical Garden although Murrill had no appointment there. It is no coincidence that Underwood (1902) published a short paper on the same fungus group.

The reason why Murrill's mycological paper was first read to the Botanical Society of America at its June, 1902 meeting in Pittsburgh, is unknown. Perhaps they expected a paper on ferns. Presciently, though, they heard proposal of several new species, a trait which would become characteristic of Murrill through the rest of his mycological life.

1902 was an active year for collecting as well. Surely the most intriguing cache was made by Mrs. Edna Murrill in the vicinity of Falls Church, Virginia. A few specimens were gathered in early April, but 30+ in late October. No specimens from Virginia bear William's name in those periods. An educated guess might be that Edna visited with her family while William remained in New York and travelled about Europe.

Fig. 3.4 The Snuff Mill. The Tobacco Mill on the Bronx River in the Lorillard estate, later called "The Stone Mill," and "The Snuff Mill." (New York Botanical Garden collection)

Murrill's field activity was also productive. During his summer in Europe, a few specimens were collected in Austria and Kew, but he returned with 25+ from the environs of Uppsala, Sweden, with all collections identified by Lars Romell, a "student" of Fries (Petersen and Knudsen 2015).

In 1902, F. S. Earle (Appendix 1) was Assistant Curator of Mycology and Plant Pathology at the Garden. He and Underwood had shared a year on faculty at Alabama Polytechnic Institute (now Auburn University) where they had published two papers, one minor, the other significant (Underwood and Earle 1896a, b, 1897). Now, with Underwood at Columbia, the two were reunited and possibly through Underwood, Murrill and Earle became friends and field mates. In June, they spent a day in Van Cortlandt Park in the Bronx and on successive Saturdays in September collected on the New Jersey side of the Hudson River in Fort Lee atop the Palisades and Edgewater just below them (MycoPortal NY). Today, such a trip would take a half-hour automobile ride from northern Manhattan, but in 1902 (30 years before the George Washington Bridge), a 45-min ferry ride was necessary just to cross the river. Later that summer, Earle was in Redding, Fairfield Co., Connecticut, for some 3 weeks, depositing a large cache of specimens, mostly higher fungi, at the Garden

(MycoPortal). Underwood owned a small farm in Redding, and Earle may have visited there.

Following a paper describing new phyllophorous fungi (Tracy and Earle 1901), for Earle, 1902's highlights were again tropical. In June, Earle spent a short time in Cuba, but a Garden excursion to Jamaica for 4 weeks in October and November produced a large gathering of specimens and a report on his trip (Earle 1903). As with other such explorations, "The trip was undertaken on the invitation of the … Director of the Jamaica Public Gardens, for the purpose of investigating certain diseases of logwood, coconuts and other economic plants. My instructions were also to secure living specimens of tree-ferns for the conservatories, and to collect fungi and other cryptogams for the herbarium. …Forty-five specimens of tree-ferns representing fourteen or fifteen species were secured and forwarded to the Garden. Six hundred and thirty herbarium specimens were collected, of which nearly five hundred were fungi, the remainder being lichens and mosses with a few fresh-water algae." For his Jamaican hosts, more important were descriptions and suggested remediations of several diseases of economically important trees. It was the first of several Jamaican expeditions by Garden staff.

It is likely that Murrill became aware that his understanding of polypore taxonomy and nomenclature expressed both in print and in person was beyond ordinary education and he felt obliged to summarize the generic names applicable to these fungi (Murrill 1903e). The paper was read before the Botanical Society in Washington in December, 1902, not long after his work on *Ganoderma* appeared. The latest offering, surely aided by his second summer in Europe, outlined every genus name chronologically and alphabetically. For one used to such things, the paper resembled such resumées of Nomina Generica Conservanda Proposita penned by Donk (1949) some 30 years later. This sophisticated exercise appeared in JOURNAL OF MYCOLOGY, William Kellerman's (APPENDIX 1) niche journal specifically for mycologists. Within the same year, five more papers in the series on North American polypores appeared (Murrill 1903a, b, c, d, e). One step further, Murrill (1903f) took less than a page to propose a new family, the Xylophagaceae.

In spite of cooperation in bringing the Botanical Garden to life, to a certain extent the Torrey Botanical Club, Columbia's Botany Department, and the Garden were rivals. Columbia's botanists had access to the Garden's facilities, which by then included Columbia's herbarium and botanical library. But Columbia's botanists also had access to the Garden's laboratories. The Torrey Botanical Club had a large membership, a thriving BULLETIN, of which Lucien Underwood at Columbia was Editor (succeeding N. L. Britton, also at Columbia, but by then Director of the nascent Garden). To the BULLETIN had been added TORREYA, with the Garden's Marshall Avery Howe as editor. Even under Underwood there was a small roster of associate editors, of which Britton, Howe and MacDougal came from the Garden, and Underwood, Curtis and Lloyd from Columbia. When John Hendley Barnhart (Garden) succeeded Underwood as Editor of the BULLETIN in 1903, Murrill's name appeared as one of the associate editors. Aside from his membership in the Club, he was unaligned with Columbia or the Garden, but his name now appeared nationwide—not bad for a high school teacher in lower Manhattan. He remained a co-editor of the BULLETIN until 1911, 2 years after he started MYCOLOGIA.

THE AMERICAN MYCOLOGICAL SOCIETY. It was the "season" for organizing independent societies to represent groups of scientists of related interests, and, although these days it seems like an ethereal ghost, the mycologists were no exception. It is tempting to conclude that this was the birth of the Mycological Society of America, but that history began in 1932, as well-described by Fitzpatrick (1932, 1934, 1944). Instead, at the annual meeting of the AAAS in Washington, DC, in December, 1902, a very small group of mycologists informally discussed the advisability of forming an organization. The sentiment "proved to be strongly in favor of such a movement, but it was felt that the matter should be carefully looked into before decisive steps were taken. Accordingly, a committee on organization, consisting of Messrs. Earle [NYBG], Shear [USDA, Washington], and [Edward] Clements [Nebraska], was appointed to correspond with the mycologists of the country, and learn their views on the project." The Committee huddled to draw up the nationwide list of mycologists, numbering twenty-seven. Response to an April letter of inquiry was enthusiastic; arrangements were made with AAAS to allow a meeting at the next conference, December 1903, in St. Louis. Mycologists there were 16, and upon election, the following slate of officers was approved: President, Roland Thaxter [Harvard]; Vice-President, Franklin Earle [NYBG], Secretary-treasurer, Edward Clements [Nebraska]. The American Mycological Society had been born, and all attending mycologists were considered charter members. A full slate of papers was read at a special session (Shear 1904). Earle's paper was entitled "The necessity of reform in the nomenclature of fungi," the effects of which would be far-reaching. The paper, as such, was not published. Murrill's (1903f) paper was coincidentally read at the self-same meeting, and, although he might have considered it a taxonomic offering, it was just as nomenclatural as Earle's

It was decided that the new Society should be affiliated with Section G of AAAS, to give it a home and a place at annual meetings. There it remained. Shear (1906) reported on the third meeting of the Society, where, in the absence of President C. H. Peck, Vice-President Earle presided. A committee was appointed to meet with two other societies with the intention of merger. "The new constitution recommended by the committee of the Botanical Society of America, the Society for Plant Morphology and Physiology, and the American Mycological Society, as a basis for the union of the three societies, was adopted and the present officers continued as a committee with power to co-operate in completion of the details of reorganization." And with this, the American Mycological Society ceased to exist. Parenthetically, Cornelius Lott Shear, one of the AMS founders in 1903, was the father of the American Phytopathological Society, which organized just 5 years later.

Atkinson (1897a, b, c, d, 1899), at Cornell, had already published the first two of his illustrated mushroom guides and now published a revised and augmented edition (Atkinson 1901a). An intriguing parallel can be seen if Murrill's interests and publications over this period are matched against those by Atkinson. As mentioned above, as a student Murrill had collected polypores for Atkinson, and now Atkinson wrote some papers on similar subjects as some by Murrill: Atkinson (1900a, 1901b, c, 1902) on wood- and tree-destroying fungi, some especially on shade trees. Atkinson's reports were of a more popular tone than Murrill's, and soon diverged into studies on "lower fungi," but a certain audacity might be detected in Murrill, the fledgling mycologist.

Murrill's friends, Underwood and Earle, were engaged in pursuits of their own. In 1903, Underwood was granted a leave from Columbia from February to October for a fern collecting tour through the West Indies, with major stops in Jamaica and Cuba as well as smaller islands. Scouting the area for future exploration was also important. Both Cuba and Puerto Rico had been Spanish colonies, but now were either under U.S. occupancy (Cuba) or a U.S. territory (Puerto Rico), but whether exploration of Jamaica and the Lesser Antilles was possible hung in the balance. The leave was emblematic of the 1895 agreement between Columbia University and the Garden—the Columbia Professor and Department Head on leave to undertake the Garden's mission. The trip was to be attenuated by a summer in Paris, Kew and Berlin for critical herbarium work. In February, Earle joined Underwood in eastern Cuba for a month to collect fungi along the same lines as those he had procured in Jamaica previously. Cuba also withstood two invasions by the Brittons during the year.

Underwood returned to New York in mid-May. "A collection of about three thousand numbers was made in Jamaica…about four hundred species of ferns were collected, some of which are known to be new to science." All the ferns, alive and dead, required curation—the living at the Garden, the preserved at Columbia.

In early May, 1903, Earle was granted a leave to aid the USDA in Puerto Rico. His mission was "…. a general examination of the island for the purpose of determining the principal diseases affecting the agricultural and horticultural products." While taxonomy was essential, Earle's Garden title in phytopathology was not only utility, but the trip probably sealed his future residency in the Caribbean tropics. It must have impressed Murrill that all this travel was made possible by funds available through the Garden, while the money for his own trips to Europe came from his pocket.

Quite parenthetically, it was notable that Curtis Gates Lloyd of Cincinnati, Ohio, paid the Garden a visit in the spring in order to study the fungi in the herbarium. At that time, he could have been interested in gasteromycetes or the ascomycetes of the Ellis and Everhart collection. Soon enough, he would show his displeasure with Murrill's brand of nomenclature, but in 1903 they probably did not know each other, although Murrill's historical lexicon of polypore generic names might have raised an eyebrow (Lloyd library 2023). Although not accomplished for another 15 years, Lloyd already envisioned a library and herbarium in Cincinnati.

THE CITY. Murrill's commute from home to work each weekday was becoming more arduous: Manhattan was importing tens of thousands each day for work downtown, especially in the financial district.

By 1903, the New York Stock Exchange had grown so large and complex that it needed a new home (again), this time into specially designed quarters on Wall Street. Always, the newest modes of communication were adopted: "ticker tape machines," typewriters, and telephones with rotary dialing devises just introduced.

From a gathering of 24 New York stockbrokers who signed the Buttonwood document in 1791 (NYSE history 2023), the number of persons employed by the financial sector and its orbit grew and grew, until thousands of employees gathered daily, with the financial district expanding geographically slowly but increasingly congested. Outside the Exchange itself, "curbstone brokers" bought and sold shares in young, upstart ventures like railroads, steel, textiles and chemicals, the very stuff of enterprising young developers (American Stock Exchange 2023).

By the end of the nineteenth century, hordes of commuters flowed into lower Manhattan each morning, and out late on each workday. While horizontal expansion was difficult, upward expansion was commencing—the start of "skyscrapers." Women were not hired for the trading floors, Blacks were absent except as laborers, and Jews were frowned upon. Certainly, this sub-city, "cheek to jowl" with the nearby neighborhoods of squalor revealed by Jacob Riis's photographs in "The Way the Other Half Lives" (Riis Photos 2023) made the City difficult to navigate as well as to govern.

MycoPortal data from the Garden for 1903 would indicate that the year was less than productive for fresh collections by Murrill. In May, Falls Church was again represented, but the summer yielded only a half-dozen European specimens. Of note, however, was one determined by Giacomo Bresadola, considered the dean of European higher fungi, and another collected by both men together (Fig. 3.5) in the "Austrian Tyrol" (perhaps the Tyrol in the neighborhood of Trento). Switzerland and Kew were also visited, but Christmas was spent between Falls Church and Blacksburg, family territory. A couple details: (1) Printed, dated specimen labels were used for selected specimens, but probably the species binomial was stamped later; (2) Murrill felt free to use pre-printed New York Botanical Garden specimen labels—technically permitted since that was the destination of the specimens, but ignoring the possibility that Murrill kept his own herbarium, or perhaps deposited specimens with Underwood or the Columbia herbarium. One way or another, the practice bespoke a certain familiarity—Murrill had become one of the New York "botanical fraternity." He was a married man of 34, spending his third successive summer in the herbaria of the capitals of Europe.

Soon after his in-person friendship with Lucien Underwood was taking shape, Underwood extended use of his microscope, library and herbarium to Murrill. Surely with mutual agreement, Murrill had begun a study of polypores. Many years later, Murrill (1945a) remembered Underwood fondly: "I collected ferns for Dr. Underwood in Virginia several years before I met him. He was responsible for my first articles on polypores and very probably my position at the Garden."

In January, 1902, Murrill wrote to Peck: "I am doing some monographic work on *Fomes*, *Polyporus* and Co. and desire all your available reports as well as lists of all your species of the Polyporaceae exclusive of *Boletus*. If this is too much trouble,

Fig. 3.5 Giacomo Bresadola & William Murrill in Italy. Formal portrait of Giacomo Bresadola and William Murrill. (Journal New York Botanical Garden, 1906)

please let me know if Scaccardo has overlooked any so that I may get all American species correctly."

By early twentieth century standards, Murrill was on the right track. Over the academic year, his daytime was spent commuting and teaching, and his evenings in study at Columbia. Was there a home life? Specimens passed under his gaze and Underwood's lenses, notes were taken, and polypore data grew. That was taxonomy, but it was soon clear that organismal names required library study as complicated as the morphological work on the specimens. Some questions could be answered in Underwood's library, but many could not, and even if only the North American species were to be covered, if the "type method" was to be obeyed, a trip to Europe—and another and another—was imperative.

By mid-1902, Murrill (1902b) felt ready to share the results of his labor. The first chapter was *Ganoderma*, which, by the way, was a genus name with a checkered past. Of the seven species treated, five were new and two were transferred. The paper was presented to the Botanical Society of America by invitation in June, before he embarked on his second European pilgrimage, and was printed in October, after he had returned and was back in the classroom. He generously acknowledged help: "A large number of the specimens examined have been collected during several years past by Professor Lucien M. Underwood, and by various contributors to his private herbarium. The extensive literature and large collections of The New York Botanical Garden and Columbia University have also been most generously placed at the writer's disposal." He went on to thank numerous mycologists in Europe for "many courtesies" extended during his time there.

The parade of genera marched on through 1903: *Pyropolyporus* in February; *Fomes* in April; *Elvingia* in May, and *Cryptoporus*, *Piptoporus*, *Scutiger* and *Porodisculus* in September (Murrill 1903a, b, c, d); with the historical review in May (Murrill 1903e).

THE HIRING. Surely, Franklin Earle's comfortability with the tropics had been displayed during his pre-Garden career, and he had come to the Garden from Alabama. During his time at the Garden, he had joined Underwood in Puerto Rico and Cuba, and now, in March, 1904, an announcement appeared that he was taking a leave at the request of the Cuban government, to aid the USDA in the establishment of a Cuban department of agriculture. Although the Republic of Cuba was mentioned in the announcement, all such developments were subject to Washington's approval. "The plan contemplates the establishment of one or more thoroughly equipped and thoroughly manned agricultural stations in Cuba, similar to those which have proved of such value in the development of the United States, together with an agricultural and botanical survey of the entire island" [J. New York Bot. Gard. 5(51 March): 63)].

Apparently, Earle's leave doubled as a prolonged job interview, for upon return, his attitude must have seemed buoyant and he must have conferred with Britton, for less than a month after his return, in late April he tendered his formal resignation to accept appointment as Director of the Experiment Station of Cuba (Anonymous 1904). No one wrote real-time commentary, but the Scientific Directors and Board of managers must have been alerted, surely Nathaniel chatted with Elizabeth. After all, Earle's move could provide an ally in Cuba, a place of great interest to the Garden Director (and his Cuban-raised wife). But just as exigent as Earle's resignation seemed, a Garden budget line existed at Assistant Curator rank, and the endowment ought not be squandered. Who should replace him and in what discipline?

From the vantage point of 40 years later (Murrill 1945a), the story was simple: "Professor Earle was then at The New York Botanical Garden working on fleshy fungi. In 1904 he was appointed Director of the Cuban Experiment Station and the Naturalist was chosen to fill his place at the Garden as Curator of the Fungi. Another great surprise!" Not an acceptance, just a surprise at being asked: an opportunity for full-time research, perhaps collecting excursions or herbarium consults, little or no teaching responsibilities. Or could such a position actually present restrictions in tasks, furnishing identifications for the public, dealing with collections poorly preserved but treasured, and a director who appeared duplex—one facet facing the public (through the press) and philanthropy, and another facing the staff? And how did Edna fit into this? Only time could tell.

In 1945, he looked back on the Garden job as fulfillment of his preparation. He wrote: "This position at the Botanical Garden was perfectly ideal. If the Naturalist had not gone to Cornell he would not have been prepared for it. If he had not taken a position in New York he would not have been in touch with the men who could help him most. If he had wasted his money during vacation instead of using it to further his scientific studies he would have been unworthy of such an opportunity.

The work at the high school was always considered temporary—only a means to an end. The work at the Garden was something he wanted to do—and at the same time he was going to get paid for it. How different life would be if all jobs were like that!" Of course, at each milestone noted, he had chosen the "high ground" and implicitly, self-congratulations were earned.

Parenthetically, he was not accurate about the title of his new appointment. Earle had been Assistant Curator, and Murrill was so-titled on at least two occasions [J. New York Bot. Gard. 5(54 June): 126; 5(57 Sept.): 191].

Surely based on interviews in late winter, the flurry of published official announcements took place in late spring through summer, and for the first summer in 4 years, in 1904, Murrill remained domestic. Herbarium labels at the Garden (MycoPortal) indicate his whereabouts. The fourth of July holiday was spent in Falls Church, Virginia, probably with Edna's family; somewhat later there appears "Crabbottom," a modern location no longer identified, but surely in Virginia, for the end of July notes Blacksburg, his family homestead.

The major summer excursion, however, was to upper East Tennessee, where the Appalachian Mountain chain merges with the Blue Ridge. This region was (and is) known to be more northern than southern, where cranberries are cultivated in true bogs. In 2 weeks, he collected numerous fungi (Murrill 1904e) published in October, the best collecting was at Unaka Springs, in Washington County, Tennessee. Here were Edna's family roots before their move to Falls Church. Just as Lynchburg, Virginia, was on the James River, so Unaka Springs was on the Nolichucky River in Tennessee. Return to New York was by way of Bedford and the Peaks of Otter, Virginia. By this time, Murrill's professional address was noted as New York Botanical Garden, and the paper included some attractive line drawings. What did Nathaniel Britton think of Murrill's first month on the job, at home with family and old friends?

For Murrill, several factors automatically shifted with his new position. His daily commute turned 180° from downtown Manhattan to Bronx; from crowded elevated train to surface train with fewer stops (one of which was Bronx Park); from urban crush to suburban quiet. The temptation was to move closer to his new work. Further, he would have an office of his own, and access to a growing library and herbarium. Access was not new (one announcement was worded: "Dr. W. A. Murrill, who has carried on extensive investigations upon the fungi at the Garden during the last four years …"), but he had had no formal connection to Columbia either. Now, his Garden "laboratory" could be exclusive or community-based, but well-equipped for his new duties. Perhaps now he could leave a book open or a specimen unaccessioned overnight (Fig. 3.6). Finally, although he had rubbed shoulders with the Brittons and the staff for the last "four years," he now could be an equal around the luncheon table (Daly 2016) with curatorial staff and others. The staff had grown since 1901, with Hollick and Howe added, and personalities had evolved as well. Burlingham, Banker, Grout, Gleason and others were students.

Soon added to the regimen of research, an announcement: "The remainder of the autumn course of lectures will be given as follows: … October 22 [Saturday], "Life

Fig. 3.6 William Murrill at his desk at the Garden. William Murrill, First Assistant, at his desk at the Garden, circa 1910. (Courtesy Jim Kimbrough 2003)

history of a fern," by Professor L. M. Underwood; October 29 [Saturday], "Botanizing in the Austrian Tyrol," by Dr. W. A. Murrill…" Thus was Underwood's research sphere stated, and Murrill's most recent trip to Europe (before his appointment) advertised.

As it turned out, Murrill's relationship with his boss was at least as complex as the one with his colleagues. Murrill (1945a) assessed Britton this way: "Others unhampered by exacting demands on their time by social duties, political contacts, appeals for funds, dealings with architects, engineers, city and corporation officials, and a thousand and one details appertaining to the establishment, financing, and development of a great institution may look with equanimity on their productive output in publication. Yet one wonders how many of them could have continued to be productive had they been faced with the multitudinal problems, large and small, that Doctor Britton met day after day, week after week, month after month, year in and year out, and still find time to produce technical paper after technical paper." His drive to make the Garden more prestigious than its closest domestic siblings—global prominence could not be far behind—was within sight. But already, the income from the endowment had shrunk somewhat with inflation, so finances were a continual thorn. The bucket could not be lowered into the well too frequently. Given the times and Nathaniel's position, even Elizabeth could not be put on staff in spite of her botanical expertise, especially in bryology. Her botanical roles were gratis, but her biopolitical skills were powerfully felt in other arenas.

The thirteen chapters in Murrill's Polyporaceae of North America series appeared in print periodically until mid-1905 (Murrill 1905a, b, c, d, i). It took 63 generic names to complete the list, the largest (*Polyporus*) with 23 accepted species, and another 39 as monotypic. Thirty-one new species were introduced and 104 epithet transfers made. His New York friends were not forgotten: two species epithets for Underwood, and a genus based on Cuban material, for Earle. Sensing that a more popular approach might be necessary, keys (Murrill 1904f, 1905e, f, g) to selected polypore groups were interspersed. Murrill, in his virgin oeuvre, had set the stage for the judgement of the mycological—and botanical—community for the rest of his career. In a biographical sense, the polypore publication chain, based on a single, large, mass of data surely gathered early in (or before) the publication process and surely subject to occasional tinkering as time passed, spanned and functioned as another of the inflection points in Murrill's life. They appeared as though done in his Garden job, but in truth, were surely compiled largely before his appointment.

CUBA. 1905. Franklin Earle's ship had hardly made berth in Havana, and the ink hardly dry on Murrill's appointment than: "During the following winter an invitation was received from Professor Earle to come down and collect fungi with him in Cuba." His report was written in a form by then routine, a memorandum addressed to Britton. The only mention of finances: "We were together during the whole expedition, sharing the expenses and dividing the specimens." "At times we were ably assisted by Mrs. Earle and Mrs. Murrill, who not only did their share of collecting but also helped to care for the plants during the tedious operations of drying and packing."

Murrill was careful to mention prior botanical survey work by Ramon de la Sagra, whose fungal collections had been arranged and published by Camille Montagne in Paris, and Charles Wright, whose fungi had been worked over by Rev. Miles Joseph Berkeley at Kew. This by way of justifying more European investigations to examine those specimens.

"Nothing could be more delightful than this quick transition from the clutches of a long and unusually severe northern winter to the fresh and balmy air of a tropical island continually swept by sea breezes. … It is safe, I think, to predict that within five years Americans will own three-fourths of all the land that can be bought in Cuba and that American capital and intellect will control to a large extent the business enterprises of the island." True in the short term, but not prescient over the next century.

Collecting was accomplished in the far west and far east of the island. Considering the dry season weather, over 2500 fungal specimens were processed, to be shared between the Agriculture Station and the Garden. Relatively few fleshy types were seen, most fungi being woody pyrenomycetes and polypores. The Garden specimens were deposited under Earle and Murrill.

CHESTNUT BLIGHT. Murrill's apparent career as a mycology researcher was interrupted in spring, 1905, by a visit from Hermann Merkel, Chief Forester at the Bronx Zoo, which shared Bronx Park with the Garden. Merkel (1906) had observed a disease of the planted American chestnut trees at the Zoo in summer, 1904, and scattered trees had died (Fig. 3.7). Not trained in phytopathology, Merkel had submitted a report and samples of diseased branches to Flora Patterson, mycologist at USDA in Washington. She had diagnosed a fungus, a species of *Cytospora*, previously unknown from chestnut. Now he came to Murrill, with phytopathological training at Cornell, and more convenient for consultation than Washington. Murrill (1906a) went to the Zoo, observed the situation, returned with specimens and set out to culture the fungus. He also sent a specimen to Charles Peck, State Botanist of New York, to find whether the fungus was previously known in the state. This activ-

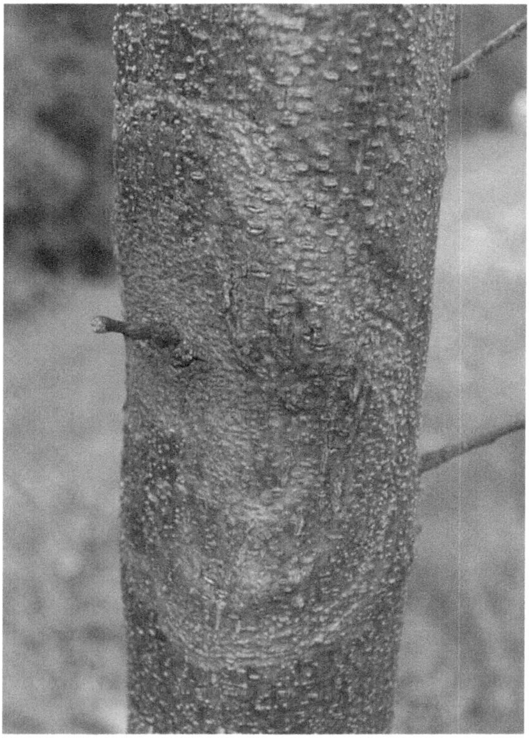

Fig. 3.7 Active Chestnut Blight infection. Prominent colony of *Endothia parasitica* on the trunk of a living chestnut tree. Note dead twig at the center of the infection. (Wikipedia)

ity took place in spring, 1905, but Murrill was scheduled to be away from the City for much of the summer and early fall and stuck to his plans (see above).

Indeed, Murrill's years at Cornell had provided a firm basis in plant pathology, both through the campus courses and the Experiment Station. There was general espousal of the "postulates" set forth by Heinrich Hermann Robert Koch (1876, 1893) in Germany just at that time, and Murrill intended to prove the cause of the chestnut disease by applying these principles. Once he was satisfied, he described *Diaporthe parasitica* (Murrill 1906b, c). Meanwhile, both Murrill and Merkel were alarmed at the prospect of the possible catastrophic loss of eastern North America's dominant hardwood forest tree (Fig. 3.8).

By mid-1906, the disease was known from New York City, surrounding New Jersey, Maryland, Virginia, Washington, DC, and perhaps some southeastern states (Murrill 1906b, c). Murrill gathered first-person observation on the train between New York and Albany to visit Peck (Letter, Murrill to Peck, June 20, 1908; NYS).

Sadly, but truthfully, no adequate remediation could be advised. It was certain that trees, both diseased and not yet symptomatic, should be sacrificed, immediately used for their wood, and all other parts burned.

Fig. 3.8 Murrill on a dead Chestnut tree. Figure thought to be William Murrill standing on the prostrate trunk of a huge harvested Chestnut tree. (Courtesy Jim Kimbrough 2003)

Much later, Murrill (1945a) spun the story with unabashed acrimony. "In Pennsylvania a plan was formulated to spend many thousands of dollars to fight the fungus which I had already shown by experiments to be a hopeless undertaking. The Washington pathologists were wholeheartedly for it because they were ignorant and furiously jealous. Only Clinton and Stewart stood by me at the Harrisburg [Pennsylvania] Convention, and we three were ridiculed beyond measure." Murrill was invited to address the Convention after its adjournment: "The address was listened to with great interest by delegates from other states where appropriations were contemplated. My knowledge of the disease appeared to them so complete and my manner so earnest that not a single other state joined Pennsylvania in its useless and wasteful scheme." The gist of this process has been recognized to this day (Brown 2022).

Although the spring had seen hyperactivity over the "chestnut canker" in New York, Murrill commenced on his field season nonetheless. His first stop seems to have been the tiny settlement of Ohio Pyle (now Ohiopyle), in far southwestern Pennsylvania, a place whose major attraction was a crossing of the Youghiogheny

River. These days, it advertises a population of 38, and acts as the gateway to the Laurel Wilderness to the north. Only a few days were spent there (July 3–8), apparently after some dry weather, but some valuable specimens were gathered. Only circumstantial evidence might indicate Edna's presence.

Possibly more attractive, however, was Veitch's Summit (now Summit-Veitch), Virginia, July 10–17. Even then, it was adjoined to Fall's Church, Edna's family's home. Herbarium labels attest that there had been recent heavy rains, and the prodigious numbers of (and names on) specimens reflect excellent collecting. Two locales were visited (perhaps more than once during the inclusive dates on herbarium labels), a spot on the south bank of the Potomac River "above" Washington, and Veitch's Summit, itself. It would not be strange if he had company from the USDA contingent in Washington—Cornelius. Shear and Flora Patterson among them.

There is no evidence to conclude time around Blacksburg—no accessioned specimens at the Garden—but otherwise there is no indication of his return to the City except a note, "upon my return in August" in one of the papers on Chestnut disease.

Although mentioned only parenthetically in his autobiography, the Murrills moved from Morningside in Manhattan to Bedford Park, Bronx, just across the New York Central tracks from the Garden. His work was a direct stroll down Bedford Park Boulevard. It was to be very transitory.

A 1905 FROLIC. he main attraction of the season, though, was a protracted excursion into the river-and-lake territory of Maine. A USDA expert on the fungi of Maine, H. W. Ricker, invited Murrill on a canoe trip of some duration. Murrill's introduction to this skill must have been strenuous, for they paddled from Oldtown on the Pennobscot, Piscataquis and Sebec Rivers into Sebec Lake, outfitted with camping and specimen-preparation materials, upstream all the way. Both Pennobscot and Piscataquis counties are in central Maine, with Sebec Lake and Sebec River at the southern end of the present-day Appalachian Mountain Club 100 Mile wilderness with Mount Katahdin to the north (Fig. 3.9). After a sojourn on Lake Sebec shore, an overland trek was made from Willamantic to Boarstone Mountain and back. They canoed downstream to Milo, hence by railroad back to Oldtown. A list of the camps, by number and name, was attached, and can be matched to herbarium label data. Murrill (1905h) wrote: "I obtained about 1,500 specimens of fungi, many of them not heretofore reported from Maine and some of them undescribed." Further: "The rivers of Maine are low at this season, both from natural causes and because the water is purposely held back in the lakes. In addition, therefore, to the usual number of dams, falls and rapids to be carried, poled or waded, one must often drag the canoe through stretches of shallow water, usually rocky and swift and abounding in holes of uncertain depth (Fig. 3.10). These 'rips' are more fatiguing going up stream, but more dangerous to canoe and cargo coming down.... This part of Maine strongly resembles certain parts of Sweden, though other portions of Sweden furnish large agarics in great abundance."

One small observation is worth relating. "It was interesting also to notice that almost all of the wild cherry trees scattered through the woods were entirely dead and their trunks covered with the sporophores of an undescribed species of *Poria*. I have

Fig. 3.9 Canoing in Maine; Etching. Classical etching of a lake in Maine, with Mount Katadhin in the background. (Maine Tourist Bureau)

Fig. 3.10 Canoing in Maine. Canoer "riding the rips" on the Maine rapids. (Maine Tourist Bureau)

never before seen the disease so abundant and virulent except in the high mountains of Virginia where in certain sections every cherry and plum tree, both wild and cultivated, appears to have succumbed to its ravages." Another deadly fungus disease to add to Chestnut blight.

> One of the objects in mind during my trip was the gaining of all possible information throwing light upon questions of geographic distribution and the history of divergence in the North American and the European fungus flora. When the collections made have been worked over, our ideas concerning the distribution of certain northern species will doubtless be considerably clearer.

Murrill (1905h, 1906d) reported the Maine experience in muted words in TORREYA and the JOURNAL OF THE NEW YORK BOTANICAL GARDEN, but the memory grew in his mind. Forty years later he narrated as follows: "The first summer collecting trip [as Garden staff]... was to central Maine in a canoe with Ricker...Five delightful weeks were spent there just at the height of the fungus season and a large collection was made. Fish, young ducks, squirrels, buckwheat cakes and mushrooms constituted the chief food....The Naturalist returned to New York in September with much first-hand knowledge and in superb physical condition. Dr. Underwood had been a powerful man and still enjoyed a little horseplay at the expense of those who were easily teased but when he tackled the Naturalist after his Maine trip he was lifted bodily from the ground and shaken as a terrier shakes a rat, until he cried 'Hold; enough!' That's what the outdoor life and paddling a canoe through rough water can do for a fellow" (Murrill 1945a).

The second move of the Murrill residence in 1905 was to Bronxwood Park, a title used these days for the landscaped Pelham Parkway running east from the Garden. He (Murrill 1945a) described it idyllically: "...Bronxwood Park seemed even more attractive, so he settled there permanently in a large house with an extensive yard facing the tennis courts and bowling alleys. Here he entertained visitors, enjoyed his many interesting neighbors, gave lawn parties, grew dahlias and peaches, took part in the local sports and walked for miles through the adjoining fields and woodlands, often accompanied by a friend or his faithful dog." The singular language employed and pointed omission of Edna—in favor of "his faithful dog"—are, as usual, telling. In the same document—his autobiography, but at the other end of the document—he revisited the sentiment, but with quite a different twist. See below.

PROMOTION? Although the calendar might have rolled over to 1906, a couple timely items were begun in late 1905. Murrill was probably hardly home from Maine when Daniel Trembly MacDougal submitted his resignation to Britton. The Andrew Carnegie desert laboratory outside Tucson was not part of the Garden orbit in location or support, but MacDougal had fallen in love with the desert flora and climate and the job of director was inviting. But his parting would leave a vacuum. Britton (1906), in his usually formal prose, put it this way: "The resignation of Dr. D. T. MacDougal as Assistant Director has brought about a reorganization of work at the Garden. Dr. W. A. Murrill, who has been serving as a Curator for parts of two years ... has been appointed First Assistant, and the duties of Mr. Percy Wilson, Administrative Assistant, have been increased." So although the word promotion would certainly not be out of place, the parameters of the job were not conveyed, except to conclude that some tasks had been added to the work of the Administrative Assistant—Britton's secretary. An interesting start for the new year, carrying with it the JOURNAL Editor's job. Work added, work transferred (Fig. 3.11).

Murrill (1945a) retrospectively discussed at length this "opportunity." Having described his qualifications (not for the first time), he wrote: "But this added honor was not without added burdens. The carefree life of the investigator was a thing of the past. Responsibility loomed large. He had to think for others as well as for himself. In other words, he was in danger of being caught in the gilded net which had blighted the prospects of so many men who could have done splendid research work if they had only been left alone. Tempted by a little prominence and a few extra dollars they handed over their birthright and traded the immortality of the original

Fig. 3.11 William Murrill, formal portrait. Formal portrait of William Murrill, upon promotion, circa 1908. (New York Botanical Garden Archive)

investigator for the temporary and hollow smugness of an ordinary executive. However," he added pensively, "there was considerable freedom at the Garden and research was esteemed and encouraged. Perhaps the danger might be averted for a time at least."

TASKS. Editor of the Journal of the New York Botanical Garden; Public lectures, often with large attendance by children; maintenance of the grounds and buildings; oversight of routine work of over a hundred laborers; eradication of fungus and insect pests; "Garden clubs asked for lectures, as did many scientific institutions; there were various celebrations and meetings where a Garden representative was expected… A public executive must be a politician in making promises but a saint in fulfilling them. His attitude should be charitable, unselfish and sympathetic, like that of a born teacher.

"In the long absences of the Director in the Tropics, the Naturalist took over all of his duties in addition to his own; such as the supervision of new construction, the handling of checks and bills with the aid of the accountant, a large and growing correspondence… To meet these extra demands on his time [the Naturalist] formed the habit of coming over very early and doing scientific work before the building was opened to the public. In the quiet of the morning, after a good night's rest, his mind worked smoothly, rapidly and efficiently. Many of his manuscripts were prepared during these early morning hours.

"He tried not to forget, too, that all his supervisory work was merely subsidiary to his real life work, which was biological research, with the accent for the time being on the fungi" (Murrill 1945a). Compare this sentence to the "gilded net" above; also the "temporary" study of mycology to later three decades describing Florida fungi.

MYCOLOGY. As though to test the elasticity of the new promotion, the second holdover was a two-page paper on a unique group of polypores (Murrill 1905i). Problems were manifold. The name of the group (Agariceae), although perhaps dictated nomenclaturally, led the reader to conclude that the paper dealt with mushrooms. Murrill tried to correct this conclusion, but the presence of *Agaricus* as a genus of polypores was doubly confusing. Moreover, using a pseudolamellate hymenophore as the leading separating character was an invitation to error. Five genera covered the lot, but this small publication, in this less-than-august mycological journal was—merely—noted by Kellerman (1906) in his JOURNAL OF MYCOLOGY. His note merely lifted words from Murrill's paper to draw attention to the paper. But another pair of eyes also saw December's TORREYA and Curtis Gates Lloyd (1906a; Fig. 3.12) was incensed, not with the taxonomy but with the nomenclature, and specifically the use of ancient names unfamiliar to most mycologists. He wrote (at length here, unfortunately, but as an introduction to his view of nomenclature): "There has been so much changing of names lately in the Polyporii that I feel it well to state our position in this regard. The most and best systematic work on Polyporus was done by Fries. His system and names have been in general use for two generations and are familiar to all. We therefore feel that no attempt should be made to change them excepting in very exceptional cases. It has become quite a fad lately to look up dates of synonyms and shuffle the names around on such evidence. There is no merit in such work, and it produces nothing but confusion. One half of the old 'synonyms' are not true or are so vague that the truth can not be ascertained, and the other half are of no importance if they are true. This, of course, applies to

Fig. 3.12 Curtis Gates Lloyd. Curtis Gates Lloyd, posed portrait, circa 1920. (Courtesy, Lloyd Library, Cincinnati)

the species considered by Fries in his latest work. There are without question many reduplications of names.... As to genera, the question is not so simple. The genus Polyporus is too large and should be broken up, but I feel that as much of the old should be retained as possible, particularly the four leading sections with which we are all familiar....

"In Europe for the last twenty years there have been three men working on dividing the polyporoids into new genera. First Karsten, then Quélet, then Patouillard. Each has proposed his own system and his own names, and neither has met with much general favor, because, in my opinion, of the vast array of new names. Mycologists in general refuse to learn a new language in order to work with old plants. I think many good ideas are expressed in their work, but they could have been better received had they been used to subdivide the old series, not to replace them. In America, Mr. Murrill is a little late in taking up the work, for most of it has been done before—at least three different ways. To rechristen the ideas of his predecessors and further add to the babel of new names, is only making a bad position worse. As the European work has mostly failed to meet with favor for this very reason, I can foresee no other fate for the American. Most of my past work on the Polyporii has been in the line of collecting specimens, and sending them to authorities in both Europe and America for names." (Lloyd did not italicize Latin names).

Probably at the same sitting, he (Lloyd 1906b) took a more jocular stab. "ARCHEOLOGY—We recently visited Rome about the same time as our fellow-countryman, Mr. Murrill, of New York. There was a rumor among the guides that the object of his visit to Rome was a search for old botanical names in the catacombs." It was not to be the last such jibe.

Lloyd (1907) continued his criticism of Murrill somewhat later, this time picking on him in considered, serious words. In describing the chestnut blight situation in Italy, Murrill had noted that diseased trees could be found with fruitbodies of *Polyporus frondosus* at the base, but that local folk did not report the diseased trees to authorities for fear that the trees would be removed, and with them, the treasured edible *Polyporus*. Lloyd's criticism was not substantive, but quarreled with Murrill's use of *Polyporus*, when in his own (Murrill) prior publications, he had used *Grifola* as the genus name. This inconsistency, Lloyd thought, pointed directly to Murrill's habit of name-juggling. His parting words: "As he has apparently abandoned 'Grifola frondosus,' let us hope that he has abandoned them all [names of split genera], for he is too good a man to be hampered by such foolishness."

Intriguingly, Murrill (1906e) found himself in Paris in early summer, 1906. He wrote: "The cryptogamic herbarium at Paris, located in the building adjoining the celebrated Jardin des Plantes, is especially rich in polypores from tropical America, including the types of Montagne and others. Dr. [Paul Auguste] Hariot, the curator, extended me every courtesy. Here I found Mr. C. G. Lloyd at work upon the Gasteromycetes. He has made Paris his headquarters for some time because of its central location: Kew, Leiden, Berlin, Trient [sic], and other places of great importance to the mycological student can be easily and quickly reached from this point." One can only wonder about how genuinely friendly their conversation could have been.

Probably referring to Murrill's first part of polypores in North American Flora, Lloyd (1908) again sharpened his pencil to issue another blast at "The result of Kuntzeism." Lloyd sought help from a German review of Murrill's paper, in which Sydow reported: "...thus about nine-tenths of all species mentioned bear the name of 'Murrill' as author." Lloyd countered: "... [Murrill] added his name to 332 out of the 409 species that he considers. ...The only effect it will have will be to fatten the already overgrown and unwieldy synonyms."

THE CRISIS OF 1906-1907. It is hard to believe that the San Francisco earthquake of April, 1906, powerful as it was, had serious consequences for Lucien Underwood and the Garden in New York. A straight line, no, but indirect circumstances surely. On the West coast, San Francisco was a center of population and wealth born of the gold rush and its aftermath a half-century previous. The local banking system, however, was not adequate to generate the dollars to underwrite earthquake property losses, leading to migration of dollars from up and down West-coast banks to central California, and once the western banks were running low, they reached eastward to the financial capitals of the country, predictably New York. The financial wealth of the United States banking system of that day was put under a strain to deliver money for rebuilding of infrastructure and food on people's plates, and every dollar migrating from east to west decreased the liquidity of eastern banks. As the need for loans increased, so did interest rates, bank to bank, bank to public. Bank profits soared, to be sure, but the larder of eastern banks drew slimmer and slimmer.

But there was an added complication. An already wealthy western mining mogul, Miles O'Brien, previously a member of the Scientific Directors of the Garden, decided to "play the market" in mining stock. He sold a copper mine for $12 million dollars, but then gambled that mining stock would decline, and thereafter, a general decline. His selling transactions on the Stock Market were copied by some New York banks, occasionally with money they did not have, but had to borrow. At least one bank was rumored to be on the brink of bankruptcy, and depositors demanded their savings—a "run on the bank." Once one bank was in perceived economic trouble, "runs" spread to others, and in the banking community of the day, bank failure followed bank failure. Soon, the banking community was threatened and even the largest banks were questioned by the public—dollars were not adequate to meet the crisis, deposits were not guaranteed—bank failures spread and many depositors lost everything. By this time, 1907 was well under way. The "crisis" grew.

WORKING HOLIDAY. Having drawn attention to the added labor of the First Assistant upon the absence of the Director, Murrill absented himself from the Garden for a good portion of 1906. Presumably, arrangements for his European trip were already in motion while his promotion was finalized.

The report of his trip (Murrill 1906e) was titled: "A summer in Europe; some foreign botanists and botanical institutions" and lived up to its advertisement. Leaving New York May 29, the Murrills arrived in Gibraltar, thence to Naples by mid-June. An extensive touristic trip of the northern half of Italy produced short visits in botanical gardens. A week was spent in the company of Abbé Bresadola in the northern, walled town of Trento at the southern extremity of the Italian Tyrol (Murrill 1906f). Bresadola had become one of the leading publishing mycologists of Europe, with the added value of attractive aquarelle illustrations of his fresh gatherings. Several Americans made pilgrimages to Trento to see Bresadola's herbarium, but while Bresadola was polyglot, he did not speak English, so French or German were the lingua franca of conversation. He was over two decades older than Murrill, but led productive day-trips during their time together.

In France, it was the time of Narcisse Patouillard, the mycologist-pharmacist of Neuilly, Paris. He had recently published his thesis (noted by Lloyd, see above), a revision of classification of the higher fungi, based fundamentally on microscopic analyses (Patouillard 1900). He had already introduced his microscopic analysis methodology in a regional monograph (Patouillard 1884). Nonetheless, this approach, resulting in a serious reorganization of the Friesian system in general use, plus the French language, plus poor availability outside the French-speaking community, had made little headway. It was another 60 years before the work was "rediscovered," and more widely appreciated. Murrill estimated Patouillard's herbarium at 30,000 specimens, accompanied by 3000 or more original colored drawings.

Murrill's next stop was Berlin, where Paul Christoph Hennings was mycologist. Murrill (1906e) correctly commented: "Although a prolific writer, Professor Hennings seems almost overwhelmed with the immense collections of new material brought in from various parts of the world. His own types are numbered by hundreds, and the collection also contains much valuable material worked over by older authors." Parenthetically, much of the Hennings herbarium was later destroyed during World War II.

Over time, Murrill's (1945a) memory of Hennings soured a bit: "Hennings was an excitable, unreliable sort of person who perspired over dusty bundles of fungi sent in from the uttermost parts of the earth in the Berlin Botanic Garden when Engler was director. He inspired my distrust when he claimed to remember the spore measurements of scores of species. Maybe he did."

Two other workers visiting Berlin were mentioned: "[Arthur Louis Arthrurovich] Jaczewski of St. Petersburg (soon to be called Leningrad before returning to its traditional name), and botany professor Paul [Wilhelm] Magnus, by then on the cusp of retirement. A walk in the new botanical garden passed the mansion of Heinrich Gustav Adolf Engler, half of the team (with K. Prantl) producing the German DIE NATÜRLICHEN PFLANZENFAMILIEN, a magnum opus prototype used as a model for NORTH AMERICAN FLORA.

A visit to Sweden and Denmark was hasty, and interestingly, no mycologist was mentioned in Upsala [sic], although Fries's herbarium was open for research. In Stockholm, however, Murrill was welcomed by Lars Romell, a worker directly in the legacy of Fries (Petersen and Knudsen 2015). Copenhagen produced no working mycologist, but Murrill was able to examine specimens from the American tropics. More important was Leiden, The Netherlands, and the herbarium of Chistiaan Henrick Persoon, only recently put in order by Dr. Johannes Paulus Lotsy. In Amsterdam, the botanical garden director and famous geneticist, Hugo de Vries, was in the U.S., with New York on his schedule.

Just 2 months after embarking, Murrill took obvious pride in representing the Garden at the meeting of the Royal Horticultural Society in London. The conference took nearly a week, with travel to local botanical gardens, both public and private. Mycologically, though, 10 days at Kew were of utmost importance. He was under the leadership of George Edward Massee, mycologist and successor to Miles Joseph Berkeley and Mordecai Cubitt Cooke. Already, the Kew Herbarium was second to none, with worldwide collections still arriving. Cooke's splendid aquarelles gave an additional dimension to preserved material. In an important statement, Murrill wrote: "Mrs. Murrill and I remember pleasant hours spent at his [George Massee] home, near the Gardens."

> Shortly before leaving London, I visited Dr. M. C. Cooke, the distinguished mycologist, at his home in Kentish Town, where he has lived for forty years. He met me at the door and gave me a hearty welcome. His cheerful voice, vigorous step, and firm round handwriting seemed hardly in accord with a busy life of eighty-one years.

The Murrills arrived back in New York from transatlantic passage on August 18, hopefully rested and fresh for the second half of the year.

THE GIRL SCOUTS. Return to their Bronxwood Park home must have been a relief. Considering the voluminous luggage not only requiring international transport but constant packing and unpacking, a period of rest and relaxation must have been welcome. Based on much later extensive conversations with him, George Weber (1961) wrote: "The Murrills occupied a large, beautiful, well-furnished home near the Garden, filled with many art treasures and rare collector's objects from abroad. Mrs. Murrill served delicious dinners to their friends and to her husband's scientific friends. There was continuous entertainment of a widely diversified guest list, but he most enjoyed children and groups of young people, particularly the Girl Scouts, with whom he spent most of his off-duty hours, although he participated in rowing, swimming, tennis, volleyball, horseshoes, skating, bowling and camping." In all probability, the Murrills had servants—full- or part-time—to help with cleaning and landscaping. From Weber's rendition, one might conclude that there was more than adequate off-duty time, but Murrill (1945a), himself, painted a more austere picture (Smith-Vikos undated; Rose 2002). MycoPortal records only a few, sporadic Murrill collections for the year, all local. Nothing from Virginia.

Especially his association with Girl Scouts should bear some explanation, but none is forthcoming. There are references to numerous gatherings, indoors and out, apparently enjoyed by both William, Edna and the girls. Much later, Murrill wrote a sizeable booklet of fictional short stories sketching his own ideas of girlish behavior leading to success. The booklet, he wrote, was distributed to 10,000 Scouts

(Murrill 1950; Smith-Vikos undated). Murrill (1945a) devoted an entire chapter of his autobiography to the subject. "It was very difficult to find men who were willing to undertake work with girls because they did not understand them, but the Naturalist had learned by experience that all young people are much alike whether they wear pants or slacks." His years of teaching at the Wesleyan Girls Seminary in Staunton contributed to his views. He did not lack for a constituency. "There were 2,800 girls in the Bronx who looked to him for counsel and instruction, and 8,000 others in the greater city who often heard him speak."

He prided himself with the memory of the house: "His large house was filled with curios from other lands and surrounded by a large shaded yard now justified its existence. There were three pianos with stacks of music for dancing and singing ... Sometimes three hundred girls were packed at once in the front room for a Christmas entertainment, a spelling match or some other party, while the polished parquet floor often felt the beat of dancing feet. ...In the summer the Scouts were taken in groups to a lake in the mountains where they occupied a large and well-appointed camp. Every week the Naturalist would join them there, occupying a little tent near the water, where he could conveniently help with the swimming."

George Small, son of J. K. Small, Curator of Museum Exhibits, recalled that Director Britton expected his staff to put in "full time" and to be on call daily. Members of the Garden staff took turns taking a full day of Sunday duty, during which they were supposed to make an inspection tour of the entire garden and check in with the guards; "it was quite a chore," Small remembered (Mickulas 2007). One must wonder which Sundays could be spent in the mountains with the Girl Scouts.

1907 AND NOMENCLATURE. The subject of botanical nomenclature is, for even most botanist readers, tedious, dense, and even bordering on repugnant, but, like a not-quite-dormant volcano, it exudes constant burps of lava, and now and then, erupts for real. Whether on the sidelines or as a participant, Murrill was entangled, if for no other reason than his association with Lucien Underwood, so nomenclature is a necessary topic for inclusion in this narrative.

So far, this narrative has traced several participating actors through the second half of the nineteenth century through the opening decade of the 20th, but botanical nomenclature has had a chronology of its own, and it is necessary to recoup some years to do it justice.

For both Columbia College and the Torrey Botanical Club, botany's crown jewel was a herbarium. The Hosack-Torrey-Britton genealogy developed herbaria in the Lyceum, the Torrey Botanical Club, and the College. Each herbarium had had a curator, whether professional or not. A well-curated collection was a source of pride.

Herbarium specimens, however, bear labels including, at the very least, the name of the plant followed by the author(s) of that name, but as the century pivoted, the very protocol of naming of plants underwent a series of events and opinions threatening systematic chaos. Botanical nomenclature had its own contemporary history and that story is also a subplot in this narrative.

Botanists of all stripes gathered together annually at the national meeting of the American Association for the Advancement of Science (AAAS), but AAAS covered all of the sciences, physical as well as natural. As outlined previously, already by 1885, a rather amorphous "Botanical Club" (not to be confused with the Torrey Botanical Club) gathered informally at the AAAS meetings, and had its own program of papers and business meetings. Botanists petitioned for a formal designation, and became "Section G" of the AAAS. Finally, the Botanical Society of America was officially founded at the AAAS meeting in summer, 1893. Pertinent to this narrative, its first slate of officers was: William Trelease, President (Missouri Botanical Garden), Nathaniel Lord Britton, Vice-President (Columbia College, New York), John D. Smith, Treasurer (Peabody Institute, Baltimore) and Charles R. Barnes, Secretary (University of Wisconsin).

Systematics remained a very significant part of botany, although morphology and physiology had begun gathering students and journal space, but the botanical communities of America and Europe were concerned by the seemingly aimless personal idiosyncrasy of the plant naming process. Of course, "nomenclature" had already become chaotic by the time Alphonse de Candolle penned his "*Lois*," accepted at a special Botanical Congress in Paris in 1867 (Hoste 2022). Candolle's document became known as the "Paris Code." But even that document was open to broad interpretation and it was necessary to know each worker's interpretation to make sense of plant names in taxonomic monographs, revisions and checklists. Some universally accepted rules remained needed to govern naming, especially in the era of worldwide exploration and its resultant "new" species and genera (Greene 1887).

In early fall of 1892, a so-called International Botanical Congress was to be convened in Genoa, Italy. In April, 1892, the Botanical Club of Washington, D.C., met to mull over the upcoming Congress and questions of nomenclature. A committee of seven men was charged to report back on the subject, and at the next meeting in early May, the Committee opined "…that nomenclature might better be represented by a body of American botanists, and not an international congress." Furthermore, "We recognize the Botanical Club of the A.A.A.S. as a representative body of American botanists and commend to that body for discussion and disposal the subject of nomenclature …" (Dewey 1892).

At the AAAS meeting that summer, in Rochester, New York, American botanists formed a nomenclature committee which presented a set of nomenclatural "canons." The "Rochester Code of Botanical Nomenclature" was ratified by the attendees, and Prof. Lucien M. Underwood was deputized to carry the Rochester document to Genoa (Anonymous 1892a). Although the Rochester document accepted the 1867 *Lois, pro forma*, it called for change or abandonment of individual articles of the *Lois* where the *Lois* and the "Rochester Code" disagreed, an adroit political maneuver to assuage European botanists partial to the *Lois*.

For his part, in 1892, William Murrill was teaching young ladies in Staunton, Virginia, and one of his students had been a prior pupil under Lucien Underwood.

The Rochester Code was not unanimous even among American botanists, and almost simultaneously, a small group of German botanists formulated and published some "recommendations," which were also discussed in Genoa (Underwood 1892b; Smith 1892).

The four-point proposition had been subjected to international judgement in a circular to 706 botanists, of whom 359 were German. Several botanists had voted for all four, including Nathaniel Britton in New York, testing his loyalty to the Rochester Code.

Lest there be some misunderstanding, Lucien Underwood, who prominently contributes to this story as a Columbia College botanist, in 1892 was working in Indiana and publishing on ferns and hepatics (Underwood 1899a, b).

Underwood faithfully carried out his mandate and reported back (Underwood 1892a, b). John Merle Coulter, editor of the BOTANICAL GAZETTE, put it this way: "A letter from Prof. L.M. Underwood, delegate from the Botanical Club of the American Association to the International Botanical Congress at Genoa, announces that the attendance was large and representative. Articles I, II and III of the Berlin recommendations were adopted, except that the American suggestion prevails and 1753 was adopted as the uniform [starting] date for genera and species. Article IV and other matters were referred to a standing international committee upon which the American representatives are N. L. Britton, J. M. Coulter and E[dward] L[ee] Greene [Berkeley, Caifornia]."

Underwood (1892b) quoted his own speech, furnishing a "first person" account to his detailed report on the Congress. The general tenor of the Genoa event was impressive, and a movement began among American botanists that their upcoming 1893 meeting should be expanded into an international congress (Underwood 1892c).

The "Rochester Code" seemed to come together with little published fanfare, but it stirred a whirlwind of opinion in its wake. Like the Paris Code, it ruled that names were judged by chronological priority—the oldest name was the correct name—and that botanical nomenclature began with Linnaeus's binomial (genus name + species epithet) system. His SPECIES PLANTARUM, 1753, was considered the starting point book (for the text of the Rochester Code, see Anonymous 1892a, b).

A well-travelled German, Otto Kuntze (1891; Fig. 3.13), though, understood that Linnaeus, himself, had employed genus names prior to SPECIES PLANTARUM in several books dating as early as 1735, and proceeded to suggest thousands of name changes based on absolute priority (Zanoni 1980).

Kuntze's publication, REVISIO GENERUM PLANTARUM, was more than controversial. Erwin Smith (1892) wrote: "There seems almost no end to the changes a persistent rummaging of old literature can bring to light, and we may be certain it will not end with Kuntze. For one, I most devoutly wish the strict law of priority were at the bottom of the sea. It does seem that it would be better to study nature more and parchments less. But the proof of a pudding is said to be the eating, and this closely printed book of a thousand pages is commended to the digestion of Messieurs, the systematists."

In his very last days, the venerable Sereno Watson (1892), at Harvard, took issue with Smith, and penned words that would be resurrected years later: "In my opinion botany is the science of plants and not the science of names. Nomenclature is only one of these tools which is necessary to botany, and this being the case, points of

Fig. 3.13 Carl Otto Kuntze. Carl Otto Kuntze, German botanist and nomenclaturalist. (Wikipedia)

nomenclature should be subordinated to science." Another Harvard product, Edward Lee Greene (1894) broke with the other Harvard botanists, becoming "the odd man out" of nomenclature, wrote in the preface of his Manual: "No botanist will be obliged to adopt the nomenclature of [Greene's] Manual of Bay Region Botany. The author is convinced, however, that the day is coming, and at a fair rate of speed, when the employed generic names which Linnaeus substituted for older ones of Micheli, Tournefort, Lobel or of Gessner, instead of such as have right of real priority, will no longer be thought of by those who name priority as a leading principle in plant nomenclature." So much for dictated priority.

Another, separate committee of the Botanical Club of AAAS had compiled a substantial check list of vascular plants of northeastern North America, and had tried to adhere to the Rochester Code (Anonymous 1893). It presented the list to the next AAAS meeting in 1893, at Madison, Wisconsin, where the botanists tweaked the "Rochester Code" to allow the genus name and species epithet to repeat each other, now (in 2024) outlawed as tautonyms, but this modification took on the appellation as the "Madison Code" (Bessey 1893; Robinson 1895a). Botanists attending numbered 37, "but few foreign botanists were able to attend…" The Madison Code was hardly universally acclaimed.

William Murrill, still in Virginia but on the cusp of leaving for Cornell and graduate study, found a specimen of a rare fern, and through it, became a published botanist and a correspondent of Lucien Underwood, who requested a herbarium collection of all the ferns of Virginia.

In New York City, Underwood, by then a professor at Columbia College, wrote a miniscule paper perhaps a harbinger of both his and Murrill's future. Underwood had only arrived in the City a year prior, but was about to assume the editorship of the Bulletin of the Torrey Botanical Club, where the one-page paper appeared. Underwood (1897a, b) took to task the hydnum genus name *Kneiffia*, which, he contended, was a homonym of a prior flowering plant genus name. The paper signaled his willingness to exercise his nomenclatural skills: it also betrayed some interest in hydnaceous fungi, which was to be expressed by his student, Howard James Banker, who compiled a Ph.D. thesis on the subject.

Lester Frank Ward, a sociologist who dabbled in botany, acted as first author on a committee paper on nomenclature to the Torrey Botanical Club (Ward et al. 1893). He traced "the nomenclature question" to the "Strickland Code," a committee document adopted by the British Association for the Advancement of Science in 1842. As supportive of name priority, Ward quoted from Asa Gray (1879): "For each plant or group there can be only one valid name and that is always the most ancient if it is tenable, consequently no new name should be given to an old plant or group except for necessity. That a name may be better is no valid reason for changing it."

Benjamin Lincoln Robinson (1895b, c, 1897, 1898), part of the Harvard group, repeatedly criticized absolute priority and Rochester Code, but provoked a rebuttal by Frederick V. Coville (1895) at USDA in Washington. Coville (1895) wrote: "He [Robinson] surely cannot expect American botanists to revert to a now discredited system of nomenclature under which they had been chafing more and more for the past fifteen years." A long list of botanists and centers of botanical research supporting the Madison Code was appended.

Lucien Underwood (1899a, b) published a bibliographic paper on the genera of ferns, in which he adhered to the Rochester Code as rigidly as he could, an effort supported by Pollard (1901; Underwood 1901a), but criticized by Fernald (1901a, b).

Orator Fuller Cook, a former Syracuse student under Underwood, emerged as an important arbiter of the type method of the Rochester Code (Cook 1898a, b, 1900). His attempt to classify "species," however, was a classic case of confusing taxonomy and nomenclature (Cook 1899).

Anticlimactically, the 1900 Botanical Congress in Paris merely pushed nomenclatural considerations to the next-planned Congress in Vienna in 1905, which appeared—to Americans—to present a gathering to which American botanists could present new amendments to bring the European taxonomists into agreement with the Americans.

In a run-up to the Vienna Congress, in January, 1903, a botanical Nomenclature Commission was formed at the AAAS meeting in Washington, DC. After some deliberation, the 23 Commission members signed off on the amended "Code of Botanical Nomenclature," which consisted of 19 "Canons" regulating the formation of plant names as well as extended portions on orthography and citation of names (Arthur et al. 1907). Published in May, 1904, comparison of this "American Code" with our latest—2023—version shows the extent of the roots of the former to the latter.

"[Nathaniel Britton] was an individual of pronounced ideas, and in the often acrid nomenclatural controversies of the last decade of the last century and the first decades of the present one, he was an outstanding champion of strict priority in publication and a strong exponent of the so-called 'American' code of botanical nomenclature as contrasted to the international rules. As a result, his own publications and most of those prepared by his associates in New York were issued under the 'American' code" (Merrill 1938). For the most part, this included William Murrill.

The words used by the Commission to accompany their "American Code" are worth quotation, for how their Code would be received by the Europeans depended on the surrounding written context within which it was delivered. "A Nomenclature Commission was appointed by the Botanical Club of the [AAAS] at a meeting held in Washington, D.C., January 2, 1903. At this meeting a series of rules for nomenclature was presented and referred to the Commission, which has carefully considered all the principles involved, and has tested the application of the principles to all kinds of cases. The Commission has found that, for purposes of more exact statement, and to reach more satisfactory results, some rearrangement and modification of the rules as proposed at this Washington meeting are advisable. We therefore recommend, and propose also to move in the Vienna Congress of 1905, that, in accordance with the provisions of the committee on organization of the Congress of 1905, the code of laws of 1867 be amended by the abandonment of all its articles and the substitution of the appended code." It would not be difficult to imagine some affront by those reading these pronouncements, in English, suggesting deletion of the esteemed Alphonse De Candolle's *Lois*.

Several major differences had developed between the European "*Lois*" and all of the American codes. One "bone of contention" became known as "the type method." The two predominant ranks (i.e. species, genus) were each to be based on a type, whether a type-species for a genus, or a type specimen for a species. European mycologists bridled on this point precisely because individual authors had individual opinions on the choice and characters of individual taxa (a term only introduced in 1950; see Donk 1949).

Inherent in the American Code were methods through which to select the "type" necessary for principles of nomenclature to be universally accepted. Rejection of the "type method" inherent in the "Paris Code" of 1900, invited the Europeans to continue their individual application of nomenclatural principles. But simultaneously, the "type method" required regulation of the process of selection of a "type," but such regulation came after the chronological strictures of this narrative. Nicolson's (1991) history of botanical nomenclature history is recommended.

By imposing 1753 as the starting point for botanical nomenclature, numerous generic (and less so specific, mostly of polynomial construction), pre-1753 names would be declared invalid, and Europeans preferred to retain such "pet" names. Conservation of specific names was voted down handily, but conservation of generic names would come to plague nomenclaturalists for another half-century.

The American Code was presented to the Vienna Congress in 1905, but there, Kuntze presented his findings, and nomenclatural debate was postponed until the next Congress, which did not materialize until 1910, at Brussels (but see Cheeseman 1907, for a good review of early Congresses, especially Vienna). The American Nomenclature Commission further refined the American Code (Arthur et al. 1907). The "type method" (Canon 15, with five infracanonical points) was spelled out with examples, this time including the "first species rule," as a means of selecting the type of a genus. In searching for a "fit" species to serve as type for a pre-existent genus name, the first species listed by the original author and not transferred to a later-published genus, was to be selected as the type. Some years later, the "first species rule" was deleted, judged to be overly mechanical and not employing scientific principles. If the first species had also been transferred, a taxon was to be chosen from the "residue" of species in the original publication, but in all cases, the type was to be a taxon included in a genus by the original author.

Both Europeans and Americans agreed to a universal starting date for botanical nomenclature, 1753 (Linnaeus' Species Plantarum, in spite of Kuntze), but over time, some cryptogamists introduced the efficacy of an even later starting date for fungi, but whether 1801, and Persoon's Synopsis Mehodica Fungorum, or 1821 and Fries' Systema Mycologicum, was under discussion (see Pfister 1981). George Atkinson (1910), Murrill's Cornell mentor, weighed in, as did Farlow (1910) at Harvard. For even the Americans, the fate of names introduced between 1753 and either 1801 or 1821 raised intricate problems (Pfister 1981; Pfister et al. 1990; Petersen 1975a, b, 1976a, b, 1977). Any starting point which departed from 1753 meant a split from "green-plant" botanists and complications in the "*Lois*."

The "American Code of Botanical Nomenclature" was formally introduced by the Nomenclature Commission appointed at Philadelphia in the April issue of the Bulletin of the Torrey Botanical Club, with Underwood's name as a member (Arthur et al. 1907). By Mid-November, he was dead.

BILTMORE. Only the barest announcement (J. NYBG 7: 256. 1907) reported: "Dr. Murrill visited the Biltmore Forest School, at Biltmore, North Carolina, in October, where he secured collections of Polyporaceae and made some observations on diseases of trees." Without these words, penned by Murrill, himself, as editor of the Journal, we would be unaware of this introductory visit. As it is, there is no report of an invitation, and if so, by whom. It would have been extraordinary if Edna had not spent this time with her family in Falls Church, but only the scantiest MycoPortal record indicates Murrill's Blue Ridge visit.

Earlier, in May, Murrill had visited with Charles Peck in Albany to study Peck's specimens of pileate polypores, but also to discuss the Chestnut blight and its rapid spread in all directions from New York. Peck did not yet possess specimens of the fungus, and Murrill proposed an exchange of his *Diaporthe parasitica* for fragments of Peck's *Poria setigera* and *P. marginella*, promptly accomplished.

CRISIS OF 1907, PART II. The owner of the largest bank in the country, John Pierpont Morgan, in his Manhattan mansion, scrutinized the banking problem. His banks showed signs of stress so although unconventional, he began "salting" his banks with his personal fortune to keep them afloat: almost sufficient to quell the emergency but not the panic. Unheard of, he convened a private conclave of several other moguls to form a "money trust" of personal wealth from which banks could borrow in order to stay in business. The "money trust" fulfilled its purpose and the financial crisis subsided, at least at the bank corporation level (Crisis of 1907 2023). Many local banks, though, were bankrupted during the year-long panic. Lucien Underwood lost much (or all) of his savings, and his Connecticut farm was threatened. Although outwardly he seemed his usual friendly, gracious self, he fell into deep depression.

At the Underwood farm in Redding, Connecticut, Helen had come home from Cornell and the family was gathered around the table for dinner on Saturday, November 16, 1907. Surely the conversation turned to the precarious Underwood finances, and the probable loss of the farm. With no warning, Lucien seized one of the sharp kitchen knives, and attacked Marie. Bewildered and instinctually defending herself, she suffered a deep wound on her hand, and screaming, protected Helen as best she could, and the two fled. Lucien seemed not to pursue.

Marie and Helen ran to a neighboring home, where immediate aid was forthcoming. Her tale must have seemed unbelievable, but a doctor was summoned as fast as possible. Time passed; Helen was treated and the men walked to the Underwood house to check on Lucien. Pounding on the door brought no response and the men entered, only to find Lucien in the bathroom, dead. By self-inflicted wounds, he had bled out. The New York Times described the entire scene.

1907 had not even drawn to a close. Lucien Underwood's life's pieces had to be gathered by Marie, Helen, Britton, Columbia, the Torrey Botanical Club, friends, fellow botanists across two continents, many holders of all the other fragments of Lucien's persona. Murrill had lost a long and personal friend, a benefactor, a model, a mentor, and a colleague. The keystone of collaboration between Columbia and the Garden was also lost.

While Underwood's name as Chairman of the Scientific Directors remained in place for the November number of the Journal of the New York Botanical Garden (his death was noted within), it was replaced by H. H. Rusby in the next issue, which included a fine necrology by Rusby (1907) himself.

Murrill of the Garden

Chapter 4
Home and Away

Lucien Underwood's suicide hung over the New York botanical community. After all, his service reached into the Torrey Botanical Club, Columbia, and the Botanical Garden. But as he would have seriously told them, "Life must go on." Nathaniel Britton was still Director at the Garden and although the legal entity was a dozen years old, site improvement and construction went on unabated. The BULLETIN OF THE NEW YORK BOTANICAL GARDEN produced report after report of dry narration—the roads, the conservatories, landscaping, library, collections, exploration, personnel, endowment—all required perpetual attention.

For Murrill (1907) the mycologist, in mid-December, 1907, volume 9(1) of NORTH AMERICAN FLORA appeared, covering a portion of the polypores. Descriptions and keys must have been written through fall. The second part of the polypores came out in three parts (Murrill 1908a), and, in the JOURNAL OF THE NEW YORK BOTANICAL GARDEN, a description of the fungal herbarium—and its growth—in the museum (Murrill 1908b). In the latter, the use of naphthalene flakes as a herbarium insecticide was introduced for the first time in the U.S.

For Murrill, life's trajectory seemed to be constantly positive. Prep school, college, teaching, advanced degree, settlement in the leading city of the nation, and a job seemingly created for a naturalist. But as time slipped by, so did the complexity of the world of science, and Murrill would find that his thinking might not keep pace. As the frontiers of science expanded knowledge of life, some of the undergirding precepts would require repeated examination. For the staff of the Garden (including its Director) taxonomy was a central occupation, so some foundations needed periodic refreshment. Seemingly occupying excessive space here, the rest of Murrill's life and activities may be seen as reflected in the following few pages.

THE DEFINITION OF "SPECIES". As outlined above, botanical nomenclature deals chiefly with two ranks of organisms, species and genus. Strangely, it was not essential that either of these terms be defined, only that the assignment of names for

R. H. Petersen, *William Alphonso Murrill: The Naturalist*, Memoirs of The New York Botanical Garden 124, https://doi.org/10.1007/978-3-031-93343-1_4

taxonomically similar gatherings be done uniformly by all workers. As time passed, botanical experimentalism had grown and philosophically broadened; workers in major subdisciplines, taxonomy, morphology, physiology and ecology, found it difficult to communicate across the scientific community. As the calendar turned over to the twentieth century, the young Botanical Society of America took it upon itself to explore the apparent disagreements.

THE 1908 SPECIES CONCEPT SYMPOSIUM. Well into the twenty-first century, an old proverb persists: "A species is what a competent taxonomist thinks it is." Only a diversion from serious consideration of the thought; but so much so that a symposium to investigate this conundrum was convened at the meeting of the Botanical Society of America in Chicago, January 1, 1908. Several of the invited papers challenged various parts of the aphorism; "species," "competent," "taxonomist," "thinks," all would come under close scrutiny.

The printed papers were not delivered verbatim to a sitting audience, but manuscripts of papers read—probably edited—were published. In the introduction by Duncan Starr Johnson, Secretary of the BSA, the lines were drawn: C. E. Bessey and N. L. Britton, the "taxonomic aspect;" J. C. Arthur and D. T. MacDougal, the "physiological aspect;" F. E. Clements and H. C. Cowles, the "ecological aspect." The order itself revealed a bias: taxonomy being basic and the most historical of the three aspects; not a modern or experimental pursuit. An edited record of closing discussion was included, and the whole collection occupied an entire number of AMERICAN NATURALIST.

The presiding personage, D. S. Johnson (1857–1937) was well-published on ecological and comparative morphological subjects, especially on maritime and littoral plants. He was Professor of Botany and director of the botany laboratory at Johns Hopkins University.

The strong influence of the University of Chicago was reflected in the attendees. Cowles was on the faculty there, J. Merle Coulter (discussant) was director of the Hull Botanical Laboratory of the University, J. C. Arthur was a partner with Coulter in the editing of BOTANICAL GAZETTE, and George H. Shull (discussant) was a product of the Chicago lab. Without any explicit declaration, the Chicago lab was a rival of the Columbia-Garden collaboration for American botanical supremacy, although facing in quite distinct directions.

Bessey (1908) summarized some major themes: "Nature produces individuals, and nothing more. She produces them in such countless numbers that we are compelled to sort them into kinds in order that we may be able to carry them in our minds. This sorting is classification—taxonomy. … So species have no actual existence in nature. They are mental concepts, and nothing more. They are conceived in order to save ourselves the labor of thinking in terms of individuals, and they must be so framed that they do save us labor. If they do not, they fail of their purpose. Here we perceive one of the principles which must control in the limitation of species."

Concerned with "competence," Bessey wrote: "It is increasingly evident that botanists must insist upon an adequate training by those who are proposing to make and describe new species. Every teacher of botany should impress his pupils with the seriousness of the work of making new species, and should train them to feel that such work should be left to the few who are masters in the subject."

Without attempting to define "species," Bessey put forward a twelve-point program mixing pedagogy, taxonomy and nomenclature.

Nathaniel Lord Britton (1908b) presented a chronological summary of "species" through writings of leading botanists from Lindley ("The determination of species is, therefore, in all respects, arbitrary, and must depend upon the discretion or experience of the botanist.") to Hugo de Vries ("Species are considered as the true units of nature by the vast majority of biologists. ... They are now to be replaced, in their turn, by smaller types, for reasons which do not rest upon comparative studies but upon direct experimental evidence.").

Britton: "As long as species were generally understood to be relatively fixed in characters, their delimitation was relatively simple, but the general understanding that all living organisms are descended from others which were different from them has greatly complicated the subject. Whether the evolution has been by imperceptible progressive modifications of structure, or by mutations, or by both methods, the result is essentially the same from the practical standpoint of taxonomy; from these standpoints, then, similarity of individuals must remain the consideration to which most weight will be given in taxonomic usage."

Arthur's (1908) brief paper centered on parasitic fungi (Uredinales) and their treatment taxonomically.

MacDougal (1908), given to lofty words and dense sentences, seemed more bent on denigrating taxonomy (and taxonomists) than contributing to some definition of species. "... botanists of academic habit will with the greatest gravity attempt to read out of existence, and estimate of no importance, hereditary groups of organisms unless they conform to certain ill-defined, arbitrary standards. ... Taxonomic thought rounds its broadest conceptions when its conclusions are based upon the aggregate of individuals within natural groups, and embody ontogenetic procedure, environmental relations, heredity, evolutionary change and comparative functional performance."

Clements (1908): "It quickly becomes evident that species so-called are widely different in rank, origin and relationship. Some are clearly species in the usual sense, while others are merely variations and forms of these. The ecologist thus comes to look with doubt upon all species. He accepts them reluctantly and provisionally until they meet successfully the test of experiment." Parenthetically: "The consequences in the form of unequal and invalid species have been almost innumerable. This is particularly true of the fungi. To one familiar with them, it seems certain that the number of valid species is considerably less than half the number published." Although not mentioned explicitly, Clements' opinion certainly fit the large papers by Murrill (1903b, c, d) and Earle (1909).

Clements: "In concluding, the ecologist will confess frankly that he does not know what a species is. On the other hand, he is certain that he knows some of the things it is not, and that the species of the descriptive botanist comprise several widely different things. Just what these are and what their relation is to species, if there are such, is a matter to be decided by experiment alone. The question of what a species is can not be answered provisionally until a sufficiently large number of experiments have been made to indicate the regular procedure in the origin of plant forms and to reveal the principles that control it."

Cowles (1908): "In the main it may be said that there are two opposing conceptions of species that are to-day struggling for mastery in the realm of biologic thought. The more prevalent idea, dating in its essence from the time of Darwin, has been that species are artificial creations, mere matters of convenience in the classification of the organic world, arbitrary concepts that have no great and enduring reality. Partisans of this view hold to the doctrine of continuity, maintaining that all species have been connected with other species by a series of intergrades, and that there is no vital distinction between variation and mutation. … In somewhat striking contrast to this concept of species stands the idea that species are entities, which arise by discontinuous variation or mutation, and which have their full specific value from the start. Nor does time change specific form by any slow gradations; the species at its death shows no essential difference from the species at its birth." Louis Agassiz would have nodded in agreement; Asa Gray would have frowned.

Appropriate as a (Cowles) closing statement: "Taxonomy must be scientific. It must require for its devotees a training as rigid as that required by professional workers in morphology, physiology or ecology. Species-making by taxonomic types must be abandoned. … In the future it must be recognized that the final test of the validity of species is experimental, and taxonomists must work no less in the herbarium, but more in the field and in the garden. If the taxonomists of the future fail in these respects, a hard but certain fate awaits them. The world of morphologists, physiologists and ecologists has borne with them patiently and long, and had deferentially abided by the species determinations of the taxonomists. The recent ebullitions of the taxonomical radicals have evoked in botanists in general successively dissatisfaction, contempt and rage. These things will not be endured much longer; a little more and the sinning taxonomists will be 'cast out into the outer darkness where there shall be wailing and gnashing of teeth.'"

George Harrison Shull of the Cold Spring Harbor Laboratory of the Carnegie Institution, still on the young side of 34, had attempted in vain to make scientific sense of the work of Luther Burbank, the popular California plant husbandman. Anonymous (1908) spoke about the ecological view. "It is my experience in the cultivation of things brought in from the field during the last 4 years, that we know nothing about the significance of variations in the field as to their bearing upon the question of the species, if by species we mean a continuously variable group, because in some instances we find that very distinct variations as seen in the field are immediately lost when the individuals possessing them are grown under uniform conditions."

The final discussant (Anonymous 1908a) was John Merle Coulter (Fig. 4.1), who opened with a pithy statement: "One thing agreed upon in all that has been said is that nature makes individuals and men make species. This means that in a certain

Fig. 4.1 John Merle Coulter. John Merle Coulter, Influential botanist at the University of Chicago and editor of *Botanical Gazette*, circa 1925. (Wikipedia)

Fig. 4.2 Thomas Jefferson Burrill. Thomas Jefferson Burrill, "elder statesman" botanist at the University of Illinois, circa 1915. (Wikipedia)

large sense the individual is the unit to be recognized. At the same time it is evident that the attempt to record such units on the basis of any present scheme is entirely out of the question."

Perhaps an ultimate sentiment was expressed by Thomas Jefferson Burrill (Fig. 4.2), venerable Professor of Botany at the University of Illinois. "I do not know that I have anything very definite to say, but I am somewhat conscious of a feeling that I am glad that I am not a taxonomist this afternoon. The taxonomist has certainly suffered this half day."

FRED SEAVER Were it not for his long, upcoming career at the Garden, Fred Jay Seaver (Fig. 4.3). Would not warrant mention at this juncture. Like several other students, this one from Iowa, he had applied for and received a fellowship to study at Columbia and the Garden in 1906, where he surely interacted with Britton, Underwood, Murrill, and the staff at the Garden before his short-lived appointment in Fargo, North Dakota on faculty of the State University.

During his time in New York, Fred Seaver had made a positive impression, and when the Garden Board of Managers sought a replacement for MacDougal as Director of Laboratories, which carried responsibility for the pedagogical Columbia-Garden cooperative program, Seaver was invited and returned to the City even before his initial year in Fargo was finished. As an added enticement, Seaver had background in plant pathology and could deal with the public seeking advice along those lines. Once on staff, Seaver served as laboratory Director for 3 years (1908–1911), Curator for 31 (1912–1943), Head Curator for 5 (1943–1948) and Curator Emeritus in retirement. His Ph.D. from Iowa State University came with his promotion at the Garden in 1912. His thesis, "Hypocreales of North America" (Seaver 1909a, b, 1910a, b, 1911) also brought him close to the chestnut blight interest of Murrill.

ANOTHER BOOST? Whatever Murrill had been doing must have met with the approval of Britton, for announcements appeared in the JOURNAL and in TORREYA of his promotion from first assistant (of the staff) to Assistant Director of the Garden. Surely a raise, but just as surely, the prospect of added tasks and responsibilities not attributable to systematic research. In subsequent listings of staff, Murrill was under "administration" (with only Britton) with no stated curatorial duties. Whether this was intended, he continued his mycological work, with frequent papers and colla-

Fig. 4.3 Fred Jay Seaver. Fred Jay Seaver, Midwestern plant pathologist, heir to *Mycologia* upon Murrill's resignation. (Wikipedia)

tion of the higher fungi. It was Seaver who henceforth was known as Curator of Fungi.

Murrill, himself, felt that he was reaching the end of the polypore project and his interests were turning toward the boletes (Murrill 1908c), poroid to be sure, but with fleshy, stipitate fruitbodies not easily preserved. They required a very different protocol than the polypores. Nonetheless, he wound down active work on polypores by preparing "a few duplicate sets of common polypores, 77 species in all, for free distribution while they lasted. The species are named in accordance with the nomenclature used in my treatment of the family in the two parts of the North American Flora." To Peck: "If you care to have a set, I will ship it to you at once…." Three sets of keys were prepared for TORREYA (Murrill 1908c, d, f).

In early March, Murrill again wrote to Peck (Letter, March 7, NYS): "I write for information regarding the collections of Boleti made by [Charles Christopher] Frost. I have written to Burlington, Vt., where his collections are at present supposed to be, but I am sure that you must have a number of his specimens which he sent to you for determination, and you can probably give me more detailed and accurate information regarding his work in this group than anyone else. It is my intention to discover what I can in a helpful way regarding his types. If these type specimens are not worth examining, then I should like to save myself the expense and trouble of examining them. If you have more recent specimens that you have compared with his types, then my work may all be done in Albany." Murrill's reference to the "type method" introduces nomenclature to his letters. Although he had grasped the idea and used it previously, he was not a leader in nomenclatural thinking or publishing. Later, he published on what he found of Frost's reliquiae (Murrill 1908g, h), chiefly at the University of Vermont.

Murrill's overlooking of Frost's "type" specimens, regardless of their state of preservation, in favor of Peck specimens exhibiting taxonomic characters more clearly would, these days, not pass muster. In 1908, however, more latitude had not been spelled out, and lecto, neo- and epitype were not in the nomenclatural lexicon.

In May, Murrill (J NYBG 9: 91.) took note of the fate of the herbarium of the respected mycologist-teacher-farmer Alfred P. Morgan, recently deceased. "Professor Morgan was one of the leading mycologists of the country. Two others, Professor Underwood and Professor Kellerman, have died during the past winter." For more on the latter two, see APPENDIX 1. Murrill's (1908f) most erudite paper combining taxonomy and nomenclature of polypores had been reserved for Kellerman's JOURNAL OF MYCOLOGY, surely to inform a more sophisticated audience, if not to put his name on the mycological marquee. The year would not pass before Murrill sought to replace Kellerman's JOURNAL.

Murrill's first professional travel of the year was noted in the Journal (J NYBG 9: 91): "Dr. Murrill visited Washington [DC] about the middle of April to examine the collection of Boleti at the Division of Vegetable Pathology, and to confer with Dr. Metcalf, of the Division of Forest Pathology, regarding the distribution of the chestnut canker" (Murrill 1908i). The blight had continued to spread over time, and it was critical to bring federal forces to bear on the problem, however futile. A new disease, this one on plane tree—*Platanus*—was detected at the Garden (Murrill 1908j).

In June, Seaver, as Director of the Laboratories, reported in the Journal on "Research facilities at the garden in cooperation with Columbia University" [J NYBG 10(114): 127–132]. Lecture and laboratory facilities were fully described, but under "Research subjects," fields of study were accompanied by the names of engaged staff. Taxonomy, morphology, physiology and ecology were listed, and as expected, green plants (as opposed to fungi) were emphasized. In mycology, however: Taxonomy of fungi—Murrill, Seaver; Morphology of fungi—Seaver; "Special taxonomy"—Murrill, Seaver (et al.); General pathology—Seaver. Such listing would have been unthinkable without prior consultation, but Murrill's name is limited to fungal taxonomy, even though he was hired as a pathologist in the shadow of Earle, and considered himself much more broadly, as a naturalist.

In mid-June, another visit to Peck in Albany, again to examine bolete specimens (Letters, Murrill to Peck: 4 June 1908; 20 June 1908, NYS). By this time, they had established a friendly relationship and Murrill looked forward to some days of field work with Peck in the fall, probably on Long Island. The Albany trip was noted in the Journal of the New York Botanical Garden (9 July 1908: 132): "Dr. W. A. Murrill visited the State Museum at Albany in June to examine the collection of fungi of the family Boletaceae, which probably contains more original specimens than any other collection of its kind in America."

Intriguingly, MycoPortal records a few specimens collected by Murrill on the "Cornell Tract" at Upper Lake in the Adirondack Mountains, with labels indicating July second (Thursday) and third (Friday). Google Earth records scores of lakes in the Adirondacks, but not an Upper Lake. Upper Lake is prominent on Long Island, but not upstate. The labels, however, are all handwritten in Murrill's hand, strengthening the confusion. With at least a day for travel in each direction, the trip took no less than 4 days, but went unnoted by any other source. A few years later, however, Murrill spent some days at Upper Regis Lake in the Adirondacks, prominent in Google Earth, and a probable accurate name of the 1908 site.

BILTMORE II. Certainly, the most anticipated fieldwork of the year was a result of the short visit by Murrill to Biltmore, North Carolina the previous fall. The invitation was repeated, and July 13–24 was spent at the Biltmore Forest School, again with school director, Carl Schenck. This time, Homer D. House accompanied Murrill and specimens at NY are attributed to both together. As expected, Edna spent the time with family in Falls Church, Virginia, where Murrill gathered several specimens on his way home to New York, notably boletes.

Fig. 4.4 George Washington Vanderbilt. George Washington Vanderbilt by John Singer Sargent, circa 1900. (Vanderbilt Mansion Archive)

By 1908, George Washington Vanderbilt (Fig. 4.4) had taken the family's place on the roster of members of The New York Botanical Garden Corporation. Aside from the stability of his fortune, under some pressure during the current "Crisis," Vanderbilt was "land poor"—his million-dollar chateau (Fig. 4.5) not far from Asheville, North Carolina, and his land holdings of some 130,000 acres of forested Blue Ridge did not contribute a sufficient "living wage" for him and his family. Although tree cutting had taken place previously, it was not on a programmed plan, and Vanderbilt approached his Forester, Carl Alvin Schenck, the imported expert on forest management from Germany, to see what portion of the forest could be cut to generate more cash, but Schenk resisted. He was in the business of forest management, not forest sacrifice. He and his predecessor, Gifford Pinchot, now involved in the formation of the U.S. Forest Service, had started the Biltmore School of Forestry in 1897 (Fig. 4.6), and already over 50 students had graduated, every one of them gainfully employed.

To avoid a misconception that the school was at Vanderbilt's "Biltmore" chateau, from April to November the "school" was located north of the tiny town of Brevard, in Transylvania County, and took in the entire watershed of Davison Creek. Notable was a side-hill aptly called "The Pink Beds," referring to the profuse flocks of wildflowers. As this was generally considered the territory of Moses Ashley Curtis and Lewis David de Schweinitz—more or less "hallowed ground" (Rogers 1981; Petersen 2001, 2019)—Murrill readily accepted an invitation to collect. Given his

Fig. 4.5 Biltmore Estate. Vanderbilt Estate near Asheville, North Carolina. Vanderbilt was a member of the Garden's Board of Managers and owner of Biltmore School of Forestry. (Vanderbilt Visitor Bureau)

Fig. 4.6 Cradle of forestry school. The reconstructed headquarters building of the Biltmore School of Forestry, near Brevard, North Carolina. This was the location of Murrill's Biltmore collecting. (Courtesy of Cradle of Forestry Office)

Fig. 4.7 Homer Doliver House. Homer Doliver House with his apparatus for production of photographs of wildflowers. House would later become State Botanist of New York, succeeding Charles Horton Peck. (Wikipedia)

distance from "the Biltmore Mansion," it is doubtful that Murrill met Vanderbilt while on his property.

In his report (Murrill 1908k), he set the scene: "When I reached the valley, on July 13, a season of wet weather had brought out quantities of fleshy fungi, which, with the assistance of Dr. H. D. House (Fig. 4.7), were collected in abundance. Many of the thinner forms dried readily in the sun, but the more fleshy agarics and all of the *Boleti* had to be dried by artificial heat…." In spite of facilities provided for this purpose, it must be expected that a large minority of material did not survive self-hydrolysis or destruction by mold or insect larvae. As it was, over 200 specimens are cataloged in the fungal herbarium of the Garden. Represented were the majority of agaric genera, but whether any of the mushrooms were prepared for the table was not reported.

Murrill gave special attention to chestnut blight. "The chestnut canker was not seen in North Carolina. The chestnut tree is of immense importance in the Biltmore Forest, being cut in great quantities annually for lumber and for tannin extract. It is said that the successful employment of chestnut wood pulp, now a waste product from the extract factory, for the making of paper, would increase the value of Pisgah Forest, with its wealth of chestnut timber, 100%. This tree is very sensitive and is dying in many parts of the forest from the effects of the chestnut borer and the disturbance of the natural forest conditions, but, fortunately, the new disease, so abundant and destructive about New York, has not yet been introduced at Biltmore."

Whether Schenck passed these remarks to Vanderbilt cannot be attested, but no increase in chestnut cutting took place until the blight reached the area. Within 20 years, the Blue Ridge chestnut forest component was decimated. Only the trees at highest elevation were spared, but chestnut root sprouts continued to be common for many years.

In the August issue of the Journal of the New York Botanical Garden (9: 146), the following appeared, surely penned by Murrill. "Dr. Homer D. House, who was connected with the Garden and Columbia University in 1902–04 and again in 1907–08, has recently accepted the position of associate director in the Biltmore Forest School."

House would go on to become an assistant to Charles Peck in Albany, substitute for Peck upon Peck's incapacitation, and, upon Peck's retirement (and death), House would become State Botanist of New York for the rest of his career.

Mention was also made of an encounter with Miss Gertrude Simmons Burlingham (1872–1952), camped out in the Pink Beds with a student from Vassar College. Burlingham had been associated with the Garden for some years, and just in the spring of 1908, had received her Ph.D. from Columbia University, for her studies of *Lactarius* and *Russula* (Burlingham 1908) largely under Lucien Underwood.

An apocryphal story also became associated. Soon after the two ladies pitched camp, they met a large rattlesnake, well-known as venomous. They had dispatched the reptile, but a ranger from the forestry school advised them that rattlesnake meat was quite tasty when cooked, resulting in their seeking out unfortunate visitors. An invitation to share a meal was tendered to Murrill and company to join them when possible. While Murrill was interested in their mycological discoveries, his epicurean participation was not recorded.

Far from the Blue Ridge outdoors and the plethora of fresh fungi, Murrill's responsibilities in mycology persisted at the Garden. An example: a Garden employee had been dispatched to Panama to survey the possibilities of formal expeditions by Garden personnel. He reported back (Williams 1908), but in the gatherings were some fungi. A few pages after Williams' report, appeared: "Probably between 900 and 1000 species were collected on the entire trip. Of the lower forms, Dr. Murrill has already examined the small collection of fungi and reports some 24 species. Of these, 17 are common tropical or world-wide species. Of the remaining 7, 6 species are interesting or rare, while one is perhaps undescribed, although collected before in Cuba" (written by Murrill; J NYBG September 9: 158).

MYCOLOGIA. The death of William Ashbrook Kellerman (Appendix 1) in March had brought his Journal of Mycology to an abrupt halt. It could have been only a few weeks thereafter that conversations began between Murrill and Britton about the need for a mycological journal and the responsibility of the Garden to take up Kellerman's mission. Murrill would be editor, of course. Britton, not an acolyte of mycology, would have been dubious about spending the Garden's money and staff talent on such an endeavor. Murrill, conversely, saw layers of service for the

Garden, the mycological and botanical communities, and probably least important, his own association with the effort.

Two important questions required answers. First, was there sufficient interest from the scientific community (most importantly, mycologists)? Whether before or after discussions with Britton (and others) is not known, but Murrill took two steps to identify a response. There was a need to identify a strong cohort of mycologists willing to serve as associate editors, partly to communicate the seriousness of the endeavor. Thirteen were willing to have their names made public, of whom five were foreigners. Significantly, Peck demurred. Second, a form letter was sent broadly. Its content showed considerable forethought. In its entirety (dated October 15): "Dear Sir: A number of mycologists and other persons interested are considering the publication of a journal devoted to the subject of mycology, including lichenology, and they desire to ascertain how such a journal would be received and to what extent it would be supported."

The proposed journal would be published quarterly, with a total of from 200 to 300 pages and a subscription price of three dollars a year. Considerable attention would be given to features of general interest, such as personal items, notes, comments, and reviews of current mycological literature. A colored plate and several illustrations in black and white would accompany each number.

"Would you subscribe to such a journal in case its publication were assured? An early reply is requested. Very truly yours, WA Murrill. Assistant Director." No record was kept of the results.

The second important question: there had been strong rumors that Kellerman had shouldered much of the financial support for his JOURNAL: where could such funds be found for a successor?

The answer was the Lydig Fund. A large portion of the land beneath the Garden had come through the Lydig family. One of the founders of the Garden was Judge Charles Daly. Maria Lydig was married to Judge Daly, and David Lydig was part of her father's lineage. Maria had an interest in the founding of the Garden, and Judge Daly acknowledged his wife's interest in his (Daly's) will by establishing a fund in Lydig's name, but bequeathed by Daly. The awkward citation appeared in the "front matter" of Garden publications, notably MYCOLOGIA: "The David Lydig Fund, bequeathed by Charles P. Daly."

MYCOLOGIA, to be sure, was an organ of The New York Botanical Garden and remained so into the twenty-first century. The Mycological Society of America was still two decades away. Volume 1, number 1 was issued February 27, 1909.

Murrill (1909a, b) occupied the first two offerings, the first of which, "Illustrations of fungi," over the years ran to more than 30 entries. The second, however ("The Boletaceae of North America") was a floristic monograph of the family (as he understood it) with keys to species. The third paper was by Fred Seaver (1909c), by this time on staff at the Garden as a plant pathologist and specialist on minute, parasitic ascomycetes.

In his annual report of the Assistant Director, Murrill (1910a) noted: "This bimonthly periodical, devoted to fungi, was inaugurated in January and has appeared on alternate months throughout the year, making a volume of 290 pages with 16 plates and 5 figures. A feature of this publication is the illustration of fungi in their natural colors."

Murrill plunged into the boletes with fervor. The Journal of the New York Botanical Garden reported: "The entire collection of Boletaceae in the herbarium of Cornell University has recently been sent to the garden for critical examination. This collection contains valuable material from Alabama, North Carolina, the Adirondacks, the Cayuga Lake Basin, Nova Scotia, Ohio, Michigan, and elsewhere, mostly collected by Professor Atkinson and his associates" (J NYBG 9: 182. 1908). Next: "Dr. W. A. Murrill visited Harvard University November 7 to examine types of certain Boletaceae in the Farlow collection" (J NYBG 9: 201. 1908). But also, his knowledge of the fleshy fungi in general, revitalized at Biltmore, was put to work with a fall lecture and a published paper (Murrill 1908l).

Meanwhile, at Cornell, Atkinson (1908) was kept busy with material sent from non-mycologists, but occasionally with detailed notes and even fresh material, including a poisonous *Tricholoma*. Another, *Amanita calyptroderma* from California, was introduced three times in three journals (Atkinson 1909a, b, c), together with Latin diagnosis, excellent photos and description of edibility.

Murrill, of course, although perhaps temporarily a mycologist, considered himself a naturalist, and the breadth of skills were put to work on an exploratory expedition to Jamaica, south of Cuba in the Caribbean Sea. He was not the first of the New York botanists to assess the situation; Underwood had reported back, but numerous destinations remained. William and Edna sailed from New York harbor on December 5, 1908, skirting the Bahamas, Cuba and Haiti before putting in at Kingston, Jamaica. A fine hotel was set as overall headquarters before fanning out. In all, 13 locations were explored for fungi; in several, gathering a rich number of specimens (herbarium labels; MycoPortal). In six locations specimens are noted as W. A. and Edna Murrill, and Edna "obtained" over 200 colored illustrations (Murrill 2023: NYBG Library finding guide:). Christmas was celebrated at Cinchona, Jamaica and collecting particularly noteworthy, but no special mention was made of establishing a Garden tropical station. They docked at New York on January 27 of the new year, 1909.

EARLE'S "GENERA". When Franklin Earle (Appendix 1) came to the Garden in 1901, it was specifically for his experience in plant pathology, and perhaps his potential as an agent for tropical studies. When he left in 1904, these expectations were fulfilled when he became director of the U.S. Experiment Station in Cuba. Since then, he had revisited the Garden, but his specific mission was not reported. Surely Murrill, a good friend and mycologist, must have anticipated a major publication, but Earle's departure from his established norm was colossal.

Released early in 1909, "The genera of the North American gill fungi" combined unexpected erudition in taxonomy of this large group as well as exhaustive library research to satisfy nomenclatural requirements (Earle 1909).

Considering the opinions of the botanical community summarized in the 1908 "species concept symposium" (see above), Earle's contribution could have been seen as name-generating, both taxonomically and nomenclaturally. He explained the nomenclature conundrum: "The most advanced thinkers among the taxonomists have become convinced that the only method for preventing the most unfortunate shifting of names is to insist that a generic name shall always be retained for the species or group of species to which it was first applied, or in other words, that a generic name to be valid must always be inseparably associated with some type species. This is, in reality, only the strict recognition of the law of priority, but it is known as the 'method of types.' Unfortunately, this principle has not been widely accepted in Europe and it was voted down at the recent International Botanical Congress held in Vienna during the summer of 1905. It is, however, recognized by the great majority of American systematists and has been incorporated in the code of botanical nomenclature formulated by the nomenclature commission appointed by the Botanical Club of the American Association for the Advancement of Science at its Washington meeting, January 2, 1903. This code was published in the BULLETIN OF THE TORREY BOTANICAL CLUB, May, 1904, and a revision, employing a few slight changes, in the same journal for April, 1907.

"This code has been followed in the selection of the generic names which have been adopted. … This is necessarily so, since the idea of the type of a genus held by its author was a mental concept and not a concrete species. It is believed, however, that the provisions of the code are so clear that in the great majority of cases the same result would necessarily be reached by any conscientious worker. … [The list of available generic names] includes the names that have been applied to genera of the gill fungi since 1753…

"… By referring to the two most recent and authoritative general works on the fungi we find that Saccardo, in his *Sylloge Fungorum*, recognizes 82 genera of gill fungi, but of these only 50 per cent. are to be found in Engler & Prantl's *Pflanzenfamilien*. Of the 54 genera found in the latter work 28 per cent. have names not used by Saccardo."

Earle listed 236 generic names and their origins, most with a carefully chosen type species. It was a huge nomenclatural effort.

But Earle (1909) also provided keys to the genera for North America, clearly a taxonomic exercise. 147 genera were included, based on characters selected by Earle. The primary choice, "gymnophilli" versus "cryptophylli," was as abstruse then as it is today, and doomed the key from serious usage.

What Earle could not know; only one year later, at the Brussels Congress in 1910, a later starting point for nomenclature of most fungi would be set as 1821 and Fries's SYSTEMA (Fries 1821). "The goalposts were moved," essentially by American delegates over lunch. Earle's taxonomic contribution was relegated to obscurity, but his nomenclature—especially his care in selecting type species—was unflinchingly accurate to this day (e.g., Petersen and Hughes 2021).

CURTIS GATES LLOYD. A lesson learned from C. G. Lloyd; taxonomic publications spawned correspondents and specimens. A note in the Journal of the New York Botanical Garden for May [10(113): 122] noted receipt of a collection of Japanese polypores. At the May staff/student conference [J NYBG 10(114): 136], Murrill "spoke briefly of our meager knowledge of Japanese fungi and the desirability of further material from that region, as throwing light on the distribution of temperate species. The collection consists of 72 numbers, containing 44 species, 26 of which occur throughout the North Temperate zone, 2 are Eurasian, 8 are oriental, 3 are known Japanese species, and 5 are undescribed." Murrill's (1909d) assessment was duly published and the specimens joined the fungus herbarium.

By early July, 1909, Murrill was surely ready for his formal 2-week vacation. As could be expected, he made for western Virginia, probably dropping Edna with her family on the way. His destination was Mountain Lake, reached through Blacksburg, where his family and friends still lived. His trip was reported in the Journal of the New York Botanical Garden [1909 (117): 224] and in his annual report of the Assistant Director (Murrill 1910a). The Lake (Fig. 4.8) was found at about 4000', a "moist and heavily wooded" area, and yielded "nearly a thousand specimens, mostly large, fleshy species." He wrote: "… its fungous flora has been up to this time practically unknown, although by inference closely related to that of the high mountains of West Virginia and North Carolina. On my return, the local species claimed my attention far into November;" and a lecture in the fall schedule. Allowing for poor drying conditions for "large fleshy species" from their collection, note-taking,

Fig. 4.8 Mountain Lake. A modern photo behind the dam at Mountain Lake in western Virginia, a favorite destination of William Murrill. (Virginia Tourist Bureau)

packing, shipment and storage until examined, MycoPortal records a couple hundred specimens, not nearly a thousand. The Blacksburg-Mountain Lake area would be a repeated destination (Murrill 1920a, b, 1923a), and the future location of the biological station of the University of Virginia (UVA Station 2023).

In early fall, 1909, Mrs. Murrill and Mrs. Rider, a friend of the couple, spent nearly a month together at the Rider home in Chappaqua, in northern Westchester County, New York. Printed herbarium labels carry both names, and aside from a very few specimens from the family environs in Virginia, comprise the only domestic collections from Edna thus far. She apparently also executed aquarelles of some collections.

Murrill (1910a) considered: "My most important scientific work of the year [1909] has been the completion of a monograph of the Boletaceae of North America, which will appear in Volume 9, part 3, of 'North American Flora' (Murrill 1909a). A preliminary arrangement of most of the North American species of this family was published in Mycologia earlier in the year" (Murrill 1909b, e). The species of Cantharellaceae, the lowest tribe of the gill-fungi, were also monographed for Volume 9, part 3, of "North American Flora" (Murrill 1910b).

CHRISTMAS IN MEXICO. Considering the plans necessary for tropical travel, even within the continent, conversations about a second Christmas season "on the road" must have begun in the summer of 1909. Considering a destination in tropical Mexico, an area hardly known botanically (especially true of fungi), special plans were necessary, and to include Edna, complications were compounded. At the very least, she must have been willing after Jamaica the previous winter, to again calculate a winter in New York versus a tropical climate, even if non-English-speaking, and approved of the trip. Parallel talks must have taken place between Murrill and Britton (and the Board of Managers). Whose idea was this? Was an endowment in hand for an excursion, or was the nebulous mycological tropical expedition searching for the funds to support it?

Murrill, remember, considered himself a naturalist, and this time in Mexico exercised all such attributes, and added significant sociological observations as well. He observed the living conditions of the natives, the major crops being produced on various geological formations; the commodities raised in scattered locales; elevations noted for their botanical profiles; selected products—chicle, pulque, sisal, sugar—dominated landscapes; local customs for the Christmas season were all experienced. And amid all this geographical report, a few days of rich mycological plunder stood out. The tropics, of course, are especially susceptible to temperature and moisture, and from Murrill's (1910c, 1945a) descriptions, much of covered territory was in "the dry season." Under such conditions, polypores on rotten wood were most common, with fleshy agarics found only in locally moist situations.

Briefly, the itinerary began with a few days in Yucatan, where henequen, or sisal, was the dominant crop and chief export. The countryside was unrelieved, and huge plantations were owned or controlled by foreign investors. Rail travel brought them to Veracruz on the Caribbean coast, during the rainy seasons a sweltering coastal plain, but Jalapa was 80 miles inland and 4000 feet higher.

Even higher in the neighborhood of Orizaba was coffee country, and more plentiful fungi. On to Mexico City, already a metropolis with little collecting. The trip continued southwest to Cuernavaca, capital of the state of Morelos. Over several days, forays were made in all major directions. The area was botanically rich, judged by the confusion of wild flowering plants, including some very interesting cacti (one of Britton's targets). One day started on horseback at dawn up the slopes of Vulcan Popocatepetl, then by foot for 4 hours, collecting along the way. Murrill's use of "we" is to be judged. The plunder was 175 numbers upon return to Cuernavaca.

Return to Mexico City was followed by rail through Guadalajara to Colima, not far from the Pacific Ocean, where collecting was spotty and the major crop sugar cane. After some days, the excursion began its return to Mexico City, Orizaba, Córdoba to Veracruz. From Córdoba, a single day foray to Mortzarongo, some 25 miles southeast in a secluded valley where "I secured the largest collection obtained during any single day while in Mexico." Forty-eight hours later a similar day was spent around Xúchiles, with another long but productive day. From Córdoba to Veracruz, and the voyage back to New York, arriving January 29, 1910, the exploration was "routine." A lengthy report was composed (Murrill 1910e) and remarkable memories of the excursion were made for his autobiography (Murrill 1945a). He estimated the mycological success as 1250 numbers comprising over 3000 specimens, of which upon examination, 50–100 were undescribed. In spite of dubious collecting conditions, a rich deposition for the herbarium: Mycoportal search reflects similar results. Elsewhere (Murrill 1910e) "… with full field notes on fleshy species, and over 200 illustrations of the more interesting ones made by Mrs. Murrill."

According to Murrill (1911g): "Returning from Mexico late in January, I at once began to study and arrange the large collections of tropical gill fungi obtained by various exploring excursions sent out by the Garden in recent years, and later in the year, specimens of most of these were taken to Europe and compared with type material." Thus, Murrill's research track seemed to have switched from boletes to tropical American gill-fungi. One result was intercalation of his polypore indices against his Mexican collections (Murrill 1912h).

But in what appeared to be another research direction: "During the spring and summer, my studies of local edible and poisonous species of gill-fungi were continued, and a number of colored plates were prepared under my direction for publication in future numbers of Mycologia. This feature has proved very attractive and helpful to students. Professor Giles [consultant chemist for the Garden] and his assistants have cooperated with me in the study of mushroom poisons, obtaining valuable results by physiological and chemical methods."

The debut of Mycologia in January, 1909, soon necessitated shuffles in Garden administrative duties. Editorship of the Journal of the New York Botanical Garden was passed from Murrill to Percy Wilson, Administrative Assistant. Over ensuing months, however, Wilson's responsibilities were re-assessed by Britton and the Managers, and his title was raised to Docent, a position of wide duties mostly over the well-being of the Garden operations. Fred Seaver, listed among the co-editors of MYCOLOGIA, helped Murrill with editing duties but also provided papers for content.

Murrill had complained to Peck that the work of McIlvaine (McIlvaine and Macadam 1900) was obscure in descriptions and poorly illustrated, so the very purpose of the book was diminished. MYCOLOGIA volume 1 number 1, as promised, carried a full page, full color plate of five fungi, four gilled, one bolete. Special note was taken: "The photographic work for these illustrations is being done by Mr. H. C. Berte and the color work by Mr. E. C. Volkeart," the latter continuing for some years. Even a century later, colors are rendered well. All five fungi were judged edible, although with variable enthusiasm.

TROPICAL MUSHROOMS. To arrive at correct names for collected specimens, whether tropical or temperate, edible or poisonous, whether old names or new, Garden taxonomists were obliged to find or establish type collections. This principle, the type method, was an American phenomenon (see above). A type, in this case a specimen, was almost certainly not the most beautiful or well-preserved herbarium example; in the case of tropical fungi it might well be the most poorly dried, wizened scrap imaginable, often fraught with mold.

Murrill gathered all tropical specimens from the Garden herbarium, starting, of course, with his own from Cuba (with Earle), Jamaica and Mexico, but also by otherwise occupied Garden collectors, or in purchased collections or exsiccati from Europe. Given the conditions of such specimens, varying from poor to atrocious, gathering "likes" into units could be shear happenstance. The exercise was pure taxonomy, for when all these specimens had been "pigeonholed," they still had no accurate names. After all, collections from numerous previous collectors had made their way to European herbaria, where they had been bestowed names by European workers (Murrill 1911e).

Even more chancy, identifying type specimens of tropical American gill-fungi in European herbaria, where the type method was anything but uniformly accepted, was an even dimmer possibility of applying science. But such were science and American standards, and Murrill was obliged to conform as best he could.

His answer was, if anything, unorthodox. Instead of borrowing specimens from Europe and staying in New York, he decided to tour the appropriate European herbaria, bringing a selection of his own (and Garden's) collections, notes and aquarelles with him, to compare dried material.

Also, as a secondary objective, for the edible/poisonous temperate fungi, he would seek to collect with the resident mycologists who could furnish authentic names, in an attempt to make matches with his own after his return.

So discussion took place at home and at work. Britton may have considered that Murrill would represent the Garden's mycology interests to Europe, and since he was a good ambassador, it would be worth the money to send him away. His absence, however, would increase the Director's workload.

Hopefully by mutual agreement, it was decided that William would go alone. In all probability, Edna would spend the time with her family in Virginia. An itinerary was plotted, funds were secured, and the trip took shape. Italy would be excised: it had no New World empire and therefore any collections of edibles for Murrill were miniscule. Saccardo, after all, although becoming famous, was performing taxonomy with fungal names but had not amassed a large herbarium.

Analysis of Murrill's tour of European herbaria in 1910, includes not only the trip itself, but the resulting series of papers in Mycologia, as well as a bookmark of his career at the Garden. It might appear over-lengthy. Real-time writing is available (Murrill 1911d, e) as well as remarkably detailed memory 40 years later (Murrill 1945a). Murrill's schedule at the Garden was already busy, certainly Mycologia, but including five public lectures during the summer (one of which: "The botanical gardens of Europe," delivered September 24, just BEFORE his departure). Parenthetically, two others—edible and poisonous mushrooms—signaled his interest in this subject.

Whether anyone saw him off in late September is unknown [J NYBG 11(130 October 1910): 245]. Murrill had covered the ground of his trip before, in fact more than once, so the cities, the herbaria, botanical gardens, and mycologists were familiar to him. Only his new mission had been better sharpened. Docking at Hamburg, his first destination was Copenhagen, Denmark. A Dane, Anders Oersted, had traveled to Central America, collecting and illustrating several fungi. Upon his return, he sent the mycological specimens and copies of their illustrations to Fries (1821), who published on them. The specimens stayed with Fries, but the original illustrations were in Copenhagen. Some of Oersted's fungi found their way into the twenty-first century (Cifuentes et al. 2003).

In Copenhagen, he was introduced to two of the "younger" mycologists. Why he didn't meet Ove Rostrup or Jakob Lange is unknown, but they were not located immediately at the Museum in the city. Lange, at least, could have been of use in Murrill's interest in temperate edible/poisonous mushrooms (Petersen and Knudsen 2022). Although Danish bonhomie thrived for Murrill, he was soon on his way to Uppsala, Sweden.

As it had been before, Fries's Uppsala herbarium was disappointing: not many specimens, many poorly curated, but Stockholm was soon visited. There he again met with Lars Romell, a student of Fries, and a prior friendly acquaintance. Edible specimens were all laid out in advance, and they were able to take an afternoon to collect.

Soon Murrill was off to Berlin, its Dahlem Botanical Garden, and former friends. Murrill wrote of his disappointment for tropical North American specimens, but he was impressed by the collections from Africa (the subject of Paul Hennings) and tropical Southeast Asia.

The last two stops made up for the first four. Paris, its Jardin des Plantes, and Paul Hariot, made a fine combination and collections named by Camille Montagne and those of Antoine Duss published by Patouillard were a gold mine. The magnificent Galerie d'Evolution (Fig. 4.9) and planted gardens must have reminded him of the Garden back home (even if the building's official title had only recently been changed from Galerie du Zoologie). A day spent with Narcisse Patouillard himself was edifying.

Fig. 4.9 Jardin des Plantes, Paris. Grand Galerie d'Evolution of the Jardin des Plantes, Paris, originally completed in 1897 as Galerie du Zoologie, but renamed in 1904. (Google Images)

The best was saved for last. George Massee and Arthur Cotton had arranged all the eligible collections, and the Royal Botanic Gardens at Kew provided names for fresh specimens as Murrill strolled the grounds. The gigantic conservatory (Fig. 4.10) was obviously a model for the construction at the Garden in New York, although his work was in the herbarium building of the times (Fig. 4.11). Three weeks were spent before pushing off for home. He docked in New York harbor on Sunday, November 20 [J NYBG 11(131 November 1910): 360]: whether welcomed dockside seems lost in history.

From his return to New York to appearance of his first paper in Mycologia on January 31, 1911, gave Murrill very little time to incorporate his new European findings with his pre-trip data on tropical fungi, especially considering in-press time with the printer in Pennsylvania. Nonetheless, the introduction to the paper gave some pertinent information about the project.

The first fungus considered in this first tropical agarics paper was *Plicatura obliqua* (Berk. & Curt.) [Murrill]. Contrary to accepted practice his name was not included, although he was transferring the species from *Marasmius* to *Plicatura*. Moreover, under the binomial, five taxonomic synonyms were listed, each with the geographic location from which the specimen was collected. American nomenclature dictated that the type of a species epithet was a specimen, not some geographic indication, overlooked by Murrill (1911a). Such individual specimens were not cited.

Fig. 4.10 Royal Botanic Gardens, Kew. Glass conservatory at the Royal Botanic Gardens, Kew, England. Note similarity to the conservatory at the Garden. (Google Images)

Fig. 4.11 The old herbarium building, Kew. The old herbarium building, Royal Botanic Gardens, Kew, circ. 1935. (Google Images)

In the Chanterellaceae, there appeared *Chloroneuron* nom. nov. (with Murrill understood). Murrill noted that this genus name replaced *Chlorophyllum* Murrill (North American Flora 9: 172. 1910) which was a (later) homonym of *Chlorophyllum* Massee (Kew Bulletin 138: 135. 1898). Perhaps Murrill learned of his mistake from Massee himself in conversation.

So far, it is obvious that Murrill intended to conform to nomenclatural rules, although not totally fastidiously.

During the summer before his European trip, an International Botanical Congress had been held in Brussels, Belgium. Numerous nomenclatural topics had been earlier repeatedly postponed, but delegates at Brussels were resolute to take them up. One dealt with name priority: until Brussels all plant names were based on priority which started with Linnaeus, 1753 (see above for more). Cryptogamists, especially mycologists, were uncomfortable with this, since Linnaeus knew little of the cryptogams. The result, reached over lunch, was adoption of January 1, 1821, "legislated" as the appearance of Fries's Systema Mycologicum for most fungi (with Persoon, 1801, for the rest). Atkinson (1910) previewed the Congress; Farlow and Atkinson (1910) reported its results to Americans, so Murrill had every chance to be informed.

In the second polypore instalment (Murrill 1911b), the first genus taken up was *Leucomyces* Battarra (Fung. Hist. 2, 1753), a name prior to *Venenarius* Earle (1909), in obvious violation of the brand new mandate. Murrill further wrote: "The genus includes the species usually known under the name *Amanita*, which name is properly a synonym of *Agaricus*, as pointed out by Earle." Even more confusing, *Agaricus* had been used by Murrill for a genus of polypores. Such terpsichore must have irritated C. G. Lloyd, who was not alone in his judgement of Murrill.

In the third paper of the tropical series (Murrill 1911c), types of new species were cited correctly. For example: *Hygrophorus subpratensis*. "Type collected on lawns at Santiago de las Vegas, Cuba, June 1, 1905, F. S. Earle 373." Missing is citation to the herbarium holding the type, in this case, NY. The last two papers were taxonomic (Murrill 1911c, d, 1912h), and not nomenclaturallly notable. Murrill could have easily concluded that he had summarized the North American tropical agarics, but his many new species from Florida much later would show that in 1911, "he didn't know what he didn't know."

In January, 1911, Murrill presided over the monthly conference of students and staff. The Garden Journal reported: "Dr. W. A. Murrill exhibited specimens of 'Shiitake,' an edible mushroom cultivated for centuries in Japan on decaying trunks of oak and hornbeam sprinkled at intervals with rice water, which species has been found to have several scientific names. Specimens obtained from shops in Japan by the Challenger Expedition in 1875 were named *Agaricus* (*Armillaria*) *edodes* by Berkeley (Journ. Linn. Soc. Bot. 16: 50. 1878), this being the first name applied and the best classification of the species. Excellent specimens have been sent to the Garden by Professor Kusano of Tokyo, during the past year, and others have been bought in the Chinese shops of New York City." It would be a century later that shiitake became a cottage industry in the U.S. and available in many supermarkets.

Over time, Murrill would draw attention to the public request for advice or identification about edible (or edibility) or poison (or toxicity) mushrooms, but it was Britton who sounded a general alarm early in 1912. He wrote: "The increased size of the collections and the increased and steadily increasing demands on the Garden for information and advice are now taxing the time of the staff to an extent which leaves most of us little opportunity for original investigation, though some is accomplished outside of regular hours of attendance. There is need for two more trained botanists, but present available funds are not sufficient to obtain them" (B NYBG 1912 8(27): 14).

As Murrill's publication record would attest, he remined busy with his augmented Mexican collections, but he saved time for forward progress in research. In early May, 1912, he visited Peck in Albany to study type specimens. Among scattered field days, two field partners were noteworthy. In August, it was E. C. Volkeart. In the run-up for MYCOLOGIA, Murrill had proposed an illustrated colored plate for each number. The first was a montage of colored photographs (Fig. 4.12), but before long, original art replaced photos (Fig. 4.13). Numerous plates featured the art work of Volkeart, who had handled the color for the opening plate. In the GARDEN BULLETIN (1912; vol. 8: 22), Murrill wrote: "A number of colored illustrations of fleshy fungi for a series of articles appearing in Mycologia have been prepared under my direction, together with careful descriptive notes from fresh specimens. Twenty-three species have been published in this series during the year (1911)."

In early fall, a field companion was William Gilman Thompson (1856–1927), physician, dietician, and administrator of medical schools and clinics in New York City. Long-time member of the Garden corporation, he served as President of the Garden (and its Board of Managers) from 1913 to 1922. Their collecting was in Stockbridge, Berkshire County, Massachusetts, a neighborhood rife with the homes of wealthy donors to the Garden—probably on Thompson property (Mycoportal data).

The mushroom season in the vicinity of the Garden presented adequate material, and supported Murrill's growing interest in edible/poisonous taxa. Definitions, as always, were difficult. Many of the gill fungi disqualified themselves by size, odor, texture, or some other feature and so remained untested, and "unpalatable" was neither edible nor poisonous. But in the case of poisonous species, were the toxins all the same or were they telltale chemical substances which could act as characters in taxonomy? An adage had it: "There are bold mycophagists, and there are old mycophagists, but an old AND bold mycophagist is hard to find."

In the middle of a hot, summer of 1912, the New York botanical community welcomed the arrival of Dr. Robert Almer Harper (Fig. 4.14), lately from Wisconsin, to assume the Torrey Professorship in Botany at Columbia University, after 4 years at last a replacement for Lucien Underwood. Almost immediately, he was inundated with activities. A note in the December, 1912, issue of the JOURNAL OF THE NEW YORK BOTANICAL GARDEN: "Professor R. A. Harper is conducting a series of

Fig. 4.12 Photo illustration for Mycologia. Mycologia 1, Plate 1 of "Illustrations of fungi." (Photos by Murrill, colors by Volkeart)

seminars for the benefit of the graduate students of Columbia University, the subject considered being the reproduction of the higher fungi. The first of the series was held in the main laboratory of the Garden. Saturday, November 4, consisted of an introduction covering the earliest literature of the subject." The term "higher fungi" was not defined, but his recent work had dealt with "water molds," the Saprolegniaceae.

Fig. 4.13 Watercolor illustration for Mycologia. Mycologia 3, plate 49. (Aquarelle images by Volkeart (signed at upper right))

In early fall, another Murrill product appeared. "Owing to the abundance of fleshy fungi during the autumn and the great public demand for information regarding edible and poisonous species, a set of colored plates has been prepared under my direction for the swinging frames in the museum" [B NYBG 8: 22. 1912]. The illustrations were poster-size.

Fig. 4.14 Robert Almer Harper. Robert Almer Harper, Torrey Professor of Botany, Columbia University, and member of the Garden's Board of Scientific Directors. (Wikipedia)

MORE TRAVEL. More of Murrill's time was taken by a short article on mushroom poisoning (Murrill 1911f), and the fall and winter would include lectures on the botanical gardens of Europe, one by one. But the immediate exigency was an extensive expedition into the Pacific Northwestern U.S. Essentially, it followed the rains down the coast from Seattle, Washington, through several stops in Oregon, coastal mid-California's Santa Cruz Mountains, and a western terminus in Pasadena, California. Contacts and field companions suggested collecting locations, the weather generally cooperated, and Edna was at his side, although he seems to have collected alone during poor weather or other inconvenience. Transportation was by train, and traced the same route as did Jakob Lange and family 20 years later.

The first stop on the 5-day train trip was Chicago, where Murrill delivered a lecture at the Field Museum and surely met with Charles Frederick Millspaugh, the Museum's naturalist. The train crossed the Canadian border, skirted Banff, and descended the mountains into western Washington.

In and around Seattle, Murrill had a friendly contact in Prof. Theodore Christian Frye and his graduate student, Sanford Byron Zeller at the University of Washington. Frye, a bryologist, had a smattering of fungal collections already, of which a few were determined by Murrill, although whether in 1911 or before is not known (MycoPortal data). It would appear that the department was without literature Murrill thought important, for Murrill wrote immediately to Peck: "October 20. Will you send me here by mail c/o Professor T. C. Frye, Univ. of Washington, as many of your available bulletins (1 copy each) that you can spare." Not a miniscule request, considering Peck's productivity over the prior two decades.

MycoPortal data report hundreds of West coast collections by Murrill in 1911. In the midst of his travels, the Garden Journal could report: "Recent reports from Dr. W. A. Murrill as to the conditions for fleshy fungi on the pacific coast are most favorable and the trip will result in the addition of many interesting species to the mycological herbarium." The estimate, after preliminary sorting, was nearly 3000 specimens.

Guidance by resident naturalists led to promising locales in Oregon, throughout the Willamette Valley and east to Mt. Hood. Along the coastal range, the weather was so poor (and had been for some days) that he pursued his quarry without Edna. Although the Siskiyou Range was interesting from the train window, their first California stop was San Francisco and interaction with William Albert Setchell, thence to Stanford University. A fruitful visit to Luther Burbank's operation was apparently edifying for both men. Especially interesting were the Santa Cruz Mountains of central California, the original locale of *Amanita calyptroderma* Atkinson & Ballen (see above). The southernmost stop was Pasadena, a suburb of Los Angeles. Not a productive time for collecting, but an opportunity to see the herbarium of Alfred James McClatchie, which just a short time later would be bought by the Garden from McClathie's widow (Murrill 1912f). On the southern route, another stop was made, this time in St. Louis in order to see the Missouri Botanical Garden's library and herbarium, and to give yet another lecture. The yield of specimens was enormous, much larger than had been expected, and the Murrills were headed home (New York, that is) in time for Christmas. Moreover, just as in the wake of the Mexican excursion, several papers (Murrill 1912a, b, c, 1913a) and lectures resulted from the West coast expedition. His earlier studies on polypores and boletes were revisited: singled out for a separate Mycologia paper (Murrill 1912g), but description of the trip itself was limited to some individuals who aided his work, including Edna. Type specimens were explicitly designated as such.Interestingly, several of the phrases in Murrill's (1945a; Chapter 17) autobiographical account of the trip were obviously borrowed directly from his report in the Journal of the New York Botanical Garden (Murrill 1912e), a strong indication that he had access to the Journal long after his departure from the Garden.

A combination of Murrill's polypore interest and his work on North American tropical collections, his (Murrill 1912h) paper on Mexican polypores was a compendium, including descriptions of new species. In some cases, transfer of old epithets from one genus to another was accompanied by a citation to original literature, as usual disregarding the new date for mycological nomenclature. It was quite obvious that the best collecting of such fungi had been in the Sierra Madre Oriental in the vicinity of Orizaba. The paper bore testimony that collections of fungi of Mexico had been arranged and inserted into Murrill's database.

Under "New Combinations:" "For the benefit of those using Saccardo's nomenclature, the following new species in the above article are recombined, as follows": a concordance of Murrill's binomials to those of Saccardo is provided (Example: Murrill's *Coriolopsis caperatiformis* = Saccardo's *Polystictus caperatiformis*). This practice was to continue throughout the rest of Murrill's life.

Although Murrill's position as Assistant Director would imply constant presence at the Garden, in New York, sizeable chunks of times had been spent in Mexico, Europe, and the North American West coast. Had he given any thought to this idea, he might have mused, "So far, so good."

Chapter 5
"All That Glitters..."

In the January, 1912, issue of the Journal of the New York Botanical Garden [13(1): 16. 1912], the Murrills' 1911 Christmas was obliquely noted: "The meeting of the various scientific societies of the country [AAAS] at Washington during Christmas week was a notable one and well attended. The botanists had very full programs, as well as a dinner and a smoker, in which between one hundred and two hundred took part. The Garden was represented by Dr. N. L. Britton, Dr. W. A. Murrill, Professor R. A. Harper, and Mr. A. B. Stout. A movement to unite all American botanical associations under the Botanical Society of America was auspiciously inaugurated." It would be surprising if Edna did not accompany William and spend a day or two with family in nearby Falls Church, where conversation about the West coast trip would have been lively.

The financial condition of the Garden continued to be depressing. Britton (1913), reporting for the 1912 year, explained in some detail that the total income of the Garden—City contributions, memberships and special donations—left too little money to comfortably support research and exploration. Particularly, he concluded the following: "The tropical laboratory at Cinchona, on the Island of Jamaica, rented 10 years ago by the garden for experimental purposes and held as a convenient place for studies by botanists, has not been sufficiently utilized recently to warrant us in continuing the lease from the Jamaica Government, and it is now proposed to terminate this lease, with the hope, however, that the station may be continued as a point where investigations may be prosecuted under other auspices." The station had sentimental value, for several staff members had used and enjoyed its high-altitude tropical situation, including the Murrills. In passing: "I have been aided in administrative duties during the year [1912] by Dr. W. A. Murrill, Assistant Director, and by Mr. R. S. Williams, Administrative Assistant."

The fall-out from the Pacific coast trip continued. In the Garden staff conference in early February, Murrill used the colored illustrations by Edna to compare fleshy fungi of the western and eastern U.S. Incidentally, the entire fleshy fungal herbarium of Stanford University had arrived at the Garden for determination. Duplicates

R. H. Petersen, *William Alphonso Murrill: The Naturalist*, Memoirs of The New York Botanical Garden 124, https://doi.org/10.1007/978-3-031-93343-1_5

would be kept at the Garden. Simultaneously, Japanese collections from Professor A. Yasuda were in hand, and valued for probable information on distribution on either side of the Pacific Ocean. The newly purchased McClatchie herbarium of nearly 2000 fungi awaited attention. On such—not to mention MYCOLOGIA or five public lectures scattered from spring to late autumn—was time spent by the Assistant Director.

Chestnut blight (see above) drew attention, with a convention called by the Governor of Pennsylvania of representatives from all the states in which the disease had been reported. Although concern was high, control measures had proved insufficient and few protocols were suggested, and the convention was an exercise in futility (Murrill 1912i, j).

Parenthetically, a modest report in the JOURNAL OF THE NEW YORK BOTANICAL GARDEN read: "Mr. Fred J. Seaver, curator, received the degree of Doctor of Philosophy at the Iowa State University on June 12." As busy as Murrill, Seaver edited the JOURNAL OF THE NEW YORK BOTANICAL GARDEN, oversaw the laboratories for staff and students, authored several articles based on his own research, and co-edited MYCOLOGIA as well.

Murrill's current research had largely turned to gill-fungi, although polypores and boletes were not overlooked in the field. For the gill-fungi, two phenomena were of obvious interest: toxicity and distribution. The former subject has been introduced above, but the latter depended on two sources of data, comparison of fresh material from North America to that of the Orient and Europe, and the search for types to support European names. Not a paltry endeavor.

In May, Murrill visited Peck in Albany, as usual sorting through Peck's specimens. As possible, summer was spent in local—Bronx—fieldwork, largely gathering specimens for illustration by Mary Eaton for the "swinging frames," and Volkeart for MYCOLOGIA.

LAKE PLACID. MycoPortal records frequent local collecting throughout the summer, but two major efforts were launched, both to the same biome, and both with the aid of Edna. The general location was the Adirondack Mountains of northern New York State, with Lake Placid selected as the headquarters. Murrill's (1912k) report in the JOURNAL OF THE NEW YORK BOTANICAL GARDEN began by address to Britton: "Following your instructions, I spent the last half of July and the first half of October, of the season just passed, in the Adirondack Mountains making a general collection of fungi." This wording implies some sort of support from the Garden, not merely vacation time, but the specific nature of the support remains unrevealed. Murrill sought advice from Peck about a place to make his headquarters: "Can you recommend a convenient, quiet, and altogether desirable hotel among the large number listed? I should not mind paying three dollars a day, if the place were thoroughly suitable. Mrs. Murrill will be with me" (Murrill to Peck, July 1, 1912).

A bow was taken to Charles Horton Peck (1899): "It was not our expectation to add many novelties to the known fungous flora of North Elba after the long and brilliant work of Dr. Charles H. Peck, state botanist, in that region, but we hoped to increase our knowledge of this flora and to enrich the Garden herbarium with well selected and well-preserved specimens of as many species as possible." In other words, types of Peck's species would remain in Albany, but their state of preservation might be augmented by new material at the Garden, especially at hand, not 90 miles distant. From the testimony of both Peck and Murrill, the biome resembled the territory of central Sweden, including Uppsala, Fries's home during his mature years. Murrill summarized the situation: "A majority of the species are different from those found in the vicinity of New York City, belonging as they do to a more northerly phytogeographic area extending from eastern Canada and New England westward to the prairies and southward along the Alleghanies. The fungi of this area have many points in common with those of northern Europe, being largely of common origin, and the two areas must be studied together before either can be properly understood mycologically." It should be of no surprise that these words outlined Murrill's studies for the next decade. Conversely, the Adirondack excursion was hardly mentioned in Murrill's (1945a) autobiography.

Almost as an afterthought: "On our return from Lake Placid, we stopped at Albany to represent the Garden at the dedicatory exercise of the magnificent new State Education Building. The ceremonies, held October 15 to 17, were very elaborate and were attended by several hundred delegates from educational institutions located in various parts of the world." (Dedication 2024) Upon his return, he wrote to Peck: "I received your last report at Lake Placid, but was too busy to acknowledge it at the time. Also, at the dedication exercises, I saw you at a distance but did not have an opportunity to speak to you. I do not know whether any of the specimens I collected will be useful to you or not, but you are welcome to duplicates of them when I have put them in order. Possibly we may be able to make some small exchange of specimens" (Murrill to Peck: October 21, 1912, NYS). As a "down payment," a list of polypores and boletes was appended, with agarics yet to come. Murrill's estimate was about 2000 specimens, to be added to the large backlog in the herbarium.

LLOYD AND OVERHOLTS. Among the parcels of specimens piling up for determinations was one from Dr. Bruce Fink, the lichenologist at Miami University in Oxford, Ohio. The specimens were from Tennessee and Kentucky, and the parcel was anticipated, for Murrill and Fink were correspondents. Fink had informed Murrill that there was a student under his tutelage, Lee Oras Overholts, who was interested in fungi, and Fink recommended study of local polypores as a senior project. Murrill had already produced his series on polypores a decade before, and these papers had been used by Overholts as templates from which identifications could be extracted.

In 1911, a mycological paper appeared in the Ohio Naturalist, written by Overholts and purporting to be an annotated list of the polypore taxa reported to occur in Ohio. On his opening page, Overholts (1911) wrote: "The writer wishes to express his thanks to all who have aided in the preparation of the paper. Especial thanks are due to the Lloyd Brothers of Cincinnati, for free access to the literature contained in the Lloyd Library; to Mr. C. G. Lloyd for his determinations and verifications and for access to his excellent herbarium; to Mr. W. A. Murrill for determinations and verifications, and to Dr. Bruce Fink under whose direction the work has been done.

"The nomenclature followed is that of Mr. W. A. Murrill in his monograph of the family. The most generally used synonyms have been added to correlate this paper with other writings on the family." The 118 species were spread across 39 genera, a tempting target for Lloyd's pen.

Overholts' (1911) paper set off a series of derogatory comments published by Lloyd (1912; Cooke 1954). The young man was excoriated for excessive generic names, many of which were ancient. Little criticism was aimed at the number of species accepted by Overholts, but Overholts followed (as he stated) the generic outline and arrangement used by Murrill in his monograph of the family. In addition, the paper also accepted the rules of nomenclature adopted at the Brussels Botanical Congress in 1910, and Lloyd subsequently levelled his ire at the "law-makers" at that Congress, including mycological delegates whose names read like a Who's Who of the science. A few of these attendees also became Lloyd's incessant targets (Petersen 2023).

In mid-summer, 1912, a short note appeared in the Journal of the New York Botanical Garden [13(152): 136]: "Mr. L. O. Overholts, of the Department of botany of Miami University, Oxford, Ohio, has been granted a research scholarship during August and September. He will devote his attention to the preparation of a list of the Polyporaceae of Ohio, with full keys to the genera and species." Not only did Overholts have Garden endorsement for scholarship, but had a protracted opportunity to chat with Murrill.

To jump ahead of this narrative, Overholts' (1914) second paper retreated from numerous generic names previously employed, and therefore drew a more positive review by Lloyd (1915a), but Lloyd could not resist reminding his readers of his distaste for the previous paper, as follows: "...I am gratified that Mr. Overholts has gotten back to the realms of rational mycology. I presume that this is due to the conservative influence of Professor [Edward Angus] Burt, [graduate of Harvard, professor at Middlebury College and by 1915 on faculty at the Henry Shaw School of Botany at Washington University, St. Louis]. In his previous paper [Overholts] was no doubt influenced by Professor Bruce Fink, who does not know enough about fungus to form an opinion of the merits of such things, and Overholts' previous nomenclature was about as intelligible as a Chinese laundry ticket."

Even more years later, Murrill (1945a) confirmed Lloyd's surmise. He wrote: "When [Overholts] began on this group in Ohio he sent his collections to me for determination and accepted my classification with enthusiasm. When he took work under Burt for his doctorate he was much disturbed by his new teacher's conservative attitude; so he wrote to me for advice and I of course replied, 'When in Rome do as the Romans do.'" While Lloyd's diatribe on Overholts was in poor taste, his aim at Fink was unforgiveable.

Of course, Lloyd's jab was misdirected twofold. First, it was Murrill, not Fink, who was behind the nomenclature of Overholts' previous paper. Second, the number of genera considered adequate to accommodate the fungi in question, was (and is) not nomenclatural, but taxonomic. Nonetheless, given the roster of Lloyd targets (Petersen 2023), the relationship of Lloyd and Murrill remained relatively placid.

COLORS. 1912 saw the unveiling of a book which, while not mycological, would prove invaluable to the practitioners of the trade. The erstwhile American Mycological Society, formed in 1903 (see above), had raised to a place on its agenda the chaos in citation of colors in descriptions of fungi. A committee was appointed to look at the compilation of color accuracy and nomenclature, but the committee could arrive at no workable solution. There was rumor that the eminent ornithologist, Robert Ridgway (Fig. 5.1), at the Smithsonian Institution, was also concerned with this problem, and was working on a solution. The AMS committee reported back to the membership that the task might best be passed on to an authority on the subject for a solution.

Ridgway was already a celebrated artist/illustrator, concentrating, understandably, on birds. He had illustrated the five-volume *History of North American Birds* (three volumes on the land birds published in 1874, and two volumes published as *The Water Birds of North America* in 1884). In its time, the work was considered the standard on North American ornithology. While Ridgway primarily contributed illustrations to the land bird volumes, he wrote the bulk of the water bird volumes. From this and countless other illustrations, he knew the subtlety of color. In fact, in 1886, he had devised a color manual for naturalists (Ridgway 1886), but it was soon out of print, and at any rate, less than perfect for mycologists (Petersen 2016). The color blocks were painted individually—significant brush-stroke imperfections were evident, many rectangles were not fully filled, and with no evidence of cut-and-glued papers (for contents online, see Ridgway facsimile 1886). Ridgway's (1912) second attempt appeared in 1912, and although of rather limited copies, was far more useful than any other such color guide (for contents online, see Ridgway facsimile 1912). Krieger (1914) wrote a review based on his own considerable experience. Over time, it became as popular as possible, although there is no evidence that Murrill ever made use of it.

Fig. 5.1 Robert Ridgway. Robert Ridgway, compiler of "Color Standards and Color Nomenclature," as a young man, circa 1890. (Google Images)

Aside from a planned European revisit, Murrill's 1913a, b mushroom season in the local Garden area seemed spotty, perhaps only intended for the illustrations they could spawn. In spring, he visited George Atkinson at Cornell, combining a comparison of types in the Cornell herbarium and conversations with his prior professor. Atkinson had been relieved of undergraduate teaching, but administered a botany department in rivalry with the new State College of Agriculture (Petersen 2021), and his research interest had turned to comparative morphology and fruitbody development.

OVERWORKED? A virtually unanimous verdict by Garden staff for 1913 agreed that non-research duties had grown too time-consuming for research staff, Murrill in particular. The august Chairman of the Scientific Directors, not normally a policy articulator, wrote: "In addition to this very general use of the herbarium by visitors, a great amount of service has been rendered in the naming of specimens sent in to us for that purpose. In the division of Fungi, the amount of such work has nearly doubled over that of the preceding year [Bull. NYBG 8(31, April, 1914): 305]. Murrill, himself, weighed in: "Increasing administrative and curatorial duties have left me comparatively little time to devote to investigation. However, the preparation of two parts of NORTH AMERICAN FLORA is in progress and these will probably be completed within a few months" [Bull. NYBG 8(31, April 1914): 268]. In early summer, Murrill's responsibility as Assistant Director was tested when a violent storm uprooted not less than 43 trees, blocking all the roads and paths, with remediation under his direction.

BACK TO EUROPE. The 1913 European trip itself seemed less important than previous visits to the same destinations. Murrill wrote: "Several weeks were spent in Europe during the summer, mainly at Paris, Trient [sic], and London, for the purpose of comparing American specimens of fungi with original material preserved in European herbaria. Such comparisons are a necessary basis for all reliable monographic work on American fungi" [B NYBG 8(31, April, 1914): 268]. In Murrill's (1945a) autobiography, the trip was buried among other topics in a chapter titled "Miscellany." Only one event was worthy of recollection. "He spent several weeks in Trient [sic] with [Giacomo] Bresadola, who was considered the foremost mycologist in Europe so far as the larger fungi were concerned, as his many publications and colored plates will show; and he was so honest and dependable that type specimens were readily loaned to him for comparison by almost any institution. Thus [Bresdaola] was able to have at hand at the same time authentic material from several sources dealing with the same group." Murrill was able to bring numerous specimens to NY, each with carefully penned data: "Varena [on the shore of Lago di Como], Tyrol, August 1913, coll. & det. Bresadola, notes by Murrill." (MycoPortal).

Perhaps the most cogent aspects of the trip, though, were two omissions: Murrill's mention of "several weeks" with Bresadola was surely more than had been planned, and a postcard sent from Murrill to Dr. David Prain, The Director at Kew, summarized the situation. "July 25. Somewhat better. Less pain, but still very weak. Rather regret the trip to Kew is short, but depends on when I can travel. Sincerely, WA

Murrill." The picture on reverse is entitled "Castello del buen Consiglio, Trento" (JStor image). I have not found another reference to this, whether a serious accident or chronic health problem, although Weber (1961) referred to a recurring malady throughout Murrill's adult life. The postcard predates the herbarium label date cited above.

The second omission: what of Edna? Granted that her partnership in previous travel required deduction from small references by Murrill, almost accidental, or inclusion of her initials on herbarium labels, but for the 1913 trip, no mention of Edna has been found, leaving to conjecture possible reasons why she may have stayed at home, whether New York or Virginia.

WASHINGTON "SKIM." While the storm clouds of war were gathering in Europe as early as 1913, the U.S. knew to which side it would be drafted if the conflict turned ugly. Meanwhile, the "military-industrial complex" (to use a term coined almost a half-century later) was gearing up for an increase in sales to our government for possible shipment to Europe. The bills to pay industry for arms and munitions not yet in direct use were expensive. Where to get these funds? The Civil War furnished an answer: a federal income tax, this time graduated from 1% for income less than $20,000, to 7% for income over $500,000 (in 1913 dollars) (Fontinele 2023).

The federal income tax was not the first tax for New Yorkers. The city and the state already imposed taxes on goods and services, called excise taxes, the revenues from which went to a myriad of community functions, from transportation to education, from potable water to the sewer system. For the Garden, modest salaries now had a "skim," not for the city, county or state, but for Washington. As Assistant Director, Murrill heard new grumblings and felt obliged to pass the complaints to Britton.

CRACKS IN THE FACADE. Gleason (1961) sketched some incidents which pointed toward periodic fissure between Murrill and Britton, leading to Murrill's demotion in 1919, but beginning a few years previous. One such was included by Murrill (1945a) in his autobiography." The salaries at the garden were at first ridiculously low for an expensive place like New York [City] but Dr. Britton was accustomed to the earlier standards and had plenty of money himself so he refused again and again to listen to the Naturalist's appeals for more money for the Staff. When a rock wall cannot be penetrated one must go over or around it; so certain sympathetic and understanding members of the Board of Managers were approached and their assistance secured in making the desired reforms." The improvement was noted the following year in the report of the Chairman of the Scientific Directors: "The readjustment of salaries which was accomplished last year has fully justified itself and has been a source of great satisfaction to all concerned" [B NYBG 9(32): 58]. According to multiple sources (including Gleason) nothing would have nettled Britton more than a subordinate who went around him to obtain a popular result.

The "rock wall" to which Murrill referred, was described by Mickulas (2007): "Of Britton's temperament, one contemporary noted, 'the least said ... the better.' Diminutive in stature, with a stooped posture and a long beard that appeared to comprise the majority of his height, Britton nevertheless cut an intimidating figure. A contemporary botanical admirer described him as relatively ill-humored, except when engaged in botanical fieldwork, at which time his normally lugubrious face lit up and he became extremely talkative. Such transformations, apparently, occurred only sporadically."

More testimony (Murrill 1945a): "As the Naturalist had told the Director, he [Murrill as Assistant Director] was held responsible for maintenance and the men could not do their best work if worried about food and clothes for their children. The same callous attitude was exhibited when Mr. Carnegie offered $100,000.00 to start a retirement fund for the Staff, which was refused. In former years many business men [had] kept wages at a low level and made huge fortunes. Some of these lived in New York and might have influenced the Director; while the Naturalist had a background of courage and altruism."

Britton contributed: "Administrative and curatorial work required most of the time of the staff, but collectively, considerable investigation has been accomplished." [B NYBG 9(32): 1–27. 1915] In the same report, though, Britton admitted: "The detailed administration of the Garden has largely been referred to Dr. W. A. Murrill, Assistant Director, and to Mr. R. S. Williams, Administrative Assistant, under my immediate supervision."

A TIME TO REAP. Murrill—now 45, and in spite of his split time-rationing—had been sowing the seeds of research at the Garden for a decade. His collecting trips, domestic or foreign, produced large numbers of specimens with field notes, some colored drawings, descriptions and keys. The time had come to harvest these data and to share the results. The difficulty of this is evident from numerous sources. For instance, of the 2014 collections of fungi dated 1914, logged into the Garden herbarium, just 17 were collected by Murrill, and three of those were in the City with a companion; a "baker's dozen" were gathered in Virginia, perhaps on vacation. The rest originated from all over the states and globe, some were parts of exsiccati, and almost all were of "higher fungi" (MycoPortal). None of these, though, were part of Underwood's fungal collections (Murrill 1914a), numbering some 17,000, purchased from his widow along with his library, and requiring intercalation into the general Garden herbarium, identification of exsiccati and type specimens for special separation, and processing duplicates for exchange. Add in editing of MYCOLOGIA, in which "Illustrations of fungi" reached number 19.

Volume 10, Part I of NORTH AMERICAN FLORA (Murrill 1914b), released July 28, 1914, was a floristic monograph of white-spored agarics. Like all of the series, editing was laborious and fastidious, and design equally detailed. Literature and illustration citations, topotype locales, and ecological associations were all included, added to descriptions and keys.

In a matter of 6 months, five new putative floristic monographs were produced by Murrill, using essentially the data extracted for NORTH AMERICAN FLORA. All were hard-bound, printed by the same press as MYCOLOGIA, NORTH AMERICAN FLORA and the other Garden publications, but copyrighted by Murrill, not the Garden. Surely all were edited by Murrill, but whether approved (or encouraged) by Britton or not is unknown. Priced each at $1, orders by mail could be sent to Murrill in Bronxwood Park, his home address. On the title page of each, the Garden positions held by Murrill were listed, lending additional gravitas.

Three were dated in 1914:" Northern polypores" (Murrill 1914c) with preface dated October 13, "American boletes" (Murrill 1914d) with preface dated October 15, and "Southern polypores" (Murrill 1915c) with preface dated December 1. A fourth, "Western polypores," (Murrill 1915a) was signed January 15, 1915, and the last, "Tropical polypores" (Murrill 1915d) after a short pause, signed March 15, 1915. The last depended on data not readily available to the prior four and this may have caused the slight hesitation.

In all five manuals, numerous species binomials bore Murrill's name solely, but whether to be considered as new or previously published was not disclosed, for no literature citations were included. In "Northern polypores" 43 generic names were used to accommodate the wide taxonomic variation, and priority was absolute (i.e. *Ceriomyces* Batt[arra] for the largest genus of polypores), despite the direct imposition of later stating points mandated in the Brussels Code of 1910.

"Western polypores" (Murrill 1915a), with preface borrowed from "Northern polypores," but signed January 15, 1915, was somewhat leaner, but cut from the same cloth. In these five manuals was the ammunition spent by C. G. Lloyd in Murrill's direction (see above). A remaining question, however; to what audience were they addressed? Not the encyclopedic taxonomy community that was North American Flora, mycologically almost miniscule. They were also not for the rank amateur. In his autobiographical retrospective, Murrill (1945a) was most concerned with their use of the "American Code" versus the "International Code" of nomenclature, and therefore at risk of rebuff by potential readership.

A WAR TO END ALL WARS. The political assassination in Serbia in July, 1914, struck a spark among the countries and alliances of Europe, already rife with bellicose rhetoric, military mobilization, enlargement, and popular unrest. Several interlocking mutual-aid treaties locked together Germany, Italy, Austria-Hungary (Triple Alliance) and the remnants of the Ottoman Empire against Russia, Britain and France (Triple Entente).

The earliest offensive was from the French into Alsace. By late 1914, the front had enlarged northward along the French-German border. Simultaneously, northern offensives from Germany were intended to pass through Belgium and Luxembourg into northern France toward Paris. By mid-1915, war involved all of Europe, either in open warfare or mobilization.

Transnational travel on the continent, of course, was drastically reduced, so botanists from the European mainland could not come to London for a planned botanical congress. Some workers were insecure at their own institutions, and nationalism would have made the London meeting uneasy. At the same time, threats to Atlantic shipping made transatlantic travel risky. The London Botanical Congress was postponed until some date post-war (Nicolson 1991). A number of Americans who had planned to attend cancelled their plans, and with them, any American attempt to negotiate with the Europeans over the rules of nomenclature. The "International Code" and "American Code" would remain distinct for the time-being.

At the Garden, at the March 1915 staff Conference, Dr. Fred Seaver bemoaned the nomenclature situation. "Seaver spoke of the difficulty of typifying the genus *Peziza* in accordance with the rules of the AMERICAN CODE OF NOMENCLATURE, which rules are followed in our work on NORTH AMERICAN FLORA. ... Unless some of the special provisions in the rules could be found to apply to this case, the type of *Peziza* ought to be selected arbitrarily [instead of by the 'first-species' rule]; in order to keep the name of the large cup-fungi as it has been used even in remote times." It had been hoped that knots like these could have been loosened when the Americans and Europeans got together. Still unresolved were conservation of generic names; generic type species resting on type specimens; the "first-species" rule; and later nomenclature starting points for cryptogamic plants.

A case in point [J NYBG 16(186): 117. 1915]: "Volume 9, Part 4, of NORTH AMERICAN FLORA, by Gertrude S. Burlingham, William A. Murrill, and Leigh H. Pennington, appeared April 30, 1915. It contained descriptions of 348 species of gill-fungi belonging to *Russula*, *Marasmius*, *Lentinus*, etc., of which 92 species were described as new. Two new genera were included." NAF, like all Garden publications, followed the American Code, limiting use—and respect—by some Europeans.

But the prosaic mindset in the U.S.: the very next item in this issue of the JOURNAL OF THE NEW YORK BOTANICAL GARDEN; "A botanical field meeting of the Brooklyn Institute of Arts and Sciences was conducted by Dr. W. A. Murrill at The New York Botanical Garden on the afternoon of May 22. Eighteen persons were present. After a tour of inspection of conservatory range 1 and a portion of the grounds, the visitors went in a body to the museum building to attend the lecture by Dr. M. A. Howe, on 'Sea Gardens of the Tropics.'" The explicitly "devil-may-care" attitude would not last forever.

Several sources of Murrill collections are easily identified (MycoPortal). In early June, specimens were gathered in the neighborhood of Falls Church, Virginia, Edna's hometown. A visit with her family and friends was the obvious aim. In July and August, a few collections were picked up around the grounds of the Garden, and in September there was a two-day collecting trip to Tom's River, New Jersey, gateway to the "Pine Barrens," perhaps with the Torrey Botanical Club.

The most productive trip of all, however, was a 10-day period on the shore of Upper St. Regis Lake in the Adirondack Mountains (Fig. 5.2). Murrill (1915b) effusively mentioned his gratitude to Mr. and Mrs. N. H. Luttrell, owners and operators of the Kanosa Camp, a full-service resort, by photo and description. In his report in the JOURNAL OF THE NEW YORK BOTANICAL GARDEN, Murrill wrote: "With the assistance of Mr. and Mrs. N. H. Luttrell and their guests he secured about 500 numbers, many of them of special scientific interest. Of the 300 species obtained during the week, 35 were used as food, several methods of cooking being tried. One dish was entirely spoiled by a few plants of *Collybia [Rhodocollybia] maculata*, which proved to be very bitter even after thorough cooking. A dish of escalloped

Fig. 5.2 Camp Kanosa in the Adirondack Mountains. "Camp Kanosa with St. Regis Mountain in the distance," a Murrill collecting trip to the Adirondacks. (Mycologia)

Clavaria [sic] was much enjoyed" (J NYBG 16(189 Sept.): 198). A list of fungi used for food was attached.

In his adopted interest on inedible and poisonous agarics, Murrill came close to Atkinson's and McIlvaine's endeavors, earlier explored by Moses Ashley Curtis, the amateur mycologist, for a food source during the Civil War (Petersen 1980, 2019).

It must be noted that nowhere in Murrill's reports for 1915 is Edna's name mentioned. Surely, no trip to Falls Church would have been undertaken without her participation, or suggestion. But whether she returned to New York is unknown: the usual printed herbarium labels from the Upper St. Regis time do not include her name. Weber (1961) wrote: "His wife, with a feeling of neglect and with physical ailments resulting in inefficiency and moodiness…." Moreover, Luttrell being Edna's maiden family name, a suggestion is raised of whether the Luttrells of Kanosa Camp were part of her extended family, in which case her absence would appear even more problematic.

At this particular time, though, Murrill's personality, even in the midst of differing scientific views, shines through. Lloyd (1915b) wrote: "You can travel the museums of the world over and you will not meet a pleasanter, more affable gentleman than W. A. Murrill, assistant director of The New York Botanical Garden, in charge of the mycological department. Together we have enjoyed several visits, and I am much pleased to be able to publish his photograph in our series of prominent mycologists."

Fig. 5.3 William Chambers Coker. William Chambers Coker as a young professor at University of North Carolina. (NCPedia)

FRATERNAL TRIPLETS. A casual reading of William Chambers Coker's (Fig. 5.3) obituary (Couch and Matthews 1954) might bring to mind William Alphonso Murrill.

Coker, a Southerner (born in Hartsville, South Carolina, the first state to secede from the Union), 3 years younger than Murrill, attended boarding school, and earned a degree from a Southern university (University of South Carolina). Unlike Murrill, though, Coker was born of a notable and somewhat aristocratic family.

Coincidentally, Coker entered Johns Hopkins for graduate work in 1897, just as Murrill enrolled at Cornell, having chosen it over Johns Hopkins—else they might have been classmates—and for his dissertation, Coker wrote "On the gametophytes and embryo of *Taxodium*." Unlike Murrill, however, straight from Johns Hopkins and after a semester in Europe, in 1902 Coker went on faculty at the University of North Carolina, where he spent the rest of his career. Murrill was teaching high school in Manhattan.

Couch and Matthews (1954) wrote: "Dr. Coker arrived in Chapel Hill in the fall of 1902 with a copy of Atkinson's 'Mushrooms' and, from the records of his collections that appear on the margins of numerous pages, it is evident that he began using it almost immediately." This overlooks a migration from leafy plants to watermolds: notes in the Journal of the New York Botanical Garden observed Coker visiting the Garden to use the library on this fungal group. But almost abruptly, in January, 1915, a note in Mycologia (written, of course, by Murrill): "Professor W.C. Coker has sent in a valuable collection of gill-fungi from Chapel Hill, North Carolina, containing a splendid representation of several difficult species" (Mycologia 7(1): 45. 1915). By that fall, another note: "Professor William

Fig. 5.4 Calvin Henry Kauffman. Formal portrait of Calvin Henry Kauffman at the University of Michigan, circa 1920. (Wikipedia)

C. Coker spent a part of August at the garden in the study of the literature and herbarium material relating to the higher fungi of North Carolina" (J NYBG 16(189): 198. Sept.). There was ample time for Murrill and Coker to find commonality of interest and background.

Just 2 years later, Coker's (1917) floristic monograph, "The Amanitas of the eastern United States" appeared. Different nomenclature was immediately evident, with Coker' use of *Amanita* versus Murrill's older (1914) *Venenarius*. Coker seemed unperturbed. Had these two mycologists not been Southern gentlemen, a rivalry could have developed. As it was, Murrill continued his praise as Coker's subsequent floristic monographs appeared. Another example: "The collybias of North Carolina," by Coker and Beardslee (1921), "... a splendid illustrated paper" (MYCOLOGIA 14(1): 93. 1922).

Yet another budding mycologist had common roots with Murrill and Coker. Calvin Henry Kauffman (Fig. 5.4) was a country boy, brighter than most, and spent a couple years right after high school teaching in the grade schools of the Midwest—Pennsylvania, Indiana and Illinois. In 1900, he acquired Atkinson's "Mushrooms" and "spent many happy hours studying the agarics of the vicinity. As a result, he decided to specialize in science and spent the following year (1901–1902) at the University of Wisconsin where he took courses in botany and chemistry. There he came under the influence of Professor R. A. Harper, and through him became interested in and started his studies of the Saprolegniaceae [the water molds]" (Mains 1932). Harper would become head of the Botany Department at Columbia University, and the Chairman of the Scientific Directors at the Garden.

The following 2 years (1902–1904) were spent at Cornell University where he was a personal assistant to George Atkinson and just missed Murrill. Kauffman seized the opportunity to study agarics, developing a protocol similar to Atkinson's. Encouraged by Atkinson, Kauffman devoted special attention to the difficult genus, *Cortinarius* (Mains 1932), and like Atkinson, gathered photo images of the fleshy fungi. At least once, he visited the Garden to examine herbarium material. On a personal level, he was offered, and accepted, a faculty position at the University of Michigan.

In his most extensive and outstanding publication (Kauffman 1918) "The Agaricaceae of Michigan," he described 884 species of agarics illustrated with 172 plates. It has been regarded as remarkable for its clear descriptions and crisp photos. He joined Atkinson and Coker in using Friesian "tribal" names to summarize the agarics, in contrast to the system of Murrill. Counting Atkinson, there now were four mycological systematists, each at a respectable institution with deep inventory of herbarium and library materials, all concentrating on the fleshy fungi. Only one of them, however, slavishly followed the American Code of Nomenclature, and pre-Linnaean name priority.

ILLUSTRATIONS AGAIN. Murrill was not the first (nor was he the last) to see the value of illustrations in both the identification of agarics and the separation of edibles from others, and he contributed on at least two fronts. First, in MYCOLOGIA, even before its actuality, he promised a series of illustrations, and by 1916, had delivered (Murrill 1916a).

Repeating the sentiments of some prior authors (Gibson 1895; Underwood 1899a, b; Petersen 2020b), he proffered (MYCOLOGIA 8(3): 183. 1916): "The subject of edible mushrooms has become very popular of late and many inquiries have been received regarding reliable literature. It can now be stated that there is in press and very soon to appear a large colored chart and a descriptive handbook containing the principal edible and poisonous species of fungi found in America. It is hoped that much safe and useful information will be disseminated by means of the long desired publication "(Murrill 1916b). It represented his bid for a piece of the popular mushroom book market, with the added gravitas of his position at the Garden and the colored chart depicting numerous mushrooms from both sides of the edibility choice. If this announcement was an exercise in advertisement, a much more blatant announcement came a couple months later (MYCOLOGIA 8(4): 224. 1916).

A case in point was a mushroom poisoning in Pelham Manor, just a mile or so northeast of the Garden. A family had obtained what was assumed as reliable spawn of the market mushroom, had spread it in a "mushroom bed," and harvested the fruitbodies that erupted. These were fed to five people, who almost immediately suffered paralysis, followed shortly by sickness. Their condition was alleviated only through prompt medical attention (J NYBG 17(197): 78. 1916). The culprit, soon described by Murrill (1916c) was a *Panaeolus*, aptly named *P. venenosus*. The case for a good illustration was made.

SACCARDO. On another familiar front, Murrill (1916h) announced an addition to NORTH AMERICAN FLORA consisting of several genera of agaric fungi. Appended was a long list of recombined binomials from "Murrill genera" to "Saccardo genera." His habitual inclusion later opened the question of "alternative names," awkward, but a direct answer to Lloyd's criticism of Murrill's blizzard of generic names. Another complication was more intricate: each of the "Saccardo names" would be followed by Murrill's name as "(Murrill) Murrill," if Murrill's (1916e) paper were published prior to Saccardo's appropriate volume, but if Saccardo expropriated the names BEFORE Murrill's MYCOLOGIA paper, all those names would be cited as "(Murrill) Saccardo." An intricacy left for future mycological systematists.

A CATSKILL VACATION. During William's official vacation in August, 1916, he and Edna spent some days in and around Arkville, New York, at the southern end of the Catskill Mountains. At the local Pakawatan Inn, they were afforded all the accommodations needed to process specimens. In fact, several of the other guests became interested in their activities and joined in the hunt. Over 200 collections were processed, but at least as important, the forests and geology differed from previous stations in the Adirondacks, such as the Upper St. Regis Lake the preceding year. The time of year was also different, the Adirondack collecting having taken place in October. Whether this was the rationale for the 1916 location, it was at least noted and reported (J NYBG 18(205): 18. 1917). Also, whether Edna illustrated any specimens was not reported, but her very presence (she was not mentioned for the preceding year) was noteworthy. Murrill annotated his report with a list of fungi collected (Murrill 1916f). One way or another, it was a "working vacation."

BLUE RIDGE FUNGI. In particular, his fieldwork in the mountains of western Virginia in October, 1916, exemplified both his stamina and use of publication outlets to justify his trip. During a brief vacation "In October [22], 1916, I left New York at 9:30 P.M. on the 'Memphis Special' and soon after breakfast next morning was gazing with delight on the splendid range of mountains north of Bedford, Virginia, which includes the celebrated Peaks of Otter so aptly characterized by the early settlers as the 'Blue Ridge'" (Murrill 1916g). From a broad appreciation of the natural history of the place, phanerogams were summarized in TORREYA (Murrill 1916i), but the fungi were singled out for a report in MYCOLOGIA (Murrill 1917a). Even in his autobiography (Murrill 1945a), his trip to Apple Orchard Mountain, "18 miles north of Bedford, Virginia," although only 4 days in the field, was remembered with pleasure. All evidence indicates that he was alone. Two additional reports appeared in Garden publications (Murrill 1916g, 1916h).

Past experience in this biome: "Large collections of fungi had previously been obtained by me from the regions about Blacksburg and Mountain Lake, in the southwestern part of Virginia and from Falls Church, near Washington, but my knowledge of the Blue Ridge fungi had been chiefly gained at Biltmore and the Pink Beds in western North Carolina" (Murrill 1917a).

As winter descended on the City in late 1916, wartime headlines were everywhere. The annual meeting of the AAAS, this time in New York, was somewhat sparse—not everyone could find the time and money to reach the East coast; even the railroads were bringing conscripts east for training and eventual assignment to Europe.

A MISSED CHANCE. Forty-eight years old, at the prime of his career, Murrill, like so many others, was moved to justify his accomplishments. For the venue of this sermon, he chose an invitation to address the meeting of the Botanical Society of America in New York, during the holiday season and the U.S. at war. From its title, "The taxonomy of the Agaricaceae" (Murrill 1917b), an attendee might have been persuaded that Murrill would put forward some overview of "mushroomoids," and his reputation as a fine public speaker might move an otherwise unoccupied delegate down the corridor to hear the talk, but if the botanists could be counted in double digits, mycologists would surely be less than a handful.

Among the introductory paragraphs, he exhorted: "The injection of the personal element only impedes progress; the mind should be kept open and unbiased, ready for any new light. What crimes have been committed, both in Europe and in this country, in the name of Science! Crimes of ignorance, of carelessness, of egotism, of petty selfishness, of border antagonisms and national hates! But that is human nature, and, after all, the scientists have more than their share of piety." To whom was this aimed; American, European?

"Not so long ago, most American botanists were taxonomists. Now the pendulum has swung the other way. ... National and state aid, also, has been largely in the direction of physiology, morphology, and pathology, which is only natural and proper; but it means that taxonomy and the study of plants not strictly economic must be emphasized in institutions properly endowed and equipped for this purpose ..." (Hurrah for the NYBG!)

Among the misguided topics embedded in the speech were the following. How can a botanist in a small institution make his raison d'etra was put forward in four numbered dicta. As the old hymn goes: "Brighten the corner, where you are." Another subject was Murrill's use of naphthalene flakes as an insecticide for herbarium specimens of fleshy fungi, ranking only with how to construct packets to hold specimens without crushing. As usual, taxonomy was liberally confused with nomenclature. "The very men who criticize most loudly the multiplication of species would probably lack the courage, ability, and patience to publish new ones." Could C. G. Lloyd be on Murrill's mind?

The promised topic was not even included in the paper body. Murrill's perceived taxonomy of the Agaricaceae was presented with lantern slide illustrations, only AFTER the close of the lecture. Only those still in attendance were exposed to the very title material of the paper.

AMERICAN TROPICALS. Murrill's series of papers on tropical American fungi was based on more than his own sorties in Cuba, Jamaica and Mexico. The Britton winter migrations had picked up scraps of fungi, often in the dry season, occasional packets from Small's annual time in Florida, and scattered specimens in the herbarium. The geographic coverage did not fit NORTH AMERICAN FLORA, so a separate series was written for MYCOLOGIA. Most of the series (Murrill 1909c, 1911a, b, c, d, 1912h, l, 1913b, 1918c, d) were announced in the GARDEN JOURNAL, and when thought realistic, a summary appeared, surely also composed by Murrill: "At a recent meeting of the Torrey Botanical Club, Dr. W. A. Murrill spoke of his investigations of the gill-fungi of tropical North America, which he just completed, the final paper on this subject appearing in the March number of *Mycologia*. Some of the larger genera were mentioned and the number of species in them compared with those of temperate regions. Of the 525 tropical species recognized by Dr. Murrill in his studies, 300 have been described by him as new" (MYCOLOGIA 19(3): 170. 1918). Such statistics were nearly universal in summaries of staff publications, especially in NORTH AMERICAN FLORA, satisfying Britton's penchant that the botanical world should appreciate the Garden by its output of species descriptions, especially "new."

THE MUISHROOM FRATERNITY. All botanical systematic monographs are compiled between two guard rails, taxonomy and nomenclature (in that order). While the taxonomy of each monograph in Nathaniel Britton's NORTH AMERICAN FLORA was left to the author, the FLORA slavishly followed the AMERICAN CODE OF NOMENCLATURE. In much of the fungi, the author was Murrill. It was apparent early that he preferred to gather "species" into small groups considered closely related. The result was a plethora of generic names, many comprising only a few "species," while some others, such as *Polyporus*, based on taxonomic characters and selection of its type "species," were well-populated. This protocol violated mycological history, in which the number of genera was low, and the number of "species" in most genera was large.

Put simply, Murrill belonged to the Splitters' school, while most other mycologists, American and European, belonged to the Lumpers'. The splitters did not have an evangelist, although they had a preacher (Murrill), while the lumpers had Curtis Gates Lloyd, the self-taught mycologist. Murrill, with NORTH AMERICAN FLORA and MYCOLOGIA at his fingertips, made New York City the splitters' denominational headquarters, while Lloyd, with his own proprietary MYCOLOGICAL NOTES, was itinerant throughout Europe but hailed from Cincinnati, Ohio.

Taking note of Lloyd's propensity to scold the splitters in print (Petersen 2023), one might think there could be ill will between the two taxonomists, but the evidence is quite the opposite. Not only did Lloyd find Murrill quite friendly, but Lloyd visited the Garden repeatedly to make sure his research was on sure footing. What is more, on at least three occasions, Lloyd sent large collections of specimens, once of polypores, once of puffballs, once of discomycetes, to the Garden for determination, with enough duplicates for the Garden herbarium.

A CASE IN POINT. William Coker had visited the Garden several times during his research on water molds, and upon migration to the fleshy fungi (for Murrill = "gill-fungi"), continued to do so. In 1917, he published the first installment in what would become an illustrious floristic monograph. Murrill, using his pulpit as editor of MYCOLOGIA, wrote the following: "A very handsome and beautifully illustrated handbook of the Amanitas of the eastern United States, by W.C. Coker [Coker 1917], appeared last summer [1917] as a double number of the JOURNAL OF THE ELISHA MITCHELL SCIENTIFIC SOCIETY, published at Chapel Hill, North Carolina. This handbook contains the results of the work of years in the vicinity of Chapel Hill and other parts of eastern North America. Dr. Coker and his assistants have collected a great many specimens and made excellent notes and photographs of them. Seven species of *Amanitopsis* are recognized and nearly thirty species of *Amanita*, most of them illustrated with halftones. Although some mycologists may not entirely agree with all of Dr. Coker's conclusions [i.e. *Amanita* instead of *Venenarius*], they cannot question his scientific activity and the quality of his photographs" (Mycologia 10(2): 101. 1918). Later in the same year, Murrill (Mycologia 10(5): 265, 267. 1918) announced two more major works on North Carolina agarics, one by Coker (1918) on *Lactarius*, the other by Beardslee (1918) on *Russula*. Suddenly, the Journal of the Elisha Mitchell Scientific Society had become a reputable agaricological journal.

Buried in the details of this narrative, the Garden, now just twenty years old, had grown and matured into an institution with the largest botanical library and herbarium, more systematics Ph.D. botanists, and more living plant collections than any other institution in the U.S. As such, it had become a center visited by numerous botanists, both domestic and foreign. In mycology, for instance, for a matter of several days during August, 1916, Murrill, Seaver and Burlingham were joined by Flora Patterson (USDA, Washington), William Coker (University of North Carolina), and Lee Overholts (Pennsylvania State College), who were able to gather around the lunch table and chat. In fact, within a few days thereafter, Calvin Kauffman (University of Michigan), Leigh Humboldt Pennington (Syracuse University; working on *Marasmius* for NAF), and Edward Angus Burt (Shaw institute of Botany, and Missouri Botanical Garden, St. Louis) also visited. The mushroom fraternity remained cordial.

THE WAR YEARS. The United States, under President Woodrow Wilson, resolved to stay out of the European war. As provider of much raw material for the Allies, though, German submarines were making American trans-Atlantic voyages risky, and the plight of our European friends was increasingly threatened. In early 1917, Wilson asked Congress to declare war on Germany, and broad conscription soon followed. Over 300,000 men were drafted and although this was meager compared to the French and British armies, the Americans were fresh, confident, equipped with tanks and artillery, and joined the trenches in the fall. Concurrently, the 1917 double revolution in Russia led to its withdrawal from the conflict.

The U.S. mainland remained uninvaded, and life went on, although the Garden felt the effects of war time in selected areas, mostly financial. The city's coffers funneled money in war-time directions, foreign travel was eliminated, Britton did not loosen his reins on staff, all of which impinged personally on Murrill. He added almost nothing to the fungus herbarium: a few donations in 1917 from the Delaware Water Gap, none in 1918, none from Virginia in either year. His publication output did not diminish in bulk, but did not plow new ground—polypores, edible and poisonous mushrooms, new species from old trips. For a man who incessantly referred to himself as The Naturalist, he had less and less time for research of any kind except examination and identification of collections by others, which rated as public service.

For many years after armed conflict ended, vacant-eyed, weeping, jobless men could be seen walking aimlessly on the streets of the U.S., victims of gas warfare. Windows of homes displayed a gold star commemorating a family death in "the Great War." In its aftermath, four empires disappeared: the German, Austro-Hungarian, Ottoman, and Russian (World War I Wiki 2023). Numerous nations regained their former independence, and new ones were created. Four dynasties, together with their ancillary aristocracies, fell: the Romanovs, the Hohenzollerns, the Habsburgs, and the Ottomans. Germany was not only defeated but humiliated by the terms of the Paris armistice, sowing the seeds for yet another confrontation just over a decade later.

In his annual report as Assistant Director, for 1917, Murrill wrote: "Very little field work has been possible. During a vacation of two weeks at Delaware Water Gap, I obtained about 300 specimens and sufficient information for a preliminary report on the fungi of that locality" (B NYBG 9(36): 440. 1918). On the positive side: "Two parts of North American Flora, containing descriptions of 606 species of higher fungi, 183 of which were new, were published during the year." No supported exploration could be reported, and even the usual statement of shared responsibilities by Britton was not included. Murrill's name appeared on a list of staff buying Liberty Bonds. But in a list of six "scholars" supported by the Garden, five were in mycology. Notable were Overholts and Harry M. Fitzpatrick, later of Cornell University.

Tellingly, in Britton's report as Secretary and Director-in-Chief, he summarized that the general endowment stayed steady at $557,000 in spite of repeated entreaties for growth. Furthermore, he wrote: "Owing to disturbed conditions due to the world-wide war, it is impossible to accurately estimate [financial] requirements for 1918." All in all, it was a glum report.

SACCARDO AGAIN. In the January, 1918, number of MYCOLOGIA, under Notes and Brief Articles, there appeared:" Since the completion of Volume 9 of *North American Flora* [sic], many requests have come in for parts 1 and 2, dealing with the polypores, which were exhausted in [reprint form] several years ago. In order to supply the lack of these parts, Dr. Murrill published his series of books, ... He is now preparing a small pamphlet containing the equivalents of his names in Saccardo's nomenclature, so that those using Saccardo's work may not be confused by the changes he has made. The pamphlet will appear in February, and may be obtained direct from the author for 20 cents by any reader of Mycologia" (MYCOLOGIA 10(1): 47. 1918) (Murrill 1918e).

And as promised, March, 1918, brought the "pamphlet" to the public, requiring 31 pages to present two comparisons. The title page (Fig. 5.5) carefully listed Murrill's appointment ("Assistant Director of The New York Botanical Garden"), his editorships ("Associate Editor of North American Flora, Editor of Mycologia"), followed by authorship of six publications (the five geographical polypore books discussed above as well as the chart and pamphlet on edible and poisonous mushrooms, all for sale from his personal address. These credentials surely seemed innocent enough.

The first discrepancy: the comparison had been offered to readers of MYCOLOGIA, a Garden publication, for 20 cents (see above), but in a catalog of Murrill books (Fig. 5.6) appended to the comparison, the price listed was 35 cents; thus a special price to MYCOLOGIA readers. Bearing in mind that the copyright was Murrill's, not the Garden's, the "advertisement" in MYCOLOGIA was open to question, especially since the NORTH AMERICAN FLORA separates, the "comparison" and the MYCOLOGIA

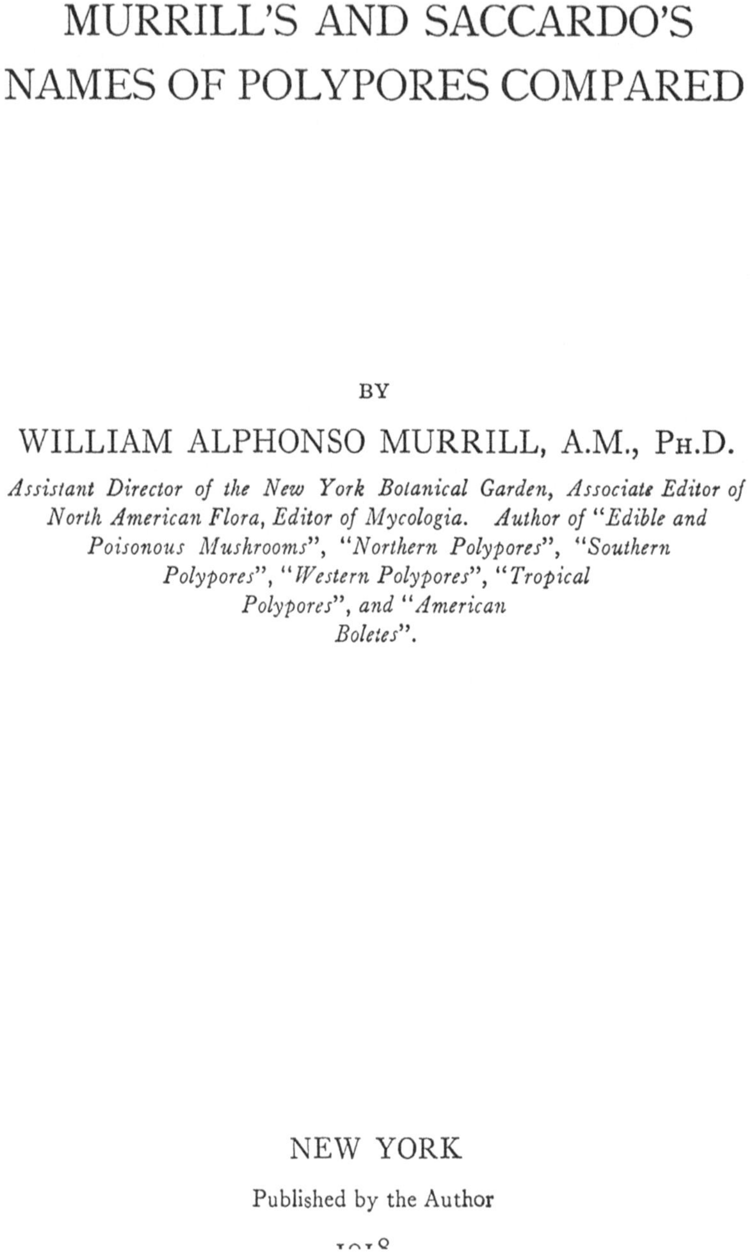

MURRILL'S AND SACCARDO'S
NAMES OF POLYPORES COMPARED

BY

WILLIAM ALPHONSO MURRILL, A.M., PH.D.

Assistant Director of the New York Botanical Garden, Associate Editor of North American Flora, Editor of Mycologia. Author of "Edible and Poisonous Mushrooms", "Northern Polypores", "Southern Polypores", "Western Polypores", "Tropical Polypores", and "American Boletes".

NEW YORK

Published by the Author

1918

Fig. 5.5 Murrill/Saccardo Concordance; Title Page. Title page of Murrill's concordance with Saccardo's generic names. Note the list of Murrill's employment, book-length publications, and publishing responsibility. (Scan)

Books on Fleshy and Woody Fungi

By William A. Murrill, A.M., Ph.D., Assistant Director of the New York Botanical Garden, Editor of Mycologia, and Assistant Editor of North American Flora.

Edible and Poisonous Mushrooms $2.00
A large colored chart and a descriptive handbook containing the chief edible and poisonous species in America.

American Boletes, *issued December 8, 1914* **. . . $1.00**
Including all the species found in temperate and tropical North America, both on the mainland and on the islands, south to South America.

Northern Polypores, *issued December 8, 1914* **. . $1.00**
Including species found in Canada and the United States south to Virginia and west to the Rockies.

Southern Polypores, *issued January 30, 1915* **. . $1.00**
Including species found in the United States from North Carolina to Florida and west to Texas.

Western Polypores, *issued March 25, 1915* **. . . $1.00**
Including species found in the states on the Pacific coast from California to Alaska.

Tropical Polypores, *issued June 15, 1915* **. . . $1.50**
Including species found in Mexico, Central America, southern Florida, the West Indies, and other islands between North America and South America.

Murrill's and Saccardo's Names of Polypores Compared, *issued March, 1918* **. 35**
This also serves as a check list of North American polypores and boletes.

The above prices include prepaid postage, even to foreign countries. *No reduction is made to anyone, dealers excepted.* The author regrets that, owing to the small editions, no copies can be distributed for examination, but a free desk copy will be supplied, if requested, with an order for ten copies of the same book sent to one address.

Remit by Postoffice or Express Money Order, or if by Check, please add Exchange.

W. A. MURRILL
Bronxwood Park
NEW YORK CITY

Fig. 5.6 Murrill/Saccardo Concordance; Inside Back Cover. List of Murrill book-length books, and generic name concordance. (Note price of Saccardo concordance)

advertisement were all produced as part of Murrill's appointment, on "Garden time." Director Britton's opinion of such an arrangement might be interesting but remains buried.

The comparisons themselves were quite straightforward; Murrill to Saccardo, and Saccardo to Murrill. Only names for poroid fungi were compared (polypores and boletes—no other fungi, including agarics), making the catalog further limited in audience, but the two nomenclatural schedules, and Lloyd's opinion of Murrill's, hopefully brought wonderment (and some relief) to the few workers who cared.

In the first (Fig. 5.7), Murrill species epithets were listed by genus names he accepted (left column, in this case five generic names) compared to their generic placement by Saccardo (right column, in this case a single genus). The second (Fig. 5.8) reversed the comparison; Saccardo placement in left column (in this case, three genera), and Murrill's in the right column (in this case nine genera).

The implications were easily revealed. In these fungi, mycologists across two continents preferred to use taxonomic characters resulting in large groups of "species," while Murrill's system used many of the same characters at genus rank, resulting in many genera but with smaller numbers of species per genus. That the mycological community seemed unified in its preference for Saccardo's outcome left Murrill alone. The progression of Overholts/Murrill versus Lloyd, followed by Overholts/Burt/Lloyd versus Murrill, discussed above, was atypical only because Lloyd's pen made it so. Murrill could not yet realize that years later he would be in exactly the position as Saccardo, versus the systematics of agarics by Rolf Singer.

One way or another, Murrill's venture into individual capitalism while employed under the directorship of Nathaniel Lord Britton might have been increasingly strained. It was to become even more so before long.

2 MURRILL AND SACCARDO COMPARED

MURRILL	SACCARDO
Boletellus	
Ananas	Boletus Ananas
Boletinellus	
castanellus	Boletinus castanellus
merulioides	Boletinus porosus
paluster	Boletinus paluster
Boletinus	
appendiculatus	Boletinus appendiculatus
Berkeleyi	Boletinus decipiens
cavipes	Boletinus cavipes
grisellus	Boletinus grisellus
pictus	Boletinus pictus
spectabilis	Boletinus spectabilis
Boletus	
amabilis	Boletus amabilis
Clintonianus	Boletus Clintonianus
Lakei	Boletus Lakei
luteus	Boletus luteus
sphaerosporus	Boletus sphaerosporus
Ceriomyces	
affinis	Boletus affinis
alabamensis	Boletus alabamensis
albellus	Boletus albellus
Atkinsonianus	Boletus Atkinsonianus
auriflammeus	Boletus auriflammeus
auriporus	Boletus auriporus
Betula	Boletus Betula
bicolor	Boletus bicolor
chromapes	Boletus chromapes
communis	Boletus chrysenteron
conicus	Boletus conicus
crassus	Boletus edulis
Curtisii	Boletus Curtisii
eximius	Boletus eximius
ferruginatus	Boletus piperatus
flaviporus	Boletus flaviporus
frustulosus	Boletus frustulosus
fumosipes	Boletus fumosipes
griseoroseus	Boletus griseoroseus
griseus	Boletus griseus
guadelupensis	Boletus guadelupensis
hemichrysus	Boletus hemichrysus
Housei	Boletus Housei
illudens	Boletus illudens
inflexus	Boletus inflexus
jalapensis	Boletus jalapensis
Maxoni	Suillus Maxoni
miniato-olivaceus	Boletus miniato-olivaceus

Fig. 5.7 Murrill/Saccardo Concondance; Sample Page. Inner page of Murrill-to-Saccardo concordance: Murrill names in left column, Saccardo names at right

MURRILL AND SACCARDO COMPARED 17

Saccardo	Murrill
Boletus	
Peckii	Ceriomyces Peckii
piperatus	Ceriomyces ferruginatus
punctipes	Rostkovites granulatus
radicans	Suillellus Morrisii
Ravenelii	Pulveroboletus Ravenelii
retipes	Ceriomyces retipes
robustus	Ceriomyces eximius
roseotinctus	Ceriomyces Peckii
Roxanae	Ceriomyces Roxanae
rubinellus	Suillellus rubinellus
rugosiceps	Ceriomyces subglabripes
Russellii	Ceriomyces Russellii
salmonicolor	Boletus luteus
scaber	Ceriomyces viscidus
scabripes	Ceriomyces scabripes
sensibilis	Ceriomyces miniato-olivaceus
separans	Ceriomyces crassus
serotinus	Boletus Clintonianus
sordidus	Ceriomyces sordidus
speciosus	Ceriomyces speciosus
sphaerosporus	Boletus sphaerosporus
Spraguei	Boletinus pictus
subaureus	Rostkovites subaureus
subglabripes	Ceriomyces subglabripes
subluteus	Boletus luteus
subpallidus	Ceriomyces subpallidus
subpunctipes	Tylopilus felleus
subsanguineus	Ceriomyces subsanguineus
subtomentosus	Ceriomyces subtomentosus
subvelutipes	Suillellus luridus
Sullivantii	Suillellus luridus
tabacinus	Ceriomyces tabacinus
tomentipes	Ceriomyces tomentipes
umbrosus	Ceriomyces communis
Underwoodii	Suillellus luridus
Vanderbiltianus	Ceriomyces Vanderbiltianus
variipes	Ceriomyces crassus
vermiculosus	Suillellus luridus
viridarius	Boletus Clintonianus
viscidus	Ceriomyces viscidus
Zelleri	Ceriomyces Zelleri
Cyclomyces	
Greenei	Cycloporus Greenei
Daedalea	
Berkeleyi	Gloeophyllum Berkeleyi
Burserae	Cerrenella farinacea
confragosa	Daedalea confragosa

Fig. 5.8 Murrill/Saccardo Concordance; Sample Page. Inner page of Murrill-to-Saccardo concordance: Saccardo names in left column, Murrill names at right

Chapter 6
The Final Straw

The preface to "Billy, the Boy Naturalist" (Murrill 1918a) was dated February 14, 1918, and began: "In November, 1917, when the author was confined to his room for a few days, he decided to try the experiment of recalling incidents of his early youth, forty years and more ago [49 to be exact]. No diary or other record had been kept and there was no one near to prompt him; neither had there been any occasion prior to this to refresh his memory of most of these incidents." In short, he was about to deliver stored memories and was proud of his ability to do so.

He went on: "The experiment was a surprise. Instead of requiring an effort, hosts of memories came flooding over him and he only had time to jot down a word or two of each as a reminder .... At his leisure during the days that followed [his confinement], with the pictures still clearly in mind, he endeavored to record these memories in simple language with scientific accuracy just as they came back to him."

An interesting observation reflected his almost two decades living in New York City. "… there may be a yearning to open the child's mind to the sweet influence of Nature and lead the starved little souls of our crowded cities to the blessings of rural life, and the charms of our native woodlands."

Murrill may merely have innocently overlooked her, but the questions arise; where was Edna? What was the cause for confinement—surely illness. Perhaps unjustified, what medications might he have ingested, with unexplained side effects? How long did it last? Whatever clues can be uncovered are ephemeral and third-hand at most.

In the last pages of the book, appears: "All that is best and noblest in the world has been developed from small beginnings by faith and patience and untiring effort. I think this is true of the world, it is equally true of each individual life. If one learns the truth and does right and works hard, he is sure to win. Billy wants his motto to be the motto of all his boy and girl friends: 'Virtus, Veritas, et Labor triumphant.' To those who follow this motto, the world will remain bright through life, and the 'clouds of glory' will accompany them from infancy to old age."

R. H. Petersen, *William Alphonso Murrill: The Naturalist*, Memoirs of The New York Botanical Garden 124, https://doi.org/10.1007/978-3-031-93343-1_6

To mirror these sentiments against his work at the Garden is difficult, especially if his protestations of too little time for research are taken into account. He recalled, allegorically, his conversation with "Mother Oak" back in Virginia: "Do not forget me, Billy, and what I have told you. When you get old, like men, if you need a friend, come back and lay your head on my breast and I will comfort you; if you wish to sleep and rest, I will take you to my heart and bid my trees cover you with their leaves and the winds sing you a soft lullaby." How prophetic, although he could not know it yet (Fig. 6.1).

"Billy, the boy Naturalist" blatantly autobiographic, related scores of stories of rural young and adolescent childhood based on reality—his own youth, embellished to be sure, but anchored in real life. Less than a year later, his second "character" book masqueraded as fiction—the plot a figment of his imagination, seasoned with real-life young people—caricatures surely.

His efforts at storytelling were purgative, perhaps in his mind. An hour spent imagining was an hour away from weighty decisions on Garden affairs and even the genera of polypores. But he also hoped for some affirmation from children (or their parents), which hopefully could spawn subsequent efforts.

On December 20, 1918, he signed the preface of "Three Young Crusoes" (Murrill 1918b) which took three shipwrecked teenagers through experiences of observation and handiwork, all based on his own (unwritten) exploration of Jamaica, years before. That two of the children were Edna (age 13) and William (age 14), may have reminded readers of Murrill and his Edna. Especially writing the final events of the

Fig. 6.1 Panorama with Apple Orchard Mountain. Farm country in western Virginia. Apple Orchard Mountain, one of Murrill's favorite destinations, in the foreground. (Google Images)

book, in which William goes to college, marries Edna, and takes her to New York City for the Christmas holidays. Unabashed photos featuring William and Edna are the only such still extant.

Murrill's books for young people joined a torrent of just such fiction for adolescents on the market. A large source of the genre was Edward Stratemeyer, whose syndicate hired ghost writers, edited and illustrated their produce, bound and sold the volumes. Series such as "Tom Swift," "The Rover Boys," "Nancy Drew," "The Hardy Boys," "The Bobbsey Twins," "Camp Fire Girls," and "Pee Wee Harris" churned out volume after volume, each with better publicity and distribution than those of a botanist from The New York Botanical Garden, regardless of how many degrees and cachets followed his name. William Murrill, though, was not seeking fame or cash, but made an attempt to bring young people to nature at its most exciting and reassuring.

During what times of day (or night) this writing was done is unknown, but the long list of his Garden's positions and other books on the title pages of these books must have reached Nathaniel Britton. One wonders about his reaction—the Garden's cachet employed for fanciful musings and personal financial gain. That sickness may have kept Murrill away from his office desk already challenged Britton's work ethic, but it obviously surely shouldn't have allowed him to pen such drivel.

Seemingly parenthetically, in the proceedings of the Torrey Botanical Club for March, 1918 (TORREYA 18(3): 59–64. 1918), there appears a lengthy discussion of how the Club (and by association, the Garden) could (or should) consider numerous programs to include grade schools. Neither of the organizations had the offices or personnel to devise or handle such a complex program; but Murrill's exhortation "... there may be a yearning to open the child's mind to the sweet influence of Nature and lead the starved little souls of our crowded cities to the blessings of rural life, and the charms of our native woodlands..." would find endorsement also to be found in his "character" books.

THE "DISAPPEARANCE." And then, there was the alleged "disappearance" of Murrill for a protracted period in 1918. The theory was summarized by Rose (2002). "The Garden had authorized Murrill's first three European journeys in 1906, 1910, and 1913, but it was the fourth, in 1918, that ultimately led to a falling out with Nathaniel Britton and a tragic rupture in his career. On this particular voyage, Murrill failed to return to New York as planned, and after an alarming amount of time with no word from him, he was considered missing."

Because this "disappearance" was identified as the cause of his demotion from Assistant Director to the newly minted Supervisor of Public Instruction, it seems necessary to dig into the report itself. A reconstructed Murrill timeline for the year follows.

As a standing routine, the staff of the Garden (including students) convened at a monthly "conference" to hear and share in new Garden developments. At the January, 1918, conference, Murrill discussed the "chestnut canker," its spread, cost and lack of effective treatment. Next, the preface to "Billy the boy naturalist" was signed on February 14. In the April number of the JOURNAL OF THE NEW YORK

BOTANICAL GARDEN, it was reported that "At a recent meeting of the Torrey Botanical Club," Murrill reported on his studies of the gill-fungi of the U.S. In the May BULLETIN OF THE NEW YORK BOTANICAL GARDEN, Murrill submitted his annual report as Assistant Director. MycoPortal and Murrill's later publications summarize numerous specimens collected in Arkville, New York, by William and Edna Murrill in August. In late September and early October, Bronx collections are reported by Mycoportal. On December 10, 1918, Murrill signed the preface for "Three Young Crusoes."

On the negative side, no European collections were deposited by Murrill for 1918. Finally, while in spite of detailed schedules normally announced in the JOURNAL OF THE NEW YORK BOTANICAL GARDEN, no such can be found for a 1918 Murrill trip. (Of course, this omission has been part of the "mystery.")

A seemingly neutral coincidence took place at the December, 1918, staff conference: "Dr. H. A. Gleason, associate professor of botany at the university of Michigan and a former student of the Garden, was present and reported on his studies of the genus *Vernonia*." (J NYBG 20(229): 15. 1919). Innocuous to be sure.

All told, it would seem that no protracted time period for a "disappearance" in 1918 can be found, or for that matter, for European travel. Such a "disappearance," however, could be proposed for 1917 (see above).

The final nail was driven in the following number of the JOURNAL OF THE NEW YORK BOTANICAL GARDEN; a short note by Nathaniel Britton (1919) told the story. "Dr. Henry Allan Gleason has been appointed the First Assistant of the Director-in-Chief, succeeding Dr. W. A. Murrill, who has been transferred to the new position of Supervisor of Public Instruction."

In actuality, the position of First Assistant was not Murrill's latest position, but the position from which he had been promoted some years previous. Also, the position of supervisor of public instruction was not only new, but a previously unrepresented facet of Garden affairs. This shuffle in title and its assumed responsibilities have been historically interpreted as a profound demotion for Murrill, but it might also be seen as a response to his recently acquired talent for popular writing: first the polypore books, and now stories for boys and girls, only tangentially involving botany. "If this is what Murrill wants to do," Nathaniel Britton might have mused, "then let him do it full time." Murrill might not have shared this perspective. After all, the new position might remove even more time away from his chosen occupation, fungal taxonomy. Of course, no mention of all this backdrop appeared: the spotlight was on Gleason, not Murrill.

"Wets" and "drys". In the midst of the winter of 1918–1919, as Murrill's personal life was in turmoil, a nationwide paroxysm was consummated in a day. On Thursday, January 16, 1919, the eighteenth amendment to the Constitution was ratified, prohibiting the production, importation, transportation, and sale of alcoholic beverages. New York City shuddered. From the Battery to the Bronx, whether immigrant ghetto and wine, common laborers and beer, to trendy nightclubs and cocktails, alcohol was a major way of life. From manufacture to consumption, an important industry was suddenly unlawful.

Far from eradicating alcohol, the City took it underground; it is estimated that 40– to 50,000 "speakeasys" moved consumption from street level to one flight down, and made "protection" payments to police a Hollywood staple.

The moral and political precursors were as old as the colonies; the Civil War had drawn political lines, and religious bigotry had added seasoning on both the "wet" and "dry" sides. One source summed it up: "Many communities introduced alcohol bans in the late nineteenth and early twentieth centuries, and enforcement of these new prohibition laws became a topic of debate. The 'drys' presented it as a battle for public morals and health. The movement was taken up by progressives in he Prohibition, Democratic and Republican parties, and gained a national grassroots base through the Woman's Christian Temperance Union. After 1900, it was coordinated by the Anti-Saloon League. Opposition from the beer industry mobilized 'wet' supporters from the wealthy Roman Catholic and German Lutheran communities, but the influence of these groups receded from 1917 following the entry of the U.S. into the First World War against Germany" (Prohibition Wiki 2023). The pivotal vote: ratification by 46 of the 48 states. Virginia, solidly "dry;" New York, solidly "wet." Murrill, the Virginian, found himself in the land of the "wets," with alcohol illegal but everywhere. His kin folk found themselves in the land of the backwoods "still," "moonshine," and the local sheriff. Prohibition had its flaws, regardless of where one stood.

Then in the spring of 1919, two more books appeared: "The Naturalist in a Boarding School" (Murrill 1919a; for figure see Chap. 1)), autobiographical but including much poetry and aphorism, and "The natural history of Staunton, Virginia" (dedicated to Edna; Murrill 1919b) a factual tour de force relating his own observations while teaching in the town. For the latter, Murrill (1919a) reported that he had a diary and notebooks to aid his memory. Murrill's name as author of both was followed by the usual cachet, led off by "In charge of Public Instruction and of Higher Fungi at The New York Botanical Garden," perhaps a bit askew but the reader got the point.

In the "Boarding School" book, an entire chapter (Murrill 1919a: 140–145) was an allegory of his young-man's overpowering love for a young maiden. Scattered facts strongly testify that it described his attraction to the female gender and to one young lady in particular, all under the pseudonym, "Sigma."

THE TROUBLES. Years later, long after Murrill's separation from the Garden and Britton's retirement and death, this self-same Henry Allan Gleason (1961: 45–46) typed a very informal series of "Thumbnail Sketches of Botanists." His entry on Murrill deals specifically with Murrill's demotion. Because the typescript is not generally available, relevant lines are quoted here in full.

"William Alphonso Murrill, tall, dark, handsome and debonair, was the victim of unfortunate circumstances. I know nothing personally of his early career, but he came to the Botanical Garden a few years before me and soon became Assistant Director. His research was on the Basidiomycetes and in his work he made both friends and enemies by breaking up the old Linnaean genera *Polyporus* and *Agaricus* [sic] into numerous smaller genera, just as the old Linnaean genera of mosses had already been split by bryologists. He wrote extensively, published books, and prepared several parts of the North American Flora.

"After more than thirty years, only two live [in 1961] who could throw any light on the cause of the trouble between him and Britton. There were many rumors. Most of the Garden staff believed it was due chiefly to a long illness which kept him from his office for several months [Presumably the same illness that resulted in the first "boy Naturalist" book?]. No matter what the cause may have been, Britton demoted him to Supervisor of Public Education [sic] and ordered him to give a considerable fraction of the public lectures himself. So Murrill stayed on at the Garden, always quiet, never complaining, and never displaying anything of the cause of his troubles.

"Then domestic troubles were added. His wife, whom he had married while she was still in her teens, went to Nevada and secured a divorce. I believe that no one at the Garden suspected the reason for her trip; in fact, I believe she was not very popular with the Garden people. Some, and I was one of them, thought she had gone because of her health. It was quite natural then, that I should ask Murrill about his wife. He replied, 'Oh, she has her ups and downs.' The tone of his voice was such that I never made another inquiry." Gleason's additional remarks are not germane here.

For the botanical community in general, especially mycologists, Murrill's demotion was nearly incredible. One such was C. G. Lloyd, who, while detesting Murrill's plethora of split genera and "new species," nonetheless held him as a personal friend. Lloyd reached out to a contact in Albany, New York, for corroboration. Stewart H. Burnham had spent over 2 years on the staff of the Garden, and now was an assistant to Homer House in the office of State Botanist. In a letter dated March 11, 1919, Burnham reported: "It is an assured fact that Dr. Murrill has been relieved, I believe January 1 as First Assistant [sic] at the Botanical Garden. … When I was in the city, I learned that Dr. Murrill had been relieved: and that talk was Gleason, from one or two side-sources. That Murrill was to become Curator of the Garden. … This is what I know. We are aware what a good appearance Dr. Murrill makes and what a charming winning natural way and manner he has; unfortunately for Murrill, I suppose sometimes that he outshines one or two others. From something dropped by the 'power behind the throne' [Elizabeth Britton], when I was talking with her [about] something quite foreign to the above matter. I in the same way gathered she might be quite on the 'outs.'

"I'm sorry this has happened to Murrill: he is growing older and not always in best of health, I suspicion. I do not quite know how far Dr. Murrill believes the name juggling: but a good deal of it must be as you say 'a matter of politics.'"

But there were surely other underlying things simmering in the relationship with the Director-in-Chief. Murrill (1945a), himself, sketched one that occurred a few years before the demotion. In the chapter of his autobiography intended to resurrect some overlooked memories, he wrote the following. "Demands for more of the common cultivated flowers at the Garden became insistent as garden clubs developed and more women took an active part in promoting and assisting in the plantings. The Director had always been interested in wild flowers [as was the Director's wife] and moreover was color blind, so he vigorously opposed the idea, claiming that there was no room for ordinary flowers even if they were desirable. While he was in tropical America one winter the Naturalist in the capacity of Acting Director sought and obtained ... 250 acres of park land adjoining the Garden on the south for this particular purpose.

"In the 'Journal' for August, 1916, there is a detailed illustrated account of the plan for the magnificent Rose Garden established on a portion of this land .... On another portion an extensive Iris Garden was developed with a Rock Garden nearby. These features, with the Dahlia Garden so splendidly cared for by Dr. Howe, attracted crowds of additional visitors and made many new friends and cordial supporters of the institution." In the eyes of the Director-in-Chief, this must have ranked not only as subterfuge, but flagrant overstepping of Murrill's title.

So two men (Murrill and Gleason) entered 1919 with new titles, even though their tasks were ill-defined. Situations did not converge for them to comment on this peculiarity. In the BULLETIN OF THE NEW YORK BOTANICAL GARDEN of 1920, both reported their 1919 activities. Based on hierarchy, Gleason's (1920) report came first. "On reporting for duty at The New York Botanical Garden [Monday] February 10, I learned that no ex officio duties were attached to my position, and I was charged by you [Britton] with various administrative and executive duties."

On the other hand, Murrill (1920c) seemed quite familiar with his new responsibilities. "Collections of living and dried plants and plant products prominently exhibited and carefully labelled; natural woodlands, fields, and meadows, with all that they contain of beauty and interest; floral exhibitions at stated intervals; a system of docentry by means of which the collections may be visited every week-day under expert guidance; lectures and demonstrations adapted especially for those interested in special phases of botany and horticulture; and regular public lectures covering a wide range of botanical and horticultural subjects." Murrill may have conceived much of this litany himself, but surely Britton couldn't complain (after all, before publication the report came to Britton, not directly to the public or corporation).

Shouldering its part of the war effort, in collaboration with the federal government, a program of gardening courses designed specifically for convalescing soldiers and sailors was announced by the Garden in spring, 1919. Under their new staff titles, the plan would seem to fall squarely into Public Instruction, and therefore, part of Murrill's portfolio. Nonetheless, in the 1919 report (B NYBG 10(39): 230. 1919), Gleason reported giving the program aid and advice, while Murrill was not mentioned. From hundreds of inquiries, 56 individuals were enrolled, and the program was enlarged somewhat later.

THE MET. For Murrill, two other events took major contributions of time and energy. The first was a collaborative effort with the Metropolitan Museum of Art (Fig. 6.2), highlighting the depiction of plants in historic artistic designs, linked to living examples of the plants involved. "There was an Acanthus Exhibit, a Lotus Exhibit, a Grape Exhibit, Persian Rugs with Pomegranate motives, Books on the subject, etc." (Murrill 1945a, Chap. 20; Murrill 1919c). The actual public program lasted about 5 weeks at the Met, with the plants and plant products furnished by the

Fig. 6.2 The Grand Salon, Metropolitan Museum of Art. The Grand Salon, Metropolitan Museum of Art, New York, c. 1900. (Courtesy Metropolitan Museum website)

Garden. Of course, much planning was required, given the synchrony of blooming and arrangement of the exhibit itself, and collaboration within the Garden including gardeners and other staff. Murrill, himself, lectured daily for the full duration of the exhibition and used its subject matter as individual lectures through the following year.

Soon, the success of the Met collaboration resulted in a new idea: "a competition in the use of plants in decorative design was held at the [Garden's] Mansion (Fig. 6.3) in April, and was participated in by 26 art students of the city" (Gleason 1919 [not in bibliography, maybe 1920?]). A detailed account of this was published in the Journal of the New York Botanical Garden (20: 124–126). The announcement came from Gleason, not Murrill.

The second time-sponge was the reorganization and moving of the thallophyte—fungi and algae—herbaria and its curators from the third floor of the museum building to the second in an attempt to allow expansion of the higher plant collections. Again, Murrill was heavily involved as *ex officio* "curator of higher fungi." Coupled with the move, of course, was the loss of his office juxtaposed to the Director's, and acquisition of "study space" with Howe and Seaver, where he would now be "odd man out," but presumably also "out of the line of fire" from the Director's office (Fig. 6.4).

Fig. 6.3 Lorillard Mansion at the garden. The Lorillard estate mansion, part of Botanical Garden property. (Journal, New York Botanical Garden)

Fig. 6.4 The Museum building completed. The Museum Building (now called Library Building) of the Garden after completion. (Courtesy, Botanical Garden website)

The ambiguity of Murrill's new charge was further demonstrated by his participation in two gatherings of plant pathologists, which normally would have been in Seaver's territory. The first, in August, 1919, took place in Connecticut, convened "to study and discuss local problems connected with crop diseases. This was attended by The Naturalist (Murrill 1945a), who took a lively interest in the discussions, collected a number of specimens and had himself photographed with the group under the giant Weathersfield Elm." The second emphasized potato diseases, and was on Long Island, again mostly a motorized tour of small farms and their products. While Seaver officially represented the Garden, Murrill was invited.

In his July, 1919, vacation, Murrill returned to the Virginia mountains. In particular, Mountain Lake, just west of Blacksburg, was again explored, but this time, even though the weather was cooperative, only a motley remnant of his collecting, almost all polypores, was eventually accessioned into the Garden herbarium. A goodly lot of the fleshy sorts was eaten *in situ* but portions were not saved. He detailed, however, that his return was by way of Lynchburg and Falls Church, the latter the family home of Edna (J NYBG 20(237) September, 1919: 185).

In September, C. G. Lloyd came through New York on his way to Europe. He spent several days at the Garden, and surely discussed Murrill's new office, the herbarium, and Murrill's attitude toward mycological research.

YAMA FARMS. The year under his new title drew down with a November lecture at Yama Farms (2023), a resort in the southern Catskill Mountains. Yama Farms Inn [yama = mountain in Japanese], established only a few years before Murrill's visit, was the product of a well-travelled couple deeply impressed with Japanese culture, architecture and gardens. Its location, in Ulster County near Ellenville, required a day-long train and auto trip from the Bronx. Although Murrill had experienced resort amenities before, Yama Farms was unique in catering to people of the intelligencia—authors, musicians, thinkers and diplomats (Yama collection 2023). To this day, Yama still is associated with the "Famous Four," Henry Ford, Thomas Edison, John Burroughs and Harvey Firestone (Harris et al. 2006), who shared concurrent vacations there, and it was said that the role of the U.S. in World War I was shaped at Yama Farms (Yama Farms 2023). Murrill had apparently been invited to lecture, and took advantage of the opportunity to do some collecting. He acknowledged his sponsor, who seemed well-acquainted with the area and accompanied him in the field (Murrill 1920d). His lecture subject was "Edible Mushrooms," and the result of his collecting was mushrooms for luncheon for 23 guests. On the menu was *Pleurotus* (*Sarcomyxa*) *serotinus*, the "slimy pleurotus." He confessed that it was not a good choice, but acted as a filler.

Later, in the Garden Journal (J NYBG 22(260): 173. August 1921) two specific cases of interest were reviewed. "The sweet-scented polypore, *Trametes suaveolens*, which is rather common on willow, was found on the large-toothed aspen, and on the same tree a splendid specimen of *Inonotus perplexus*, which latter specimens have caused considerable discussion because the type at Albany on which Peck based his description has been destroyed."

The result of his visit (Murrill 1920d) would usher in a fascination with the place, its clientele and the countryside. A year later (MYCOLOGIA 13(2): 116. 1921), Murrill could report: "The first number of the BULLETIN OF THE YAMA FARMS MYCOLOGICAL CLUB appeared in September, 1920. It contains a description of Yama Farms: the origin and purposes of the Club; plans for the future; a list of books and papers on the larger fungi; and a list of officers, including John Burroughs [a member of the "famous four"], W. A. Murrill, H. D. House (NY State Botanist), C. F. Millspaugh (Field Museum, Chicago), G. T. Moore, William Trelease (Missouri Botanical Garden), H. I. Miller (Murrill's host on his first visit), C. H. Kauffman (Univ. Michigan), Howard A. Kelly (Baltimore), Robert T. Morris, and others. "The Club intends to make Yama Farms, a vast virgin tract in the southern Catskills, an important mycological center, with facilities for collecting and studying the fungi and other interesting forms of plant and animal life. Mrs. O. B. Sarré [one of the Farms' founding couple] is permanent secretary-treasurer ..." Murrill could be proud of his accomplishment.

Murrill returned to Yama Farms again in May, 1921, this time discovering the rare moss, *Buxbaumia aphylla*. In the northeastern states, the species had been found only once, on the grounds of the Garden, nearly 20 years earlier, and then on decayed wood. Murrill wrote: "The soil at the Yama Farms locality was carefully examined under a lens and no decayed wood was found in or near it (J NYBG 23(269/270): 76. May/June, 1922). "Considerable fieldwork" at Yama was acknowledged in his report to Britton for 1921 (B NYBG 12(42): 25. 1922). October, 1921—including October 18 to be specific—saw him at Yama again, this time once more collecting the interesting polypores considered rare (Murrill 1922a). In 1922, he again returned, and acknowledged being President of the Yama Farms Mycological Club (MYCOLOGIA 15(2): 102. 1923). In September, 1923, he returned to Woodstock, with a side trip to Yama Farms (Murrill 1924a).

THE VOTE. In these days of 2024, we cast a disparaging eye toward societies in which women's roles are governed by males, but even in our country, females statistically still lag financially behind males in the work force, even though more females are entering college than males across the country. The "equal rights" amendment to the Constitution has yet to be ratified. Susan B. Anthony and Elizabeth Cady Stanton are names still prominent in U.S. historical literature, but the fruits of their labor were harvested over a half-century of campaigning, and William A. Murrill saw its progress in the City, culminating in women's suffrage across the nation in 1920.

In the wake of the Civil War, the fourteenth amendment to the Constitution guaranteed male freed slaves the right to vote in federal elections. Scattered women's rights group fought to have women included, but failed. In subsequent decades women entered the work force in great numbers, but by 1900, 60% of jobs of working women were as domestic servants (Library of Congress 2023a). By 1896, women had gained the right to vote in four states (Wyoming, Colorado, Idaho, and Utah), none in the more populated and industrialized East. Rights were acquired piecemeal: control of their own earnings, property, and in the case of divorce, custody of their children.

In economics, the turn of the twentieth century, "the golden age," also the era of "robber barons," was also known as "the progressive era." Teddy Roosevelt put the presidency behind "trust busting" as well as conservation and environment. In New York City, politics swung between conservative and reform movements, perennially against a backdrop of Tammany Hall, and always with an undertow of universal suffrage. The infamous Triangle Shirtwaist Fire in 1911, shocked the City, but pointed up the workplace conditions for many women. And through it all, there emerged a slice of the City's work-week population, largely commuting from New Jersey, Yonkers, Long Island, and Connecticut, able to afford the fare and living in a mortgaged house, perhaps with a couple kids—it became known as "the middle class" (Library of Congress 2023b).

The first attempt to gain women's suffrage was put forward 1878, but was promptly turned back. The eventual push came from World War I, when women flooded the workplace to replace conscripted men. At the end of the Great War, American men had experienced European society and had a new respect for the abilities of women. The wartime conundrum was obvious: how could the U.S. fight in Europe to preserve Democracy, when true Democracy was not practiced back home?

The actual document was written in the negative: "The right of citizens of the United States to vote shall not be denied or abridged by the United States or by any State on account of sex. Congress shall have power to enforce this article by appropriate legislation." Once proposed, Congress quickly passed it, and ratification was obtained in mid-August, 1920 (Nineteenth amendment Wiki 2023).

Although we know little about Murrill's personal circumstances at that moment, we do know that the secretaries at the Botanical Garden could now vote, and so could Edna Murrill, regardless of marital status or location within the states.

DOCENTRY. Among Murrill's new duties was "Docentry:" supervision of a group of second echelon staff acting as hosts, guides, and tour leaders for the visiting public at the Garden. In his report for the year: "My personal attention was given to at least 10,000 persons during the year; while Mr. Wilson recorded about 1,500, Mr. Williams about 1,000, and Mr. Becker 800. The total probably reached 15,000" (Murrill 1920c). A particular subset of the public was school children. Corralled by their teachers, classes ranging from second or third grade to high school were welcomed and shepherded to parts of the Garden assumed attractive, and given a talk by a staff member. An additional group was Girl Scouts (B NYBG 12(42): 24. 1922). Of all the "Public Instruction" falling under Murrill's purview, this may have been the facet to which he could relate, given his ongoing association with Girl Scouts and other young folk. In his annual report covering 1921, Britton wrote: "The increasing amount of time required for this very important program [docentry], makes it desirable that much of it be concentrated in a docent, with this as his principal occupation, which would relieve other officials for work much needed on the collections or otherwise" (B NYBG 12(42): 6. 1922). This would seem to be a partial expiation for Murrill.

LARGESSE. On the subject of public charity, Murrill related his praise for Lloyd's gift of his Kentucky house to a local organization. Murrill [2 March 1922] wrote that his way of matching Lloyd's generosity was giving the Girl Scouts of the Bronx the privilege of using most of his house. Years later (Murrill 1945a) he remembered: "[His] musical ability served him well in work with the Girl Scouts and other groups of young people. Many of his songs were composed for teenage voices. Churches always welcomed his troops of singing Scouts."

NATURE STUDY. The program of Garden docentry played directly into a nationwide pattern becoming known as the American Nature Study Movement. The Movement resembled a similar epoch that engulfed England a half-century prior. In the mid-nineteenth century British households were frolicking in their rare spare time with butterfly net in hand, or jar of water for tadpoles, or sorting through rocks to discover fossils. Darwin was one of them. Like fashion, the Nature penchant crossed the Atlantic, where it took some years to mature into an espousal by philosophers, professors, and some preachers. The "Nature Study Movement" was popular and strong enough to see both publicity and field trips. One of its strongest advocates was Liberty Hyde Bailey, through whose orchard Murrill had to cross on his way to classes at Cornell, and as conservation became part of the fad, Theodore Roosevelt and Gifford Pinchot were evangelists. As Armitage (2009) put it: "Americans flocked from their homes and their classrooms to nearby forests, prairies, rivers, and mountains to gain greater insight into the wondrous workings of nature. 'Nature study,' as enthusiasts dubbed the movement, used instruction in basic natural history, such as plant identification, animal life histories and school gardens, to promote the skills needed to succeed in industrial life and to cultivate the spiritual growth that modern life occluded."

William Murrill, strictly speaking, was the Nature Study Movement turned on its head. Instead of the secular child raised at a distance from nature, he was integrated into nature from birth. Instead of a teacher attempting to meld an appreciation of nature into the secular child, he appreciated his rural surroundings simply. When he shot a bird for its plumage, it was to integrate it into his collection of birds. He was a part of the Nature Study Movement without knowing it. Even later, when he took his girl students into the forests and fields, it was not as a pedagogical exercise, but as a sharing of his own appreciation of natural objects, coupled with praise from his charges. Although he surely contributed to the "Nature Study Movement," it was not a conscious endeavor, but a sharing of fresh air outdoors. And when he "decided" upon a future as "a Naturalist," his aim was as collector and score-keeper, not as a bland teacher of Plane Geometry, Latin declensions, beginning Chemistry, and Nature.

As a testament to his personal relationship with C. G. Lloyd, Murrill [letter 9 June 1922] asked Lloyd to send as representative a set of polypores named by him [Lloyd] for the Garden Herbarium. Lloyd [14 June 1922] replied in the affirmative, and the set was promptly donated.

Murrill's (1920c; B NYBG 12(42): 25. 1922) chronic complaint of little time for "personal investigations" was a bit hollow. By this time, he had ushered the self-chosen parts of North American Flora—polypores, boletes, and the genera of "gill-fungi" generally represented by large fruitbodies — through the press to publication. What was left in the way of major monographs was difficult to predict, and his bibliography for 1920 abounded in short, one-topic subjects. Of 24 listed papers, 14 were less than two pages long. Four were longer: two on resupinate polypores (Murrill 1920e, f), one a rationale for visiting his beloved Virginia mountains (Murrill 1920a), and one on numerous corrections and additions to his own polypore papers (Murrill 1920g). Even number 32 of the "Illustrations" series, while requiring some extra work with the colored plate, was only three pages of text (Murrill 1920h),

The trend extended into 1921. A pivotal short note (Murrill 1921a), however, may reveal more about his mental state than he might have desired. It is included here in its entirety to reflect its perceived importance. "The first three parts of this volume [of North American Flora] were issued some time ago. The manuscript for part 4, prepared by Kauffman and Overholts, will be ready for the printer within a few months. Part 5 will be chiefly devoted to *Cortinarius*, to be treated by Kauffman. Part 6 will continue the brown-spored and black-spored agarics; and part 7, the gasteromycetes and an index concluding the volume." Thus was laid out his chronological roster for upcoming but undefined months.

"Specimens of gill-fungi with brown or black spores, or any of the Gasteromycetes, will be very gladly received from mycological friends. I do not care for *Poria* at present; this group will have to wait until volume 8 [of North American Flora], containing the Thelephoraceae, Clavariaceae, Hydnaceae, etc., is well started." Although seemingly discarding *Poria*, he immediately preceded this note with another large offering (Murrill 1921b) and immediately followed the note with two small papers, all on the group (Murrill 1921c, d). The two short notes were taxonomic, but the longer paper was a discussion reporting no new studies but only the work of others, including his own prior studies in European herbaria.

His note continued: "The determination of miscellaneous collections of the higher fungi must take second place with me henceforth, as my time for scientific work is limited. I have enjoyed this kind of work intensely during the past twenty years, and a vast number of interesting things have been added to the herbarium through collections sent in from widely separated localities." Here was the declaration of the choice of occupation—the naturalist versus the administrator. Later, this would be framed as the choice between his chosen love versus the size of his salary.

"If collectors wish to deposit sets of their large fungi here without expecting reports until the various groups are worked, such specimens will be welcomed. In the case of special plants sent in for critical examination, please mention the species with which you would have them compared and also give microscopic characters, so as to facilitate comparison as much as possible." These closing paragraphs read as distinctly Lloydian—determinations in return for designated specimens—but the tedium of microscopic examination was to be accomplished by the sender (drastically reducing the number of corresonpdents), not the recipient, a hint of confrontation years in the future, especially with Rolf Singer in the 1940s.

If there were a semaphore or smoke-signal outlining Murrill's mental attitude, this short note might be such. Adjusting its date to account for time in press, it was written late in 1920 or early 1921. It is interesting to keep it in mind as time passes. The note was signed—Supervisor of Public Instruction. It could well have been "The Naturalist."

MOTHER OAK. Whatever mind-liberating field work for the Naturalist had to be relegated to vacation, given Britton's admonishment surrounding Murrill's new position. Based on what had come before, even a casual reader could predict a likely destination for Murrill's latter half of July, 1920. Much as a worker bee returns to the hive, Murrill (1920a, b) repaired to the Virginia hill country, as he had many times previous. Likely, it was time to lay his head against mother oak and take comfort in her bole.

The result was clearly a paper by a naturalist, taking notice of the biome ("mostly in oak-chestnut groves with white oak predominating"), its soils and underlying formations ("in Blacksburg, Trenton limestone, while the Brush Mountain, a mile or two to the north, the soil had its origin in Subcarboniferous shales and sandstones").

To present-day examination, Murrill's list of fungal collections is annotated, usually in the first person. Some collections on the list are not represented in the NYBG herbarium (i.e., comparison of the paper to MycoPortal). In no case is there a collection number, so collection names are apocryphal unless checked against herbarium material, but the purpose of the paper was not strictly scientific. Instead, it seems a substitute for a diary (or perhaps a distillation from one). To be a bit pedantic, the paper is not fastidiously scientific—the reader is left with no clue as to the diagnosis of a named organism except folklore.

Blacksburg specimens are at the Garden, together with those from "Montgomery County" (without more accurate location). There seem to be none from Washington, Mount Vernon, Falls Church (Edna's family home), and other locations named in the paper.

The statistics for 1921 narrowly echo those for the preceding year: mostly short papers, virtually all in Mycologia. Under "Personal Investigations," the summary in his annual report (B NYBG 10(40): 336. 1921) was lean—only a few lines long: "A large share of the time I could spare from public instruction has been devoted to work on Mycologia and two guide-books, one covering the buildings and grounds, and the other the economic museum.

"A little mycological field-work was done during my vacation in Virginia [Blacksburg] and the southern Catskills [Yama Farms]. Two articles on resupinate polypores, containing descriptions of many new species, have been prepared and published during the year [up to September, 1921]."

In fact, of the guide-books he mentioned, the first (Buildings and Grounds; Anonymous 1920) was uncredited, but required exhaustive research and familiarity with the Garden and the second, encyclopedic one on the Economic Museum was credited to H. H. Rusby (1921), the Honorary Curator of the Economic Museum, again with no acknowledgement of Murrill. In addition to public lectures and "docentry" for visitors, the workload for Murrill, therefore, included enormous editorial tasks—North American Flora, Mycologia, and the guide-books—surely mind-numbing for The Naturalist, and mirrored almost a century later, by the twentieth-century Garden mycologist (pers. comm.; Rogerson & Samuels 1996).

In his autobiography (1945a, Chap. 22), Murrill furnished a partial commentary on his visits, collecting and importance of repetitive returns to his home territory. This in addition to the basic sentiments on boyhood and education already in book form (Murrill 1919a, b; also see above). In the autobiography (Murrill 1945a), written 20+ years after his 1920 vacation visit (Murrill 1920a), and penned in a rather different world, less fettered mental state, and an amenable sociological millieu, he extolled one of his favorite haunts, the Peaks of Otter, with nearby Bedford and Apple Orchard Mountain in cameo roles. While his words are not sentimental or maudlin, they somehow convey more than simple fact. "The mountain is covered with trees entirely to its summit. Chestnut trees were large and abundant until killed by the canker. Oaks are now the chief trees; the northern hemlock is frequent in most places; sugar maple and yellow birch grow near the waterfall; while linden, dogwood, and sourwood rarely get above twenty-five hundred feet elevation. In October, the witch-hazel is abundant everywhere, with its strap-shaped, golden-yellow blossoms." These are the observations of a naturalist. Later in life, he must have been pleased to learn that much of this vista was being preserved as the route for the Blue Ridge Parkway (2023) and various government Forest Service units.

In early November, Murrill lectured the International Garden Club on "Edible wild mushrooms," but as 1921 drew to a close, Murrill was on the road. In mid-December, he observed several clusters of gill-fungi in Blacksburg, Bedford and Virginia environs (Mycologia 14(1): 44. 1922), but he was in Toronto, Canada, for the AAAS meeting between Christmas and the New Year, summarizing some papers read by plant pathologists (Murrill 1922b). It is memorable that in spite of finishing his third year as Supervisor of Public Instruction, his energy was devoted to mycological—probably more accurately, naturalist's—pursuits.

THE CLAVARIAS. According to the Journal of the New York Botanical Garden, William Coker made roughly semi-annual visits to the Garden to consult the library and herbarium. Several visits were targeted at gill-fungi, which were included in published floristic monographs. In 1920, however, his target changed to "the clavarias," a group which seemed not high on Murrill's list for work. At Missouri Botanical Garden, E. A. Burt (1922) was working on the Thelephoraceae, but had plans for the clavarias as a follow-up. In fact, he visited the Garden to check some clavarioid specimens in summer, 1921 (Mycologia 13(6): 335. 1921). When Burt's paper appeared, it was briefly reviewed by Murrill: "The North American species of *Clavaria* are treated by Burt in his usual excellent style in the Annals of the Missouri Botanical Garden for February, 1923, which appeared in August [1923]. Ninety-eight species are included, many of them being figured on the eleven closely crowded plates" (MYCOLOGIA 15(3): 148. 1923).

Murrill's high opinion of Burt was expressed again years later (Murrill 1945a). He wrote: "Edward Angus Burt … received the Ph.D. degree at Harvard in 1895 and was appointed professor of Natural History at Middlebury College, in Vermont, the same year. Later, while mycologist at the Missouri Botanical Garden, he published several important papers on the Thelephoraceae of North America, this group being his specialty. Dr. Burt was [Murrill writing in 1945] a thorough student and one of our best mycologists. I shall always be indebted to him for working so patiently on the thelephores I collected in tropical America and elsewhere."

Coker and Murrill surely chatted about the coral and club fungi (the clavarias), and Murrill sent specimens from his recent Yama Farms and Virginia collections to Chapel Hill, keeping no duplicates for the Garden. Coker's (1923) book did not appear immediately thereafter, and when it did, its title, "The clavarias of the United States and Canada," was hopelessly encompassing. Perhaps Atkinson, with his collecting trip of 1918 to the Pacific Northwest, could have advised Coker of the scores of coral fungi there, not to be seen in eastern North America. In fact, when Atkinson died in Tacoma, Washington, his availability diminished materially, but even had he lived, he held Coker's taxonomic skills wanting (Petersen 2021). Among the several workers acknowledged in Coker's introduction, Murrill appeared first—Atkinson not at all. Coker's European sources read like a Murrill (or Lloyd) itinerary, especially his esteem for Abbé Bresadola. As a gratuitous gesture, Coker sent a sample set of specimens to the Garden for use in determining future collections (J NYBG 24(288): 288. 1923).

Coker's inspection of the Garden's fungi revealed a new species, which Coker named *Clavaria murrillii.* We would know it as a *Ramaria*, collected at Unaka Springs, Tennessee, by Murrill in 1905.

For some months, Murrill had been working on a new series of papers outlining the dark-spored agarics in North America, but not as part of NORTH AMERICAN FLORA. A February visit to the Peck collections in Albany (MYCOLOGIA 14(2): 90. 1922) was rapidly followed by the first installment of the series in the March issue of MYCOLOGIA (Murrill 1922c). As usual, his large "database" was consulted for information on similar European collections and possible synonymy.

It is interesting to note that spore color is used as fundamental to the arrangement of the gill-fungi, but inspection of the keys and descriptions show that spore ornamentation was downgraded, while spore size was elevated in importance. Equally, as usual for Murrill, a topotype location was designated for each taxon, but no type specimen, and often, plural type specimens were allowed. Finally, pre-Friesian literature was freely cited, violating the newest, prevailing Code of Nomenclature. The most valuable information in the series was in the keys and descriptions as aids to recognizing the taxa. The series (Murrill 1922c, d, e, f) was not illustrated.

Another significant, albeit clerical, paper was an index to the long series "Illustrations of fungi," which had reached 33 installments (Murrill 1922g).

Statistically, 243 species had been illustrated, 213 in color on 29 plates. Interestingly, the latest chapter used black-and-white photographs taken more than a decade before publication, quite in the fashion of Atkinson and Kauffman.

A sentiment not previously emphasized followed. Murrill (1922g) wrote: "The idea of a great illustrated work on American fungi has been in my mind for many years and I have frequently spoke and written about it. Several hundred colored drawings have been prepared [several by Edna] with such a work in view. When the raising of funds for this purpose became increasingly difficult, I decided to publish in a comparatively inexpensive way as many species as possible as quickly as possible; so that collectors might be stimulated to increase our knowledge of the fungous flora and thus prepare for a more comprehensive work when its publication should become possible. From the large number of letters and specimens received during recent years, I am convinced that it was wise to publish when I did. As one writer expressed it: 'I'd much rather have a small colored figure now than a handsome folio plate after I am dead.'

"I still have faith, however, in the patriotism and generosity of American men of wealth, who will make it possible to supply nature-lovers in this country with ample, correct, and beautiful colored illustrations of our extremely varied and interesting fungous flora." Not mentioned was the Lydig Fund of the Garden which appeared on every number of every Mycologia volume. Equally, the work by Krieger for Howard Kelly and William Farlow was not cited. Surely one prominent example of the difficulty to publish fine illustrations, this of flowering plants, was Addisonia, another Garden periodical always close to running out of funds. Addisonia was an eponym of Addison Brown, the venerable founder of the Garden (see Appendix 1).

PROFESSIONAL MIGRATION? In these early years of the second decade of the twentieth century, by inspection of the Bulletin of the New York Botanical Garden, the annual report from the upper staff to the Board of Managers was in the form of subsidiary reports to the Director-in Chief incorporated into his overall summary. For purposes of this narrative, two reports have been emphasized, Britton's (1924) and Murrill's (1924b). Comparison of Britton's reports (1921 through 1923) show a growing appreciation of the Garden's efforts toward the public and therefore, with the first mention of Murrill by name, the direction is obvious. He wrote (B NYBG 12(44): 181–191. 1923) "Under direction by Dr. Murrill, our system of public instruction and information has been continued and expanded, aided by the income of the Darius Ogden Mills Fund. More public lectures have been delivered than in any previous year, through the aid of income from the Francis Lynde Stetson Fund, and there has been an increased demand for docentry service. Greater publicity has been given to the work of the institution through newspaper notices of special features and abstracts of lectures. All members of the staff have participated in this educational work, and in giving information by letters to hundreds of correspondents. Weekly radio talks on gardening topics, by staff members, with special reference to our plant collections, were broadcast through station WEAF, of the American Telephone and Telegraph Company, from April to September." All these efforts were Murrill's, who went into additional detail in his report (Bull NYBG 12(44): 199–201. 1924). Notably, under "*Personal investigations*" for the first time, Murrill noted: "The work of *my department* (Italics mine) has expanded considerably during the year, especially as regards lectures, visitors, and the giving out of information by correspondence and through the public press. Editorial and curatorial work, with the determination of an unusual number of mycological specimens, has been continued as formerly." Some details were summarized earlier in the report: "Sixty-nine illustrated public lectures were given in the Museum Building on Saturday and Sunday afternoons from April 1 to November 25, inclusive. ... A series of four illustrated lectures were given by me in early spring to the Girl Scouts of the Bronx in the Museum Building, as reported in the Journal." Under docentry: "A great many special appointments were made with visiting classes from other schools and colleges in and near New York City, such as Columbia University, Rutgers College, Hunter College, high schools in Manhattan, Brooklyn, Newark, etc., and special schools like the summer school at Cold Spring Harbor and the school of floriculture conducted by Max Schling....Many parties of Girl Scouts and Mount St. Ursula biology pupils also visited the grounds at various times during the year."

The two-decade wishes of the Torrey Botanical Club—and Murrill—were coming to fruition.

Almost as an afterthought: "My researches during the past year [1923] were confined chiefly to studies in the field. A month was spent in exploration in Florida, and all of my vacation was devoted to collecting trips to Virginia, Woodstock [NY], Interstate Park [NJ/NY], and Lake Placid [NY]; accounts of which appeared in the Journal and in Mycologia. Several taxonomic articles were prepared for Mycologia and a new series of lectures for the museum course."

Two more items, usually overlooked, are as follows. Of the $4000 from the Lydig Fund appropriated for Mycologia, the balance at the end of the year was $15.72. Moreover, the lion's share of investments of the Garden Endowment were in railroad stocks, a very few in utilities, a situation that would require change in the not-too-distant future.

SUPERVISOR?. Inspection of Murrill's timeline for 1923 is revealing. Its summary follows: February Staff Conference program: review of dark-spored agarics

March 1. Lecture: "Wood-rot fungi" to Tree Protection Institute, New Haven, CT
March 6. Gold Medal from Holland Society, at Hotel Astor; NYC
March 8–20: Month in West and East Florida + Keys.
April 8. Public lecture: "Nature talks—I. Stars"
May 12. Public lecture: "Nature talks—II. Rocks"
May 20. Public lecture: "Nature talks—III. Soil"
June 10. Public lecture: "Nature talks—IV. Ferns"
July 1. Public lecture: "Nature talks—V. Mushrooms"
July 10–20. Mountain Lake, Blacksburg, VA
July 22. Public lecture: "Nature talks—VI. Punk fungi"
July 30–31. Girl Scout camps (including lectures), Interstate Park.

August 5. Public lecture: "Nature talks—VII. Cycads and pines"
August 10. Girl Scout memorial meeting, Montclair, NJ
August 11: Walking lecture: "A walk through the Hemlock Grove"
August "early." Nearly a week in southern Catskills (Yama Farms)
September 1. Public lecture: "Nature talks—VIII. Useful trees"
September 3–5. Woodstock, NY/Yama Farms
October 19–23. Lake Placid Club, NY
December 29. Public lecture: "Cocoa and Chocolate"

The increased number of scheduled public lectures conformed to Britton's wishes, but the early subjects may not have done so. "Stars," "Rocks," "Soil," were marginal to the Garden's mission, but reflect Murrill's self-analysis as the Naturalist. Only two were mycological, and they were largely repeats of former subjects. "Ferns" took Murrill back to his first naturalistic interest even before graduate school, and his fond relationship with Lucien Underwood. Even the overall lecture title, "Nature talks," would seem indicative. Titles of other lectures might have been more appropriate for other staff members—"Cycads and pines," and "Cocoa and chocolate."

As reported above, Murrill considered the Florida, Interstate Park, and Woodstock events as his major travels for the year. All, however, are open to closer scrutiny.

Tucked in just before boarding the train for Florida in March, "Dr. Murrill was awarded a gold medal by the Holland Society of New York at its annual meeting at the Hotel Astor on March 6, in recognition of his distinguished service in the science of Mycology. After the presentation, he gave an illustrated address on 'Fungi and Their Relation to Forestry in America'" (J NYBG 24(283): 104. 1923; Mycologia 15(4): 188. 1923).

Florida: Murrill's experiences in Jamaica, Mexico and (briefly) Cuba, supported by his search for specimens from the American tropics in European herbaria, presented theretofore unsampled possibilities. Surely, the annual, wordy, meticulous, almost daily reports by John K. Small, staff member, of both general and specific botanical ecology of Florida would have raised interest. But his own narrative (Murrill 1923a, b) described much time window-shopping from trains (J NYBG 24(279) 72. 1923); strolling the neighborhoods around hotels, and spending time with well-known botanical leaders; moreover, there seem to have been no facilities or opportunities for drying specimens, which limited the kind of collecting possible. Finally, his description of Gainesville might plant a seed to his destination in "retirement."

The Interstate Park event coupled equal ingredients of science, education, and delight. In reality, it was a visit to two large Girl Scout summer camps, at which he gave talks and met with leaders and especially the girls (J NYBG 24(284): 168. 1923). The timing was in late July, and a month later, he attended a memorial Scout meeting in Montclair, New Jersey, where he arranged to have a museum constructed to exhibit natural history specimens collected by the girls in pursuit of merit badges (J NYBG 24(285): 190. 1923). His presence might have counted as representing the

Garden; it could also have counted for his enthusiastic involvement with the Girl Scouts (J NYBG 24(280): 88. 1923)—or both?

Although billed as a trip to Woodstock, NY, to give a talk to a group of folks at the summer home of a wealthy doctor, John A. Kingsbury, Murrill's host had also become a member of the Yama Farms Mycological Club. Murrill (1924a, c) described a motor trip from Woodstock to the Farms and some collecting there (J NYBG 24(287): 267–268. 1923). The locations of the relic of Yama Farms Inn and Woodstock (i.e., Google Earth) show such a motor trip improbable unless of a full day's length. The length of the trip, therefore, may be inaccurate, but it reveals Murrill's enthusiasm for Yama in a fourth successive year.

The Lake Placid trip (Murrill 1924d), again, was reported in a naturalist's vocabulary. It was at least his second visit to the resort and again, it was at the very end of the mushroom season. He doted on the differences in the northern Adirondack flora as compared to the Garden (and Yama), but one could imagine that the most common gill-fungi might have been in *Cortinarius*, the difficult genus under study by Calvin Kauffman. Even fall hunting would have been hampered by the bright colors of fallen leaves of aspen, birch, maple, mountain ash and such.

A fourth trip, though, was worthy of note—that to Blacksburg and Mountain Lake, Virginia. Home again: in mid-July, in mid-season for fleshy fungi. Having described several previous visits to the area, Murrill took the opportunity to describe Mountain Lake in natural historical terms (Murrill 1923d). In just 1 year, western Virginia would become his temporary refuge.

For the curious, though, close inspection of Murrilliana of the period shows no mentions of Edna. In prior years there would have been some mention, herbarium label or other clue indicating her presence. It might also support the very indirect idea (Gleason 1961) that she had left William and obtained a divorce sometime prior.

Of course, Murrill's daily Garden routine in docentry could not be included in his timeline, for it was only scheduled piecemeal. Omnipresent were available guides for visitors to the Garden, however casual they might be, some with appointments, others discharged from the elevated train to find themselves in a gigantic display of floral beauty, if they could only find their way to the appropriate plantings. Still others who had read a note in the newspaper—written and communicated by Murrill—about some exhibit on poison extracted by South American Indians from plants: where was the exhibit and how to get there? The library? The lily garden? The zoo? Where was the nearest restroom? For the Supervisor of Public Instruction, it all required an encyclopedic, meticulous knowledge of the property gained through constant patrolling and interviewing of gardeners and resident staff. And perhaps most important, making sure there was someone to stand in during times away from the City. Since its introduction, the docentry program had grown, and for years the Director-in-Chief had sought to establish a position to head these workers, but the budget never quite allowed such a thing, and Murrill's responsibility grew apace. Where was mycological research, and the docent to take one there?

In spite of such clutter, Murrill was able to inject a small dig at C. G. Lloyd, with whom he was friendly in spite of differences over "name juggling." Lloyd (see above) was perpetually annoyed at Murrill's constant multiple descriptions of multitudinous new or resurrected genera, species, and combinations of same, and especially Murrill's name "advertised" *ad nauseum* following all of them. In a note, Murrill recalled Lloyd's last visit: "Dr. [sic] C. G. Lloyd called at the Garden late in August [1922] on his way to Kew Gardens, England, to examine type material in certain groups of fungi. He expects to visit France and Italy also before he returns" (Mycologia 15(2): 96. 1923). Immediately following: "The polypores of South Africa were listed, keyed and described by van der Bijl in the S.A. Journal of Science for June, 1922. ... The author acknowledges the cooperation of Mr. C.G. Lloyd in this work, and his [Lloyd] name appears after a number of species not mentioned [here]" (Mycologia 15(2): 98. 1923).

As the calendar flipped to 1924, respite appeared nearby, beginning a year which would be monumental for Murrill. As much as details seem essential to tell the story of this narrative, they are sadly only minimal at pivotal points, but not at first.

A FOUR-MONTH CRUISE. "On [Saturday] January 12, 1924, the Naturalist sailed from New York on a splendid tourist ship equipped with everything one could wish for, including a gymnasium, courts for tennis and shuffleboard and a large pool into which fresh seawater was continually pumped. Fish frozen in blocks of ice in Canada were served at every meal. When the equator was crossed the usual aquatic sports took precedence over everything else for two days" (Murrill 1945a).

His trip was no secret. In a notice surely written by Murrill (J NYBG 25(289): 22. 1924): "Dr. W. A. Murrill sailed for Buenos Aires, January 12, on the S.S. 'Van Dyck,' of the Lamport and Holt Line." Proposed stops and anticipated goals were sketched, but the notice appeared after his departure: normally, purely research expeditions were announced prior to departure. In the Journal notice: "He is planning to return to New York in April."

In a long paragraph in the notice, several destinations were sketched, including one-on-one work with Carlos Spegazzini in La Plata, near Buenos Aires, Argentina, and examination of specimens in herbaria in Argentina and Brazil. What kind of planning had been required to make the trip materialize and notable for the Garden? After all, Murrill was hosted by the Governor of Trinidad at the ship's first stop, Spegazzini was generous in his treatment of Murrill, and the last stop in British Guiana was spent at a celebrated biological station. Correspondence must have been sent and answered, tickets purchased, shipment of collections (much less their drying and packing) back to the Garden all required arrangement, and luggage packed for spring, summer, fall, and winter. In the several papers resulting from the excursion after his return, nowhere was Edna mentioned.

After a short stop in Trinidad, the ship headed for Rio de Janeiro, where Murrill toured the Botanical Garden. "Everywhere there was cool shade, restful benches and pavilions, brilliant butterflies flitting about and the sound of water gurgling in brooks or leaping in cascades from the steep sides of Corcovado." It was rough duty for the Naturalist during the brief touristic stop. A similar time was spent in Montevideo, Uruguay, before making berth in Buenos Aires across the Río La Plata.

Before embarking, Murrill had done a bit of research on Carlos Spegazzini (Murrill 1924e). Before leaving Italy, Spegazzini had been a friend and colleague of Pier Saccardo at Padua, where Murrill had visited a decade before, and Murrill had assiduously cited "Saccardo's genera" in polypore and agaric papers.

Appearing in the July number of MYCOLOGIA (Murrill 1924e) as a letter written *in situ*, Murrill described his experience. "Yesterday I spent the day with Dr. Carlos Spegazzini at his house in La Plata. He lives at No. 477, Calle 53, in a fine section of the town not far from the principal places of interest and within a short distance of the railway station. His house is immense, with a large patio, or central court, filled with plants and birds." Years later, he (Murrill 1945a) recalled: "Professor Carlos Spegazzini extended a warm welcome to the Naturalist at his home in La Plata and gave him free access to all his collections with permission to take portions of any of his types." A mixture of quotes from these papers continue. "He and his son stayed with me all day long, showing me specimens, photographs, and publications; giving me anything I wanted for our herbarium; taking me out to a sportsman's club for lunch; visiting the museum; and going on a short collecting trip in the woods.

"A small green parakeet is a great pet and makes a lot of noise unless it is continually fondled. It loves especially to sit on Dr. Spegazzini's shoulder and comb out his whiskers with its curved beak." "[Spegazzini] dined with the family daily, played and sang the latest songs with the daughters, and drank Paraguay tea in the afternoon. The Professor, being somewhat advanced in age, also chewed a few dried cocaine leaves and added iodine to his milk. He seemed much pleased with the trunkful of polypores brought him from New York." The two must have had much to chat about, but in what language—English, Spanish, Italian? The answer was disclosed in a Murrill (1924e) contemporary report. "We talked French, German, English, Latin, and Spanish indiscriminately and recklessly, keeping up a continuous flow of conversation lest our ignorance in certain languages be discovered. ""I had remarked that Paraguay tea was as yet an untasted beverage so far as I was concerned; so, before leaving La Plata, the Indian girl brought me some in a cup made of horn with a tube of silver through which it is sipped. Then I came away with a large box of specimens, several books, and many pleasant memories."

Farr (1973) took 1661 pages in two volumes to gather an abbreviated compilation of Spegazzini's new taxa of fungi, but his interests (and publications) stretched from mycology to green botany, to archeology, natural history and ethnology. For fungi, he was the Charles Horton Peck of South America, but marooned by geography and language from the mycological centers of the Northern Hemisphere.

Buenos Aires being the southernmost destination, the Van Dyck turned northward, with a short stop at São Paulo, where Murrill saw the Snake Institute producing anti-venom serum. He commented on the great port of Santos, where coffee exports were loaded.

A protracted sojourn was spent at Rio (de Janeiro)—again—enough time to occupy a small chalet in the forest, from which he "roamed the woods for fungi. He also helped the boys catch enormous blue butterflies with blue nets on long bamboo poles." A side trip to Petropolis, from which he returned to Rio "loaded down with collections for the herbarium," but no mention of Abbé Johannes Rick, the local mushroom authority. Few of the Petropolis collections remain at the Garden, along with numerous collections from "Corcovado" (Mycoportal).

As time passed and temperatures ameliorated, Murrill (1945a) patched together an itinerary to bring him to [British] Guiana (today's Guyana), where he went inland to visit the biological station of Dr. William Beebe, of New York City's American Museum of Natural History. He delighted in making acquaintances with a dozen or more scientists and, two decades afterward, wrote at length about his enchantment with the natural history, the location, the well-equipped station and the good relationships there.

To affirm his proposed return to New York in April, Murrill had scheduled a public lecture, "Botanizing in Sweden," for April 26, and true to predictions, his April 9 return was announced in the JOURNAL OF THE NEW YORK BOTANICAL GARDEN [25(292): 124. 1924]. Indeed, surely by prior planning, he had scheduled four public lectures during May and June, none of them based on his South American trip. Previously unrevealed, however, was Murrill's presence at the University of Georgia in Athens, GA, from at least late April to sometime in June. A set of six plant pathological specimens there (Index Herbariorum 2023) bear "W.A.M." as at least the collector, and in two cases, identifier as well. Uniquely, the Georgia time went unmentioned in the JOURNAL OF THE NEW YORK BOTANICAL GARDEN and MYCOLOGIA.

From the July, 1924, issue of the JOURNAL OF THE NEW YORK BOTANICAL GARDEN, Murrill's next whereabouts was traceable. "The frequent rains in May and June followed by warm weather toward the end of June brought out an unusually large crop of fleshy fungi on lawns and cultivated ground. *Naucoria semiorbicularis* was never more abundant. A specimen of *Stropharia melanosperma* was picked by Dr. Murrill in front of the Museum Building (Fig. 6.5) on June 28 and the first plants of *Agaricus campester* were noticed the following day. In the woods, a yellow species of *Russula* was fairly common. Three weeks previous, he had collected *Cortinellus rutilans* at Hartsdale; but on June 29 in the Hemlock Grove were found the largest specimens ever seen here of this beautiful mushroom, which measured 15 cm. across and had a stipe 10 cm. in length and 2.5 to 3 cm. thick" (J NYBG 25(296): 209. 1924). But this July issue wasn't available for reading until mid-August.

By early July, he was able to informally address the Staff Conference. "Doctor W.A. Murrill spoke of 'the outstanding features' of his recent trip to South America. This was a general account of the journey with a display of specimens, numerous curios, and articles made of various woods. A report of the scientific work accomplished will be given by Doctor Murrill at another time" (J NYBG 25(295): 205. 1924). Instead, he spread notes in TORREYA (Murrill 1924f, 1926) and his autobiography (Murrill 1945a). But, again previously unrevealed, the latter half of July was spent at Pink Beds, Transylvania Co., North Carolina, confirmed by specimens at NY, TENN and FLAS. Some of these collections were made with Homer D. House, by then the State Botanist of New York, who had been Associate Director of the Biltmore School of Forestry at the Pink Beds some years previous. Neither Murrill's

Fig. 6.5 The Appalachian Mountains of Virginia. The Virginia hill country as envisioned by Murrill as he prepared to leave New York. (Apple Orchard Mountain in the foreground)

Fig. 6.6 An idyllic panorama of the Appalachian Mountains of western Virginia, as imagined by Murrill as he boarded the train from New York in 1924

departure nor his return were announced in the JOURNAL OF THE NEW YORK BOTANICAL GARDEN.

The next clue to his whereabouts came cryptically the following year; "During the latter half of July, 1924, the writer made frequent excursions to the fields and woodlands in and around Lynchburg…" (Murrill 1925a; Fig. 6.6).

For the first 7 months of 1924, Murrill was absent from the Garden for over five, and at least twice his whereabouts may not have been known by colleagues, and more consequently, by Nathaniel Britton. But untold must be the frame of Murrill's mental state. Aboard the cruise boat on the return voyage to New York, his mind must have spent hours comparing the natural world he had observed during his months of travel to the bureaucratic occupation awaiting him in New York; the wondrous world of "The Naturalist" versus the desk and lectures of the Supervisor of Public Instruction. How did the former morph into the latter? Absence of requited love tainted both, and his innermost stability hung in the balance.

CONNECTION SEVERED. Two additional public lectures were publicized: August 10, on "The Royal Botanic Gardens at Kew," and August 31 on "The Paris Botanical Gardens." It would have taken a very curious and close reader of the JOURNAL OF THE NEW YORK BOTANICAL GARDEN to note that Murrill's name disappeared as a public lecturer in subsequent monthly rosters.

Gleason (1961), in his thumbnail biographies, wrote "One day in 1923 [sic] Murrill worked all day as usual, closed his office door behind him, and went home. Next morning, Britton found on his office desk a note of resignation and Murrill was never seen at the Garden again."

No formal announcement or reaction came from the Director-in-Chief, although rumors must have flown through the herbaria, conservatories and gardens.

Murrill (1945a), himself, remembered the event: "In August, 1924, the Naturalist tendered his resignation and left the Garden after laboring there twenty years. The gilded net had finally so enmeshed him that little time was left for research and the long grind had worn him down." Thus, in his mind, the action was his, and his dusty premonition that money was apt to corrupt the scientist had come to pass and had to be shaken off. Indeed, to the propulsion from his circumstances must be added his next phrases of enticement. "An Aunt in Lynchburg with a big empty house wanted him to come and live with her; the Aunt who had helped nurse him when a baby; so he packed up his belongings and went back to the place of his birth, where many relatives and good friends were glad to welcome him home." In a moment searching for self-worth, his next words were: "His position at the Garden was held open for two years with the hope that he might come back, but the wild bird once free of its cage rarely returns. Fairer and sunnier fields beckoned him southward." He was the prodigal son. After almost annual trysts over two decades, "Mother oak" had seized his heart.

Britton's (1925) delayed public recognition of Murrill's resignation might be telling. The following last sentence was not needed, and might be indicative of Britton's feelings after 6 months without Murrill. "Dr. William Alphonso Murrill, who was for 20 years a member of the staff of The New York Botanical Garden, beginning with August, 1904, has severed his connection with the institution. He served as Assistant Curator from 1904 to 1905; as First Assistant to the Director-in-Chief, 1906–1919, during the last 11 years of which he was also Assistant Director; and Supervisor of Public Instruction from 1919 to August 1, 1924. Throughout the period of his service he was in charge of the collection of the larger fungi, which he built up assiduously, and his publications have been numerous. Dr. Murrill played an active and important part in the development of the Garden and his resignation has been accepted with regret."

Fred Jay Seaver, colleague mycologist at the Garden, was asked to write a few words upon Murrill's death many years later. "I recall Dr. Murrill as a tall, immaculately dressed southern gentleman. I can yet see him as he strode the halls of the Botanical Garden with deliberation, dignity and poise." (Weber 1961).

The attractions in Virginia were existential: food, drink, familiar tribe and nature, and above all else, calm. Left behind were colleagues, a large house with three pianos, parquet floors and a front room with 300 charming Girl Scouts, the largest collection of mycological specimens in the New World, the hint of possible supported travel domestic and foreign, the Yama Farms Mycological Club, MYCOLOGIA and its genealogy and finger on the nation's mycological pulse, supervision of the docentry program with constant demand for personal service, and the vibrant City with its landed aristocracy and ragged ghetto-ites. Whether his decision to withdraw from the Garden scene was abrupt or well-planned, the contents of the house and study area, with their books, correspondence, curios and art required time and action, including the packing and shipment to Virginia—to say nothing of unpaid bills. And with this pivotal and permanent denouement from "The Big Apple," this narrative could end; but there are tales yet to be told. Murrill was single and 3 months shy of 55.

The Florida Years

Chapter 7
Florida Freedom

ON THE ROAD AGAIN. The punctuation of his resignation from the Garden occurred in mid-August, 1924, and by the 24th, Murrill was in Lynchburg, Virginia, remaining in residence through at least mid-November [MycoPortal; specimens at University of Florida (FLAS)]. Contrary to some conjecture, he was not incommunicado.

A letter from Murrill to Lloyd, dated Nov. 18, 1924, replied to Lloyd's inquiry concerning his health: but to what address was Lloyd's inquiry sent? A copy of Lloyd's note [of Nov. 15, 1924] has survived, but without a postal destination. Murrill's reply is also extant (Lloyd Library Fig. 7.1). Because it is informative, it appears here.

1008 Harrison St., Lynchburg, VA.
Nov. 18 '24
Dear Mr. Lloyd
I am feeling splendid and am very happy with my relatives and friends in Old Virginia, the state where I was born. Since leaving the Garden 4 months ago, I have spent nearly every day in the woods studying the fungi within one hundred miles of Lynchburg. During the next 4 months, I expect to be following the footsteps of Schweinitz, Curtis, Ravenel, Langlois, etc. in the fair lands to the south. I may even find me a cabin in one of the pine groves of Florida and go back to nature for both my physical and mental needs. (Cf. my "Three Young Crusoes").
Ever faithfully yours
W A Murrill

Lloyd appreciated Murrill's reference to "go[ing] back to nature for both my physical and mental needs," and answered immediately [Nov. 21, 1924]. Paraphrased: "... the older he [Lloyd] gets the more he is convinced that most of our troubles and worries come from 'our vanity.' ... He continued that the happiest people he had

R. H. Petersen, *William Alphonso Murrill: The Naturalist*, Memoirs of The New York Botanical Garden 124, https://doi.org/10.1007/978-3-031-93343-1_7

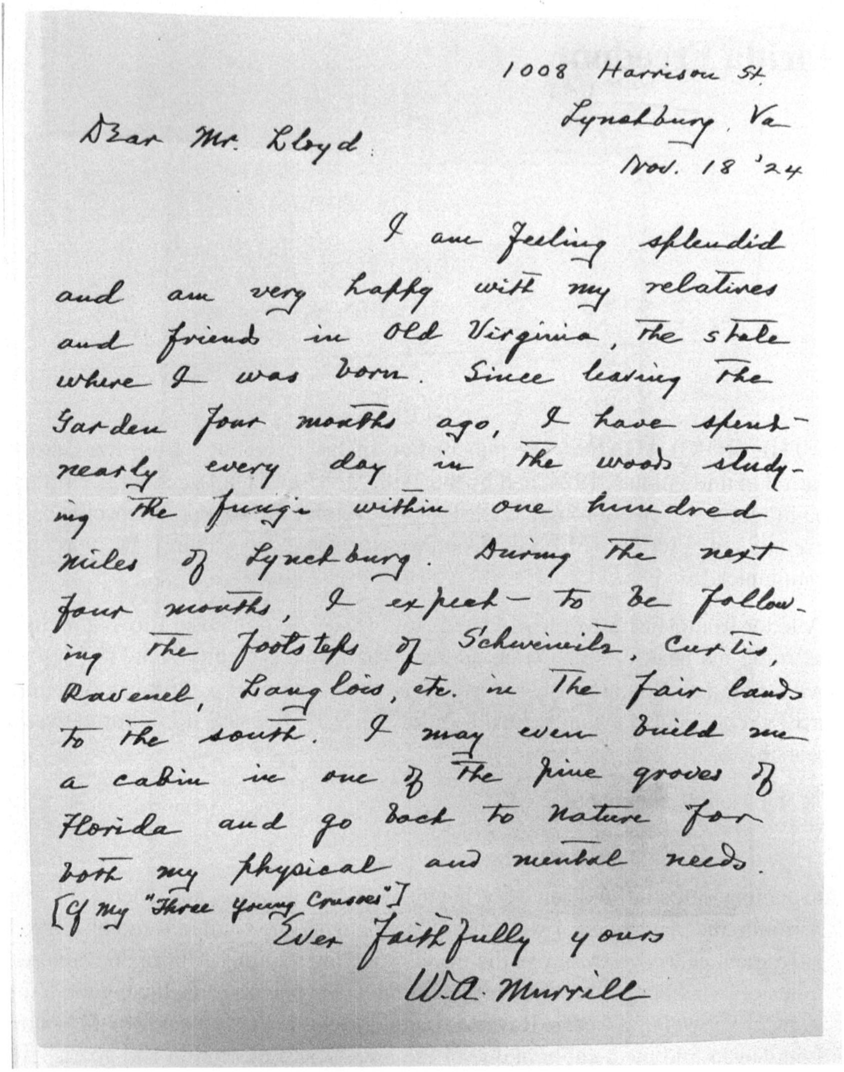

1008 Harrison St.
Lynchburg, Va.
Nov. 18 '24

Dear Mr. Lloyd:

I am feeling splendid and am very happy with my relatives and friends in Old Virginia, the state where I was born. Since leaving the Garden four months ago, I have spent nearly every day in the woods studying the fungi within one hundred miles of Lynchburg. During the next four months, I expect to be following the footsteps of Schweinitz, Curtis, Ravenel, Langlois, etc. in the fair lands to the south. I may even build me a cabin in one of the pine groves of Florida and go back to nature for both my physical and mental needs.

[Cf my "Three young Crusoes"]

Ever faithfully yours
W. A. Murrill

Fig. 7.1 Murrill-Lloyd Letter, dated 1924. Letter, Murrill to C. G. Lloyd, dated November 8, 1924. (Courtesy, Lloyd Library Archive)

ever met were the Samoans, and that if Murrill could get into that state of mind, he could be just as happy in the log cabin in Florida and free of worry than in the Bronx.

There is reason to think that the large house at 1008 Harrison St., Lynchburg, still stands (Google Earth 2023), labelled as "home of Gen. Robert Rhodes, CSA [Confederate States of America]."

This stretch of Harrison Street is lined with rather large, rectangular houses bespeaking an old, affluent neighborhood. The aunt's house, a two- or even three-story square building appearing to be brick with white trim, and with the large wing perpendicular to the road at the rear of the main structure, would have been gargantuan for a single lady. It would not have been out of reason had she run a boarding house, both for a living and for company, although this seems not the case. By summer, 1926, when ex-colleague Per Axel Rydberg— collecting in the neighborhood—dropped by for a visit, Mrs. Ford had moved to stay (temporarily, in recuperation from sickness) with her daughter near the edge of town, and Murrill was the sole occupant (Rydberg 1927). Rydberg described: "The back veranda of the house was turned over to us as a work room, where we attended to the pressing of our specimens and packed boxes for shipment."

Rydberg (1927): "Dr. Murrill entertained us with music, partly of his own composition, set to poetry relating to flowers, birds, and nature study in general, intended for Boy Scout and Girl Scout camps, etc." Murrill remained a vivacious host, especially with one of his pianos.

During the summer of 1925, in the small town of Dayton, Tennessee (a town not unlike Lynchburg, Virginia), Darwinian evolution was put on trial. John Scopes, a public-school biology teacher, had discussed Darwin's theories of gradual evolution in class, thus violating a Tennessee law which forbade teaching this doctrine. Because the subject was in vogue in those months, the lawyers, Clarence Darrow for the defense, and William Jennings Bryan for the prosecution, brought additional attention to the hot, humid courthouse. Scopes was found guilty and fined $100. The subject, though, remained controversial into the twenty-first century. Murrill did not mention the event or the controversy.

George Weber (1961) wrote much of his eulogy of Murrill based on three sources: (1) conversations with Murrill in Florida commencing not too long after Murrill's departure from New York; (2) Murrill's autobiography, written in 1945 at the urging of Murrill's circle of friends; and (3) testimony from numerous Murrill fellow mycologists in the wake of his death. None of these sources is devoid of inaccuracies and must be read "with a grain of salt." Such was Weber's (1961): 548) report that Murrill spent a year in Lynchburg before exploring Florida. In fact, he was in Lynchburg by the end of August, 1924, and early December found him (Murrill 1925c, dated Dec. 10, 1924; 1925d) in St. Augustine, Florida and De Leon Springs, Florida, for most of the winter. His opening words (Murrill 1925f) were typical: "When the chilly winds began to blow, about the first of December, and the Virginia woods no longer yielded their harvest of mushrooms, my memory turned back to quaint old St. Augustine, nestling among the live-oaks and palmettos on the seacoast forty miles or so southeast of Jacksonville. The journey from Lynchburg required only twenty-five hours, the weather meanwhile becoming notably milder, although the hotel rooms were heated upon my arrival." In fact, when he moved from De Leon Springs, in January, 1925, it was northwest to Gainesville (Murrill 1925d, penned in February; 1925e), in the middle of the peninsula in Alachua Co., where he took up residence for only a month before migrating northeast to Green Cove Springs on the St. John's River. There certainly was no time to build even a small cabin from scratch. He was truly "on the road again."

Once again, MycoPortal furnishes data on his whereabouts. Probably, in part, due to a widespread drought in 1925 (Denson 1926), the mushroom season saw fewer Murrill collections but by the writing of this narrative (2023–2024) they are now sprinkled in herbaria across the country [a total of only 39 surviving collections: generally Gasteromycetes, polypores, *Septobasidium*, a few pathogens, but very few agarics]. More fleshy types may have been collected, but given Murrill's peripatetic movements, he surely had no apparatus for drying or preparing specimens, nor a means to carry them with him in travel. There is no way to know how these specimens reached their present resting places, quite possibly by Murrill's own direction.

A survey of the Journal of the New York Botanical Garden shows no mention of Murrill's name in volume 27 (1925). It is almost like an excommunication or "shunning" by the Garden. In truth, through the decades of the Garden, resignations by curatorial staff were rare, but then always to another institution and/or elevated status. Desertion from the Garden altogether, especially by one of its most familiar faces to the managers and public alike, must have produced innumerable questions, many to Britton, and his responses would be interesting to uncover. For Murrill, a state of depression surely played a leading role, but his association of overpowering secular tasks strangling his longing for nature seems closest to the bullseye (Fig. 7.2).

Only a few months after arriving in Virginia, Murrill was contemplating a move to Gainesville in mid-Florida. He wrote (Murrill 1925b): "Now that spring is here, early in February.... Recalling other towns visited this [past] winter in central Florida, such as Crescent City, De Land, Orlando, Mount Dora, and Leesburg, where the orange was the dominant tree, Gainesville offers a decided contrast, for it is practically impossible to find an orange tree or any other form of citrus tree here except an occasional kumquat." His choice seemed more concrete: "As I was

Fig. 7.2 Virginia Blue Ridge. Panorama of Virginia mountains around Bedford, with Peaks of Otter in mid-distance. (Getty Image)

walking along the edge of a shaded swamp near Gainesville, Florida, early in February 1925, I saw a solitary frond of *Osmunda cinnamomea* that looked peculiar, so I stopped to examine it more closely" (Murrill 1925e).

1925 Chronological whereabouts (deduced from MycoPortal data:

January	No date	De Leon Springs, FL (from late 1924)
January 26—February 21		Gainesville, FL.
March 3—March 14		Green Cove Springs, FL
March 15—August 1		Unaccounted
August 2–16		Lynchburg, VA
August 23—October 1		Unaccounted
October 9—November 8		Lynchburg, VA
November 14—December 7		Green Cove Springs, FL

In early 1926, Fred Seaver noted Murrill's whereabouts (MYCOLOGIA 18(2): 94): "Early in November [1925] Dr. W. A. Murrill left his home in Virginia for Green Cove Springs, Florida, where he planned to spend several weeks collecting fungi and enjoying the mild climate of the south." A decade later (Murrill 1938a), Murrill would feel free to divulge his activity there. In a report on a new Florida agaric he called *Volvariopsis canalipes*, he wrote: "Type collected by W. A. Murrill on a sandy, exposed railway embankment at Green Cove Springs, Fla., March 3, 1926 (No. F9975). A similar plant had been observed by me on the Golf Links there in 1925 but I was too much engaged at the time to collect it. Considering my score, I wish I had." Warm winters in Florida, warm summers in Lynchburg—no seasons in New York.

In the same issue, Seaver (1926) used a paper on mycophagy to advertise Murrill's handbook on edible and poisonous mushrooms, with the colored, poster-sized chart. An interesting minor note: "This [handbook and chart] may be purchased through this institution [the Garden] or directly from the agents [with address noted]." So the funds from such purchases would come to the Garden and/or to the publisher, to be collected by Murrill at some undetermined point. Weber (1961) mentioned such sales, which would add to Murrill's "building fund."

1926 brought a new living arrangement. As sketched above, the winter seems to have been spent in Green Cove Springs in Northeast Florida. Murrill had lived there early in his post-resignation investigation of Florida, but what brought him back is unknown. By mid-May, however, he was back in Lynchburg, and spent the entire summer and fall there (May 14 through October 24). He moved to Gainesville sometime in November, and stayed there through the winter, 1926–1927.

From his 1925 chronology of two visits to Lynchburg (August 2–16; October 9– late November), he spent more time there in 1926, but it was through the summer and fall, and he therefore missed the mushroom season in Florida, although unrecognized at the time.

A grace note: in December, 1926, several Florida herbarium specimens indicate three collecting forays with George Weber, and two of the forays included Erdman West, confirmed by Murrill (1927a) a few months later. Weber has been considered to be Murrill's "discoverer" in Florida and wrote as though this were the case. Moreover, within a year before Murrill's resignation, Weber had sent a polypore specimen to the Garden, so the two men were not unknown to one another (although perhaps only by correspondence). Together with Weber, it was West who engineered Murrill's presence in the Agriculture Experiment Station at the University of Florida (FLAS). These 1926 forays are the first (that I know about) interactions of the three together, and the number of collections was rather rich. Murrill's (1927b) humorous comparison of eating *Leotia* versus pickled pigs' feet came from these collecting trips.

Like Murrill, Erdman West (1894–1965) was a newcomer to Gainesville. A Pennsylvanian by birth to a family of professional gardeners, he was brought up in the trade, and at the appropriate time, graduated from Pennsylvania State University, where Lee Overholts and Frank Dunn Kern, both with strong ties to Murrill and the Garden, were among his teachers. An inveterate collector, he amassed a large and wide herbarium, which accompanied him on his pilgrimage to Florida to accept a position as Assistant at the Experiment Station, where, aged 31, he met Murrill, almost twice his age at 56. They hit it off, and West invited Murrill to accompany him on frequent collecting trips. During their common years, West published four papers modeled on those by Murrill (West 1939a, b, 1941, 1945). In one (West 1941), he described *Ithyphallus murrillii,* a handsome stinkhorn. Possibly viewed by some as a dubious tribute, neither man considered it so.

Spending some time with Robert Hagelstein, a Long Island knowledgeable amateur collector of slime molds who came to Gainesville during several winters, materialized as a West herbarium of these inconspicuous organisms to be gradually collected, but when West (1939b) published a list of the group from Alachua County, he failed to mention Hagelstein's name.

George Weber was also included on some collecting excursions, and Murrill's necrologist also penned West's obituary (Weber 1966).

THE CABIN(S). Although some untangling of details is required, Murrill (1945a, Chap. 24) has provided a play-by-play description of his early post-resignation years through his autobiographical chronology of domicile-building. In New York, his home was large, decorated with artifacts of international travel, and full of activity, but now, having escaped the complexity of that professional life, it was time, in his words, "to return to nature." "… he wanted to live in a cabin in the wildwood; and this cabin he wanted to build with his own hands." Elsewhere he had put it differently; now he could give up thinking of the dozens (or hundreds) of demands by other people, and think of nothing but himself. Gone was the incessant transmission of the whims of the Director, and the kowtowing to the rich and influential, but concentration on the wants and needs of his own existence. So he sat for a time on his Aunt's shaded porch with books from the library, drawing plans, deciding on materials and estimating costs. "Not William Murrill, Supervisor of Public Instruction.

And here occur the first complications. "…then he became a day laborer with a lunch basket in a grove of pines not far from Randolph-Macon Women's College." The institution to which he referred was Randolph College for women in Lynchburg, now co-educational, not Randolph-Macon College in Ashland, a hundred miles away, where he had been a student. Also, did he own the land on which he wished to erect his cabin, or was he a squatter? These days, an adult, single male building a cabin within sight of a women's college and a history of camping and swimming with Girl Scouts would at very least raise an eyebrow.

A bit like the Old Testament's allegory of Noah and his ark, Murrill unfurled the saga of "The Cottage." "My first house was a little cottage on a rather steep hillside among the pines, where the wild honeysuckle grew so profusely that I called the place 'Honeysuckle Lodge.' Not a timber was lifted into place, nor a spade of earth dug, nor a nail driven except by myself. I was architect, contractor, foreman, carpenter, bricklayer, and laborer all in one. The tools used were few and simple: one saw, a hammer, a hatchet, a small square, a level, a spade, a pick, a hoe and a trowel. For the painting, I used two small brushes, and in putting on the roof an old pair of scissors were needed to cut the composition shingles." Despite the depiction by Kimbrough (2003), this was not a log cabin, although the foundation was of stout locust trunks resting on large flat rocks buried below the frost line.

"Without meaning to do so, I built my cottage about the size of Thoreau's cabin at Walden, fifteen feet long and ten feet wide. The main room was eight feet high, with a convenient attic above to keep off the sun and use as a storehouse for things that usually find their way to attics." At first, "the main room" was the only room. It served for his transition to Virginia life. "It is quite surprising how much space there is in a little house if the furniture is simple and conveniently arranged. … For once in my life, I have a bed that suits me and plenty of room for my books, which almost cover the walls." Note that although written in the present tense, this text was penned two decades after the fact.

But even as the one-room cabin took shape, he was contemplating enlargement. "By digging into the hillside and building an extension on the south side, I constructed a basement room as large as the main room and put in a thick concrete floor, which made it dry and comfortable, being cool in the summer and warm in winter. … Most of the timbers and boards were of southern pine and considerably heavier than the building law required. I hate to have a house shake when I walk across the floor." This larger house was described as "… with living rooms and a sunny music room where he installed one of his pianos and stacks of sheet music. The birds were curious until they learned to know him. At first the sound of his hammer frightened them away, but the music lured them back." So there is the sequence of events. His reference to "… one of his pianos and stacks of sheet music …" and "my books, which almost cover[ed] the walls," raises questions about the mass of personal effects shipped from New York to Lynchburg, the cost, and the accommodations temporarily required in his aunt's house.

Considering that the quotes above were written two decades after the fact, some leeway must be granted him in the very next paragraph. "In the autumns he followed the birds to the south and wandered about from one lovely spot to another until returning spring beckoned him northward again. Thus he came to lead a double life with two years as it were in one." How many autumns? According to MycoPortal data, the first unbroken construction time in Lynchburg was in 1926; mid-May through mid-October. This would seem adequate time for building a cabin, or the cabin was erected in stages over these years. Another somewhat enlarged description of his dual existence was written in one of his Girl Scout-aimed pieces from Gainesville (Murrill 1946a).

The north-south shuttle set in as a routine, and he wrote: "All this was stimulating and he began to write magazine articles about what he saw, which enabled him to see more and travel farther. With a fine new camera he took hundreds of pictures to illustrate the articles and gradually accumulated a building fund for a winter home in Florida." Unlike professional scientific writing, reward from which was anything but financial, authors of popular magazine pieces were often paid by the line, so the more articles—and the more words—the better (within reason). Murrill was free to work personal observation into a scientific subject, and as the time passed, his output increased. As he wrote: "Meanwhile, his magazine writing was continued, hundreds of articles being prepared on a wide variety of subjects. A large number were published but some have not even yet been submitted to editors [by 1945]." Many such pieces have apparently eluded Murrill's bibliographies. Weber (1961) also referred to residual money "up north" resulting from the sale of Murrill's earlier books: termed "books on fungi" (the polypore series) and "Nature and Character Books" (Murrill's early years, "Three Young Crusoes," and "The Natural History of Staunton, Virginia.")

Can Murrill's words be construed as anything less than self-praise? And if not, why should they? Here was an individual who had summarily resigned not only a respected academic position, but who singlehandedly planned perhaps the largest physical task of his life, and in his view, executed it successfully. These would not be the last such items on the itinerary of his life over what Kimbrough (2003) termed, Murrill's "twilight years."

Had Murrill assembled a journal of his last decades as was teased here from his record of time at the Garden, this narrative would be twice as long as it is. His autobiography, instead, summarized his journey in general terms, always positively and not through carefully described events. The first was the transition from the life of a migratory tourist in Florida to that of a resident.

Still, fresh memories of his South American adventure hung on, in spite of his "new life." He aimed a plucked sketch to his friends in the Torrey Botanical Club back north (Murrill 1926). In his autobiography, many pages were devoted to remembering individual sojourns, botanical observations, and the idyllic days spent with Carlos Spegazzini. It had been a naturalist's dream come true.

1927 was another year of migration. Both winters, early and later, were spent in Gainesville, but from mid-March through early September, he was in Lynchburg, perhaps cabin-building. But from the fall of 1927, Murrill (1927c) could report some new observations from Florida. Following his interest in edible versus poisonous mushrooms, he felt obliged to report on the yellow-gilled *Russula* (strangely, for Murrill, not assigned a Latin binomial). "The autumn [1926] in Florida was quite dry and few fleshy fungi were found in November. On November 27, however, I collected enough for a mess by mixing *Pluteus cervinus, Lactaria indigo, Laccaria laccata, Hygrophorus miniatus,* a species of *Helvella*, and the yellow-gilled *Russula*.

On the following day, I found a quantity of the last species in a little pine grove and used it alone." His choice of MYCOLOGIA for his experience attests to his continued friendly relationship with its editor, Fred Seaver, at the Garden.

The second paper was a typical product of "the Naturalist" (Murrill 1927d). Chronologically, it seems not to fit the mycological paper, but, in fact, it ushered in a year of silence, and when, in 1929, he resumed telling of his observations, none made an effort to be serious science (Murrill 1929a, b, c, d), and none was published in a scientific journal. Meanwhile, 1928 was again divided between Gainesville (winter/spring, fall/winter) and Lynchburg (summer). Mycoportal data report two Murrill collections in September (date unknown) from Van Cortlandt Park, Bronx, almost surely an error in data entry. Collecting totals (MycoPortal data) for 1928 and 1929, respectively, were very low: 18 and 33, surely partially blamed on inadequate facilities for drying, storage, etc.

According to Mycoportal, Murrill's 1929 mycological year saw a burst in February, when he spent several days in the field, some of them with Erdman West. Whether reflective of his interests, of the weather, or of the intrusions of other activities, the 1929 collections were dominated by polypores and Gasteromycetes, with only a smattering of agarics, probably an indication that drying facilities were not yet available.

Totally out of character, no collecting in Lynchburg was chronicled. Was he unable to relax in his hand-built cottage? Perhaps this was the year in which he built his "winter home" in Gainesville. At the least, perhaps this was the first summer he spent in Gainesville doing research on mosquitoes and spiders, but there is no evidence that he discovered the richness of the agarics during the summer and fall, for so few collections survive.

SQUARING THE CIRCLE. Of special note might be his Open Court article (Murrill 1929d). He summarized its thrust as follows: "But we have seen how remarkably well organized are the elements of nature in the atom, the molecule, the mass, the star and the universe; and it seems to me that this very beauty of form and organization eminently fits the material elements to be closely associated with and to take orders from the immaterial forces resident in living beings. Of all things existing in Nature, nothing is unworthy—except, perhaps, the man who does not appreciate and live up to his opportunities." Understanding that he had only recently entered his "second life," having shed the first, he was in no position to judge whether he was to live up to his opportunities. This theme was to return some two decades later when he (Murrill 1950) wrote "Success Stories for Girls," in which vignette after vignette revolved around girls, often orphans, who, through patience, pleasing personalities, and some overlooked skill, were "discovered" and adopted by model families.

THE GREAT SLUMP. While the macroeconomic factors leading to the onset of The Great Depression have been repeatedly summarized and argued (Romer 1993; Federal Reserve summary 2024; Eichengreen 1992), the effects on individual Americans, including our cast of characters, are more exigent. The economics of the

Garden, for example, took a negative turn; the endowment did not grow; philanthropists tightened their grips on their money. Salaries became stagnant or cut, and hand-labor workers laid off. Exploration was reduced. The great gardens were maintained, although probably not the favorite activity for the Director. On the other hand, while we have no direct data on Murrill's well-being, by the fall of 1929, he had chosen Gainesville and was settling in on year-round habitation. With no known salary, a wage could not be cut, but it could be suspected that whatever money had come from the disposal of his Bronx house, savings in banks, and (probably not) investments, might have been threatened. His thoughts on money were expressed years later in his autobiography and comparing them with his existence during the Depression is apt.

THE DEPRESSION. The first shock for the common man came with the stock market crash of October, 1929. As the talisman for the U.S.A., "the market" had gained popularity; anyone could own a piece of America. The economy had been expanding for more than a decade and a middle class was emerging, especially around the urban centers. With corporate profits reaching new heights and dividends with them, people could borrow money from a bank at an interest rate lower than the dividend on stocks—a formula fraught. The relatively new Federal Reserve System had not established a central bank role, and banks, both neighborhood and "big" New York superbanks had no "lender of last resort" with power to regulate national interest rates. As the country had seen before, small banks, short on funds, borrowed upward, but at some point, there was no "upward" left. In just 4 years, more than 5000 banks went out of business, with no protection for depositors, and therefore commensurate losses to individuals, families, and small businesses. Production flagged and by the early 1930s, unemployment spiked at more than 25% (Fig. 7.3).

Fig. 7.3 The great depression in Florida. The great depression: Florida family in front of their depression domicile. (Google images)

But Florida did not wait for the rest of the country. In 1926, and again in 1928, hurricanes pummeled the southern half of the state, and in 1929 the Mediterranean fruit fly invaded the state; citrus production was cut by about 60%. A quarantine was established, and troops set up roadblocks and checkpoints to search vehicles for any infected, contraband citrus fruit (Florida Depression report 2024). As time passed, citrus growers became caught in a vice between low prices—grapefruit was bringing five cents a box on-the-tree by the late 1930s—and handlers and shippers, who often controlled harvesting. Gainesville, chosen by Murrill for its location out of the citrus belt, was somewhat spared. The Florida Farm Bureau would come to life during the decade 1930–40 (Florida Farm Bureau history 2024).

Surely through the good offices of George Weber and Erdman West, Murrill established close relationships in the Florida State Agricultural Experiment Station, headquartered in Gainesville on the campus of the University. The Station maintained an herbarium, in which fungi were largely represented by pathogens. Murrill's magazine writings were minimally profitable, but now another avenue became available; the Florida Station, like many others across the country, produced informative outreach documents distributed to the public free or for a paltry fee. Such "extension" brochures were mimeographed and Murrill envisioned a new category which could tell the public about the environment in which it lived—in short, information about "Nature." Who better to author such document than "The Naturalist." Hopefully, the Station extension service could expand its reputation as part of the bargain.

But as charming and useful as Murrill's descriptive mimeographed materials (Appendix 2) were, they could not be expected to support the Agricultural Experiment Station at the University where Murrill gravitated.

Murrill could stand by and watch the Depression, but the Florida tourist pilgrimage, too, shrunk around him. Tourism seriously supported Florida's economy during the winter months. People drove to Florida to enjoy the warm climate. Because Floridians were facing their own home-grown economic difficulties, however, Florida State Police were stationed at Florida's border; if people did not have enough money or a job to support themselves, they were not allowed to enter the state. No record remains to relate Murrill's status. Perhaps this human embargo compromised his touristic Virginia-Florida migration (Florida and the Depression 2024).

In 1931, the Florida State Legislature created a State Racing Commission which legalized betting at horse and dog racetracks and Jai Alai frontons. A tax was levied on winnings, and while this marginally aided the state, not surprisingly, common people did not have enough money to bet!

THE TIN CAN TOURIST CAMP. The Tin Can Tourist Camp (1935); Fig. 7.4) in Gainesville was established in 1917 and existed through 1982, and Murrill surely knew it in his time. As the U.S. role in World War I drew down, the return of the "doughboys" began the serious churning of American society, including the winter migration to the South, followed by permanent residence—the Southwest for low humidity, Southeast for moist heat.

Fig. 7.4 The Tin Can tourist camp. The Tin Can Tourist Camp entrance, circa 1920. (Courtesy, Florida Tourist Bureau archive)

Kimbrough (2003): "But in the 1920s after World War I, when many folks began moving to Florida, moving meant braving uncertain lodging. Tin Can tourism (using the car and tent for lodging) was a common solution. One camp, in Gainesville, was located at the present-day site of Shands Alachua General Hospital."

As Murrill reported, however, his travel from Virginia to and from Gainesville was by train, with all the complexities of baggage-juggling.

More from Kimbrough (2003): "Interestingly, the site of the Tin Cup [sic] Tourist Court was only a few blocks from where [George] Weber and his wife Kate lived. Weber shared with me that the two of them went to the tourist court frequently to purchase jams, jellies, fruits, and vegetables that farmers from the countryside would bring to sell to the travelers. Correspondence between Murrill and J[ohn] K. Small, author of a Manual of the Southeastern Flora… reveal that Murrill wrote to Small on March 9, 1927, commenting that he [Murrill] was still at the Tourist Camp but that he was starting to build a studio to use as a storeroom while he worked on his main dwelling. Two postcards dated 1926, written by Small, were addressed to W. A. Murrill, Tourist Court, Gainesville, Florida, one indicated that he [Small] had visited Murrill in Florida recently. In Murrill's 1927 letter to Small, he [Murrill] stated "I am at the Tourist Camp in a cottage, when you came last I was off on a water snake hunt." Very likely Murrill came to the tourist camp in Gainesville in the winter of 1925.

His musical interludes at the camp might indicate his living there, not yet in his first hand-built cottage (Fig. 7.5).

A SMALL RECAPITULATION. These pieces from published sources seem to agree with the narration of the chronology above. In late 1924, he visited several different places, and concluded that Gainesville was his chief destination. He would have gone there in the late fall of 1925, but it was some time, probably about 1930, before the decision was made to make Gainesville his year-round home.

Murrill (1945a): "Gainesville was finally selected for his home because of its good location and college atmosphere. Here a small cottage was built, midway between the University and the center of the town, and he became a winter resident, travelling back and forth in shuttle fashion between Virginia and Florida with the changing season." These days the main campus and downtown have each metastasized and become nearly sympatric.

Two versions describe Murrill's physiognomy after some residence in Gainesville. Version 1 (Kimbrough 2003: 5) "It was during one of these visits [to the Tin Can Tourist Court] that Dr. Weber and his wife Kate discovered this 'tall, robust, dignified, pleasant stranger providing a piano concert for the transient tourists at their recreation hall...' Weber recalled that he immediately recognized that the gentleman was Dr. Murrill." Version 2 (Kimbrough 2003, same page): "Weber was unsure as to how Murrill arrived in Gainesville or how long he had been living in the tourist court when they first spotted him. Dr. Weber noted that Murrill looked unkempt,

Fig. 7.5 Tin Can vehicle. A tourist camping vehicle, c. 1925, reconstructed. (Courtesy, Florida Tourist Bureau)

frail, and haggard. He more or less took him under his wing and, along with Prof. West saw that he was properly dressed and fed."

The first MycoPortal records show West as Murrill's first Florida collecting companion, in January 1927, together with Weber, but West (e.g. fall, 1932) much more frequently than Weber over the years.

A FINANCIAL LIFELINE. According to Kimbrough (2003), once Weber had befriended him, Murrill asked a favor. His financial situation prompted his curiosity of whether any funds due him were in New York without a forwarding address. Weber wrote appropriate parties and "Within a short time, he [Murrill] received a check for $600.00 and a stack of unsold books and charts. This money helped buy a small lot, some tools and building supplies, and a small wood-frame house at 203 NW 9th Terrace between the University and downtown Gainesville" was the result.

Money remained on Murrill's mind. On one hand, little cash was necessary for his existence, but on the other hand, existence was somewhat fragile. As years passed, and his finances continued to be precarious, he was moved to share some advice in his autobiography (Murrill 1945a). He wrote: "It is much easier to give advice about money than to follow it, but I am going to risk a few words to young people with the hope that some of them will profit by it. If this advice, based on long experience and observation, should be generally followed, many losses, bitter tears, family quarrels and hopeless years would be averted; and lives otherwise futile turned to good account."

There follows a series of aphorisms on money. Only three are transcribed here.

"Begin to save as early in life as possible and have money working with you to make more money. There is no shortcut to wealth. Avoid all forms of gambling as you would a deadly poison.

"Invest your money only upon the advice of experts in money matters. Do not rely on your own judgement nor listen for a moment to cranks, promoters, or gamblers. They are simply playing you for a sucker, to feather their own nests. Avoid vacant lots, houses at the market price, and bonds paying a high rate of interest." This written from his Florida house, hand-built on a vacant lot.

"Always return at least one tenth to the Lord. We are simply stewards, not absolute owners, of everything we have in this world. It is right to make money if we make it honestly; it is good to enjoy all the advantages that money can bring; it is gratifying to help others wisely with what we do not need for ourselves; but it would be the basest ingratitude to forget the source of all our earthly possessions and even of life itself."

Murrill (1945a); Figs. 7.6 and 7.7): "Having built one little home in Gainesville the Naturalist tried another and another just for the fun and exercise, doing all the work himself. Then he became a landscape gardener, vegetable grower and horticulturalist, learning Florida materials and methods and filling his grounds with trees, shrubs and flowers." Weber (1961) wrote that these "extra" cottages in Gainesville were for rent, and the proceeds added to Murrill's "building fund."

Fig. 7.6 Murrill's Cottage in Gainesville. Murrill's first Gainesville "Cottage." (Courtesy Jim Kimbrough 2003)

Fig. 7.7 One of Murrill's rental cottages. A Gainesville rental cottage built by Murrill. (Courtesy Jim Kimbrough 2003)

ESTABLISHING ROOTS. Murrill's bearing, articulation, and friendly attitude helped acclimatize him in the Tin Can Camp, and purchase of his housing lot and the materials for construction of his "cottage" brought him into the resident community as well. As he was seen patrolling the neighborhood, his interests became clear and honest, and he became active in the local garden club, mostly ladies of the community, who, accepting professional advice from this older gentleman, surely buttered both sides of the bread. He was comfortable, also, in church circles, admitting that this was largely musical, but also teaching Sunday school. The thread of

Fig. 7.8 Gospel-Singing-Convention. A rural singing convention gathering in Florida, c. 1925. (Google images)

musicality had followed him from "singing school," through piano lessons in Blacksburg at age 12, through chapel and glee club at Randoph-Macon, to playing for Girl Scout games in the Bronx and the "Singing Scouts," his "music room" in the Lynchburg cabin, to "singing conventions."

In his autobiography, he reported: "In Florida he became interested in Community Singing, often driving a hundred miles to some church [surely not in 1945 when he wrote of it] and playing the piano most of the day for convention singing, solos, quartets, etc. Since he could transpose music to any key, he was much in demand for specials. In Gainesville he organized a Singing Club, which he attended for many years." [Written in 1945 — his elapsed time in Florida was then 20 years.]

Again, Murrill's vocabulary begs explanation. Out of countless congregations, almost all rural, "special music" was developed by individuals or quartets—voices and instruments, often performed while the collection was being gathered. In Murrill's youth the instrument was usually guitar or perhaps piano, but by the time of his autobiography often blended into a more popular genre, "country music," or even jazz. A "singin' convention" was a gathering of "special music" from a large geographic area as a day (or afternoon) series of performances to an as large a congregation as attracted (Fig. 7.8), a mode gradually eclipsed by similar secular performance music. Even now (2024) occasional "conventions" take place, audience

Fig. 7.9 Sacred Harp. A lady leading a "Sacred Harp" singing, probably using shaped notes. (Google images)

admission tickets unnecessary, and with the option of buying records before, during and after the convention intermission.

"In composing songs, the Naturalist always wrote or borrowed the lyrics first and let them suggest the melodies. He has about six hundred of these compositions in his collection, most of which will be deposited in some library. After the paper shortage [a relic of World War II] is over a few may be published." To my knowledge, only a few scraps of such material may still exist, and even that existence is apocryphal.

Murrill (1945a) again: "Dancing and swimming are two of the best exercises for women, giving pleasure, grace, co-ordination and control, as well as health and physical development. Dancing is dependent on music. When too old to dance, if this ever happens, both men and women can get exercise and genuine enjoyment out of choral or part singing (Fig. 7.9), either at some neighbor's house or in a school or church. A little light refreshment always adds in the pleasure of the occasion, whether it improves the voice or not."

According to Murrill, himself (1945a): "His first piece of research which was distinctly Floridian was 'A Biological Survey of Alachua County.' This connoted not only a study of the biota but also the geology, soils and ecology of the Gainesville area." I cannot find such a citation in Murrill's bibliography, only several on more specific topics. Perhaps it was an early mimeographed document from the Experiment Station (see Murrill 1937).

The Garden Club, by its very nature, was concerned with planted, tended—if not manicured—plants for either beauty or utility, or both, traits for which Murrill, trading on his years at the Botanical Garden, could make suggestions of a professional bent. But native plants in nature, their biotic communities, their relationship with their soils and subsoil rock formations, required a zeal of a different kind. Hiking boots, binoculars, walking stick and mosquito netting were prerequisites to the Nature Club, for which Murrill was leader. One day a week, nature enthusiasts gathered to take a hike in some nearby woodlot, swamp, hammock or pine grove to observe—and perhaps photograph—spiders, algae, snakes, occasional deer, mushrooms, mosses, wildflowers, soil structure and pH, with one of the group, often Murrill, acting as guide. For Murrill, one advantage was transportation to spots outside his walking radius. As time passed, he found that the more informed members could act as guides to others, in turn associated with the uninformed; especially pertinent were Boy Scout, Girl Scout, and 4-H leaders.

RETROSPECTIVE. 1929 had been a year away from "hard science." It had not been unique; perhaps it reflected a man not yet at home with his surroundings, yet also not totally withdrawn from his chosen rhythm of life. Perhaps his precarious health only compounded his loneliness—alleviated, perhaps, by writing—in his small cottage in Gainesville.

Known 1930s publications, certainly not verbose or numerous, reflected such a pensiveness. The passing of his botanical idol, Giacomo Bresadola, Murrill's senior by 22 years, busy with the affairs of the Church but sampling nature every chance he had, served as a reminder that Murrill had chosen the societally golden path while sacrificing his naturalist's marriage to the natural world. But Murrill had been reminded during his Bresadola days in the Tyrol, not once but twice. Murrill sent his heartfelt note to MYCOLOGIA, the product of his former "gilded cage," but now under the editorship of Fred Seaver, with whom he remained close. His manuscript appeared nearly a year after Bresadola's death, but Murrill's (1930a) sentiment aimed at a chosen audience. "I wished only to say a word to the young naturalist who is struggling along under disadvantages. Bresadola began at his doorstep and by careful, devoted study, gradually became acquainted with certain plants in his own neighborhood. He developed and used all his natural powers, and the circle of his acquaintance widened until it embraced the world!" Quite possibly the sentiment of a man looking back over his shoulder and pondering whether HE had lived up to his opportunities (Murrill 1930a).

A second gathering of words reflected his earlier thesis on "The dignity of matter" (Murrill 1929d). This time he mused about what sorts of organic combinations could be found in other worlds—literally.

Read alone, with no insight into the man, the question would be instantaneous: is this humorous or semiserious? The reader is engulfed in Murrill's rationalizations, but his deductions are based on only the information at hand and stand in danger of misinterpretation from our present time and knowledge. His final thought, as opposed to vast worlds with vast living forms: "I prefer to think rather of some favored spot, somewhere in space, where knowledge is so perfect that men have no fear of the natural forces, and their religion is unmixed with superstition and dread of the unknown—a religion of love and mercy, requiring neither burnt offerings nor punishment by fire. Such a place, I think, would be very near Heaven." (Murrill 1930b).

Written, of course, by an aging man, living alone, with a high regard for his own breadth of knowledge, but also anchored by the religion of the Virginia hill country.

The two remaining offerings for the year were showpieces of popular naturalistic observations, one on the lowly-regarded plane tree, the sycamore (Murrill 1930c), and the other the rare flowering hedge, *Pinckneya* (Murrill 1930d).

And with those published words, Murrill fell silent for 5 years. He had explored taxonomic mycology, popular observations of his surroundings, and even book-length fiction for young readers, but now, introspection and perhaps local social activities occupied his time and, apparently, his pen.

As though helping him to collect his thoughts, Fred Seaver (1931) dedicated the first general index for MYCOLOGIA (Vols. 1–24) to Murrill: "Dedicated to William Alphonso Murrill. Instigator and first editor of Mycologia; formerly Assistant Director of The New York Botanical Garden." In the introduction, Seaver wrote further. "Dr. Murrill continued to edit the journal until his retirement from The New York Botanical Garden in 1924. At this time the Board of Managers seriously considered discontinuing the publication as a matter of economy. A strong plea for its continuation was made and the writer was permitted to handle its publication as a part of his curatorial duties." So Murrill's resignation exposed the precarious state of the journal, and now he could read its aftermath.

And with this, the curtain was drawn on the first act of Murrill's personal drama. The Great Depression introduced a new and sinister theme, only exacerbating Murrill's modest economic situation. New location, new acquaintances, and a new life in which the rules were revealed existentially—promised but yet to be fulfilled.

Chapter 8
Poking and Proding

In 1932, after 2 years of very sparse collecting, Murrill (Fig. 8.1) suddenly had a banner year. In the winter and spring, collecting was spotty (or collections did not survive), but from July through December, collecting was frequent and rich.

In September through mid-November, he often accompanied Erdman West from the Agriculture Station. Significantly, all of his collecting was in and around Gainesville, with no demonstrable time spent in Lynchburg or even other Florida locations. Equally important, collections were dominated by agarics.

Once connected with the Experiment Station through Weber and West, located on the greater University campus, Murrill put his writing and observational skills together, "publishing" numerous mimeographed guides through the Station (Appendix 2), for sale to the public for a pittance (if the public could discover them). If the Station was generous with its time and resources (mimeograph machine, paper, transcription, bookkeeping, storage of extra copies, etc.), the proceeds could come to Murrill, and perhaps offer another opportunity to make friends with some secretaries. On the other hand, the Station could use the publicity, and take credit for these homey descriptions of surrounding locales. The first of these may have been in 1937, encapsulating the flora of Alachua County in 39 pages (Murrill 1937).

Apparent mutual satisfaction prompted three more in 1938 and a dozen in 1939. Size ranged from one page (aka "leaf") on Alachua goldenrods to 10 leaves for Trees of the County. Each installment brought Murrill's name to new audiences and growing acceptance in the local Gainesville society. As complete as known, these mimeographed guides are listed as Appendix 2.

Kimbrough (2003): [no dates provided] "Weber says that Murrill had every intention of returning to Virginia in the summer as was his practice, but he was hospitalized in the University Infirmary with the kidney problem." His stay in the infirmary lasted for several weeks, and the mushrooms [by mid-summer] were in abundance. "Weber recalled that this was the first 'spark of life' he had seen in the 'old man's eyes' during the months that he had been in Gainesville.

R. H. Petersen, *William Alphonso Murrill: The Naturalist*, Memoirs of The New York Botanical Garden 124, https://doi.org/10.1007/978-3-031-93343-1_8

Fig. 8.1 Murrill 1942. A poor photo of a clean-shaven Murrill in a second-growth Florida woods, perhaps circa 1942. (Courtesy Jim Kimbrough 2003)

"Murrill was excited; this was the first time he had seen the summer flushes of mushrooms in Florida. He needed collection supplies, a desk, and a microscope. Crowded for teaching, research, and herbarium space, they [Weber and West] provided him with only a desk, a student-grade microscope and lamp on the third floor [stairway] landing in Rolf's Hall, close to the Herbarium and to the offices of Weber and West."

The mid-1930s, according to Mycoportal data, were lean for Murrill's mycological collecting in Florida, although reasons for this paucity might be several. 1938, conversely, would seem on the surface to be quite rich, with Mycoportal listing 1670 collections, all of them from the vicinity of Gainesville. In fact, without intense correlative search, several points stand out.

While a few Murrill collections do not bear infra-Gainesville locations, those that do so (mostly at FLAS) show collecting grounds formerly unnoted by Murrill. A short list: Alligator Sink, Arredondo, Buckley's Grove, Cary Memorial Forest (probably aka Austin Cary Forest), Devil's Millhopper, east of Flying Field, Gold Head Branch State Park, Grove Park, Hatchett Creek, Hogtown Creek, Hunter's Station, Jenning's Lake, Jonah's Pond, Kanapaka Sink, Kelley's Hammock, LaCrosse, Magnesia Springs, Newman's Lake, Orange Heights, Palatka Road, Payne's Prairie, Pine Grove, Planera Hammock, Pocono Lake, Prairie Creek Hammock, Royal Palms Estates, Sanchez Hammock, Santa Fe, Seminole Hammock, South Planera Hammock, Sugarfoot Hammock, Swan Lake (in Putnam County), Tung Oil Mill (sometimes spelled Jung) on Newberry Road, and UF Campus ("south of infirmary").

Care must be exercised in distinguishing specimens from collections. A collection includes all the individual basidiomata (= "fruitbodies") collected under one number in one gathering. A specimen is a single fruitbody, often (and hopefully) several composing a single collection. Thus, a single collection of numerous individual basidiomata may be split into several specimens, some or all of which may be distributed to separate herbaria. In the case of 1938 collecting, Mycoportal data would list all the collections listed under one date and collecting ground, deposited to all participating herbaria. A more serious data search is necessary to ascertain how many separate collections were gathered by Murrill in 1938 and split into portions sent to separate herbaria.

A hypothetical example: one Murrill collection from Magnesia Springs dates from August 15, 1937. The collection included 14 specimens (basidiomata), which were separated into four packets, each packet dispatched to a separate herbarium. MycoPortal would list all four portions (under separate herbaria), and only a meticulous comparison of the data would reveal that all four portions represent a single collection under a single collection number.

Interestingly, 1937's collecting had started poorly, one collection in January, one in March. From mid-July through the fall, however, collecting was often, and all in the neighborhood of Gainesville. It is possible, considering the facts of his paucity of publications and collecting patterns, that it was 1937 to which Weber referred when describing Murrill's time in the infirmary during spring and early summer, and his "awakening" to the fungal bonanza upon his ability to observe it. After collecting became frequent in the latter half of 1937, it persisted at a roar in 1938, and equally throughout the year.

Both Murrill (1945a) and Weber (1961), though, portray a very different tableau for Murrill's apparent awakening. Both renditions are based on mosquitoes. According to Murrill (1945a), his observations on Alachua natural history required a summer assaying the local mosquitoes and spiders. But collection of mosquitoes was best in the summer evenings, dictating his presence for an entire summer in Gainesville. Ergo, natural history was the cause of the hiatus in Murrill's Lynchburg-Gainesville migrant routine. Some years thereafter, Murrill (1945a) reported that the time period from autumn 1935 to autumn 1936, was spent, in part, studying the local spiders, and if so, then the spring/early summer, 1935, may have been the period of his hospitalization and recovery during the summer flush of agarics.

Weber (1961), on the other hand, referred to a mosquito eradication program by the local authorities. If Murrill were to stay one step in front of local mosquito extirpation, an entire summer (in fact, probably more than one) had to be endured. Ergo, political and social machinery were the cause of the hiatus in Murrill's migrant routine.

What is sure, however, is that once alerted to the summer mycological bonanza, Murrill was in the field almost every week in 1938, often more than 1 day a week. He collected with Erdman West relatively frequently, with others named occasionally. Weber's name does not appear. Whether it was this "mother lode" of material which begged for space, microscope, and a few reagents, or whether this amount of material could not have been processed without these items already present cannot be deciphered, but the two circumstances are difficult to separate.

INCURSION OF WORLD AFFAIRS (AGAIN). Just as Murrill seemed to ignore World War I in his writings, so the unrest in Europe in the second half of the 1930s seemed far away from the Naturalist.

Overt belligerent acts over Europe were mere headlines in American newspapers, easily overlooked. In 1936, German troops invaded the Rhineland; Italy, Austria and Hungary signed the Rome pact, and Italy and Germany made the Berlin-Rome treaty; the Spanish Civil War erupted; Great Brittain, France and Poland signed a mutual assistance pact; and at home Franklin Roosevelt was re-elected for a second term in 1937, Roosevelt signed the neutrality act; Poland refused to return Danzig to Germany; Japan invaded China and the Korean peninsula; riots in the Sudetenland of Czechoslovakia begged for German occupation and Germany complied with invasion; Germany guaranteed inviolability of Belgium; Italy withdrew from the League of Nations; a Royal Commission on Palestine recommended establishment of Jewish and Arab states. Back home, a Stock Market decline signaled a possible serious recession in U.S.; and FDR wrote a letter to Hitler and Mussolini requesting peaceable solutions to European problems. Perhaps most melodramatic, U.S. and Germany withdrew their respective ambassadors.

Even as William Murrill was being ferried around Gainesville to collect mushrooms in 1938, Hitler was solidifying his power in Europe. In 1938, facing a mutual cooperative agreement among France, Poland and Great Brittain, the German army invaded Poland and thereby triggered a response from the other Allies, and European war.

At the other end of the world, the Empire of Japan was also showing unmistakable restiveness, and by 1938 had invaded small countries of Southeast Asia as well as China. A pact of solidarity was signed between Japan and Germany, and "the Axis" was formed. President Roosevelt signed the Neutrality Act, seemingly isolating the U.S. from the world's turmoil.

MURRILL'S WORLD. Based on titles alone, Murrill's 1938 publications would appear mundane, quite as though he was up to his old self. Closer inspection, though, reveals some "new wrinkles."

The four major papers appeared relatively evenly distributed throughout the year. If manuscript writing were superimposed on collecting dates, and again against social activities, housekeeping and other "background," Murrill must have been the proverbial "whirling dervish."

The first paper of the year (Murrill 1938b), probably written during the winter of 1937–38, was submitted to the Journal of the Elisha Mitchell Scientific Society, a decidedly Southern journal, at least so considered by non-Southerners. Avoided previously by Murrill, the Garden's (his own) Mycologia was handy and organs of the Torrey Botanical Club equally so. In fact, one of Murrill's earliest Florida papers was a travelogue in Torreya for "the folks back home."

Murrill (1938b) took up the mushroom genus *Agaricus* in central Florida, noting Coker's work on the genus farther north. Several of the Florida taxa were variations, he thought, while others seemed distinct. Quite unusual, however, was the language used. Words and phrases were jocular, almost jaunty. Examples: under *A. floridanus*: "I photographed some of the sporophores, then dried them, and ate the rest." Under *A. sylvicola* Vittadini: "Peck's *A. abruptibulbus* is more descriptive, unless Vittadini had in mind some fair wood nymph." Under *A. alachuanus* sp. nov. "Some of my readers might like to know that Gainesville is situated in Alachua County, an Indian word more difficult to pronounce than to spell correctly."

The paper concluded with a concordance of names (aka transfers) as they appeared in the paper versus their form "For those using a different nomenclature" (as *Psalliota*), a habit already established, but continued, to the consternation of future mycological nomenclaturalists.

Two papers appeared sequentially, the first in MYCOLOGIA. For the first, Fred Seaver, the editor, furnished the opening pages. In it, Murrill (1938a) treated Florida white spored agarics following his own arrangement in NORTH AMERICAN FLORA, by then over 30 years previous, and other prior publications.

The paper opened with acknowledgement to Erdman West for help in examining spores, followed by a naturalist's description of phenology: some genera were found over the winter and early spring, and others were seen in summer.

The second paper came out in the BULLETIN OF THE TORREY BOTANICAL CLUB. In it, Murrill (1938c) returned to one of his favorite themes, the polypores. Several examples of Northern polypores replaced by subtropical taxa in mid-Florida were described. Equally, "new" species were to be found, and stages of basidiome formation not previously observed. He was lecturing to his old comrades up north.

A more detailed acknowledgement appeared early in the paper. "Specimens cited in this article have been deposited in the herbarium of the Florida Agricultural Experiment Station. Mr. Erdman West, mycologist of this station, has kindly made microscopic studies of the spores. Dr. F. J. Seaver and Dr. J. H. Barnhart of The New York Botanical Garden, have arranged for comparison with some southern types and for checking of some names." Conversely, species named by others lacked citations to their origins.

The other mycological "publications" of 1938 were mimeographed pamphlets issued by the Experiment Station. In one (Murrill 1938d) he stated: "The nomenclature followed is that used by the author in North American Flora, older names have been placed in parentheses." Lists of species were organized by genus. Keys were provided for species of *Lactaria* (= *Lactarius*), *Russula*, *Venenarius* (= *Amanita*), and *Agaricus*. These keys are not the same as those gathered by Kimbrough (1972), which date from only as early as 1943.

Just as important, these mimeographed documents were never intended for immortality. In fact, by this writing (2024) they have degenerated into brittle parchment, requiring scanning in order to survive.

In his autobiography, Murrill (1945a) acknowledged that compiling lists could be cataloged under the title of "hobbies," long an activity of many naturalists, developed into comprehensive summaries of the fauna and flora of collecting locales, especially if cross-checked against edaphic factors (e.g., mushrooms and rainfall). These intercalated lists were (and remain) basal to ecological knowledge of the county and of central Florida. In his autobiography, Murrill (1945a) singled out trees, flowers, mosses, insects (especially mosquitoes, butterflies, moths), spiders, shells, frogs, turtles, lizards, snakes, birds (and their nests). Some (selected lizards, snakes, turtles etc.), were established as pets, "to study their behavior." His self-assigned title, The Naturalist, came hand in hand as am "inventoryialist."

Ancillary (or preceding) to lists were collections of objects: scrapbooks—collections of everything from postcards to stamps to replicas of sculpture—to say nothing of his music and accompanying lyrics. Hence, the herbarium of fungi was only a significant portion of an undergirding framework of collecting, a leading trait of selected psychological profiles.

In another mimeographed offering (Murrill 1938e), the rest of the basidiomycetes was covered (i.e., not the agarics or boletes). The only key was to the species of *Ceriomyces*.

Murrill (1938e): "This list is based upon collections in the herbarium of the Florida Agricultural Experimental Station at Gainesville. Some species known to be new are still unpublished and much of the mycological material remains to be determined. 'The Genera of Fungi' by Clements and Shear, published in 1931, has been followed for some of the groups; but Burt's names are used for the Thelephoraceae; my own for the Polyporaceae; and those of Coker and Couch for the Lycoperdales, or Gasteromycetes."

Of all the mimeographed pamphlets over a few years, an additional item might be mentioned. Having been bound to the "American Code of Nomenclature" for his years at the Garden, Murrill continued to use it out of habit—a testimony to routine—even though a more recent Code (Cambridge Code 1935) had succeeded in marrying the American and "International" Codes for the first time. In addition, adoption of the recent rules would have necessitated a long, involved, meticulous research effort into the facts of the new arrangements, decidedly against the current of the new Murrill.

Through the final pages of his paper, Kimbrough (2003) rarely reported a date for events, but sketched Murrill's last years (Murrill was almost 60 when he settled on Gainesville as his permanent abode). According to Kimbrough, Murrill did not drive, and it seems that at the least he did not own a car. He was, therefore, bound to his home, much of the campus, and other collecting areas provided by associates—mainly Experiment Station faculty. Kimbrough (2003): "He became a common fixture on campus with his basket and walking cane, poking around shrubs and hedges for mushrooms. Students and faculty were amazed with him and enjoyed joining him on excursions around and near the campus. He was a great story-teller and he fascinated everyone with his tales of travels to exotic places, entertaining royal families in various European countries, and his familiarity with every plant or mushroom species around. The nurses in the infirmary and secretaries around campus all 'fell in love' with him and delighted with the opportunities to chat with him and learn of his escapades."

THE EARLY BIRD AND THE WORM. Kimbrough (2003): "Murrill had strange work habits, rising after midmorning, getting his basket and collecting gear together, and striking out around campus or favorite woodlots in the suburbs. …He would return to his desk on the Rolf's Hall landing and proceed to sort, identify, and prepare his collections for drying. Most of the specimen labels were hand-written [Note: The labels proper were standard mimeographed Experiment Station papers, but the data were hand-written.] and carefully sorted. During the mushroom season, he seldom returned to his home which was 12 or 15 blocks from campus. Instead, he would work late at his desk in Rolf's Hall and in the early morning hours cross the street and make his way into the lobby of the Student Union Building, there to find very soft couches on which to sleep the remainder of the night. As students came to the Union cafeteria for breakfast, they would wake Murrill and invite him to breakfast, enjoying every moment of their meal with the old gentleman. Weber recounted that seldom did Murrill ever have to buy or make his own breakfast."

It might be remembered that in his early years at the Garden, Murrill also worked during the early morning hours before walking some blocks to his Garden desk.

Although the 1939 total roster of specimens (447 included by MycoPortal) is only one tenth of those listed in 1938, this could be accounted for by less splitting of individual collections. Frequency of collecting remained almost daily from January through late August, but September saw only a single report. October through December returned to a more frenetic schedule, but somewhat less than the earlier months.

Murrill's addition of new collecting grounds remained monumental, surely brokered by friends and Experiment Station personnel. A roster of new locations: Altamonte Springs, Bronson (Levy Co.), Buzzard's Roost, Camp O'Leno (Columbia Co.; Fig. 8.2), Citra (Marion Co.), Crown Point Hammock, Doctor's Inlet (Jackson Co.), Ellaville, Fruitland (Putnam Co.), Gainesville Courthouse, Gold Head Branch State Park (Clay Co.), Gulf Hammock (Levy Co.), Hart Springs (Gilchrist Co.), High Hammock, High Springs (Columbia Co.), Juniper Springs (Marion Co.), Keystone Heights (Clay Co.), Lake Butler, Lake Rosa (Putnam Co.), Lake Wauburg of Watson's Orange Grove, Micanopy Forest (Fig. 8.3), Nyssa Swamp, Paradise, Penney Farms (Clay Co.), Providence (Union Co), River Sink, Walker's Pine Grove, Worthington Springs, and Zellwood (Orange Co.). His radius, indicated by the new counties visited, bespeaks new drivers.

A number of Murrill collections appear in the herbarium of Erdman West, but other West mycological herbarium holdings do not indicate Murrill at all.

1938 had given the impression of a new colt, clicking up its heels in youthful exuberance. But there came a realization that this new treasure trove ought to be brought to the attention of a larger audience—the New York cabal, for instance—perhaps as a justification of his abrupt farewell. No journal gave better coverage of this group than the Bulletin of the Torrey Botanical Club, and so a series was started for that purpose, the first of which was simple enough (Murrill 1939a).

The opening paragraph told the story: "This series will include new species and notes on interesting fungi that are picked up from time to time in Florida. … Already, about two hundred Florida novelties have either been published or are soon to appear in this and other periodicals."

Fig. 8.2 Camp O'leno State Park. Camp O'Leno State Park, one of Murrill's collecting destinations; a walking path over a bridge. Current. (Florida Tourist Bureau)

Fig. 8.3 Micanopy. Micanopy, Florida, with live oak, Spanish moss, and palmetto. Current. (Florida Tourist Bureau)

A second paper, not part of the series, shared Murrill's (1939b) theory on why the Florida summer mushrooms were so unique. The "Oligocene Island" was a palaeogeological phenomenon and helped hypothesize the origins of the fungi of the region. The two papers could be paired and the results set the stage for some serious taxonomic activities; the same format as usual, now supported by the mandatory

Latin diagnosis. Not all species were based on his own collections; Erdman West was prominent as co-collector.

This time, it took two opening paragraphs to tell the story (Murrill 1939b). "Since I have been spending my summers in Florida a new mycological world has been opened up to me, with many strange and wonderful forms never heard of or even dreamed of, before. As I write these lines, I note on a blotter beside me that three hundred and twelve new species have been described [not necessarily published] from this region during the past year. Why are they here in the first place; and why have they been overlooked for so long? I believe I have the answer to both of these questions."

"The 'Oligocene Island' extends from near the south bank of the Suwannee River to the south end of St. Petersburg Bay and Alachua County, the special field of my recent explorations, lies well within its boundaries. All this region was once a real island and retained its insular character continuously over a period of at least twelve million years before being connected with the mainland. Its early climate being tropical, its original species were naturally those of adjacent lands in tropical America, while its insular character favored the development of many new species from the original ones." His experiences on Caribbean Islands and tropical Mexico were part of his personal database.

The mushroom story continued in the Journal of the Elisha Mitchell Scientific Society (Murrill 1939c), but the bulk of the year's writing was in the informal Experiment Station mimeographed material, some twelve in number, outlining natural groups of flowering plants of the county and only one (the amanitas) on mushrooms (see Appendix 2).

Over the year, both of his professional friends, Erdman West and George Weber, were remembered with eponymous fungi.

1940 was another year of consummate field work, most including Erdman West, but Murrill obviously perceived that in the joys of fresh air and penciling fascinating new data on the area's ecotones in his notebooks, he was not keeping up with describing the many variations of basidiome form, color, odor, and color changes in drying and especially in spores (but never methodically describing spore sizes, shapes and ornamentation, which may have been relegated to Erdman West, who also could not keep pace), and that such variants would be informative to non-Floridian readers. He embarked on a series of five papers in the Bulletin of the Torrey Botanical Club (Murrill 1939a, 1940a, b, c, d). Eighty-four new taxa were proposed, not only agarics, but polypores and hydnums as well. A single species was sneaked in from Lynchburg.

Sprinkled in were some eponymic names, one of which might be interesting. Together with co-collectors Lillian Arnold, Erdman West, George Weber, and Arnold S. Rhoades was Wilmon Newell, retired Director of the Experiment Station, for whom Murrill named a species of agaric. Perhaps a nod to patronage, for Kimbrough (2003) related that West was able to provide Murrill with an income of $50 per month from the Station, surely with the approval of Newell as Director, a reward for the numerous collections Murrill was adding to the herbarium.

Living up to his self-bestowed title of Naturalist, Murrill was a close observer of habitat as well as taxonomy, and was especially impressed by the distribution of "native" violets in central Florida, so much so that he wrote two papers on the subject, one from the taxonomic viewpoint (Murrill 1940e), the other the geological (Murrill 1940f).

Similarly, among the half-dozen mimeographed products of the year was one on the local *Crataegus* (hawthorn) species (Murrill 1940g). This common-language piece seemed to set the stage for two such papers in scientific journals 2 years later (Murrill 1942b; Murrill et al. 1942). As Murrill (1945a) wrote: "Hawthorns were widely collected, their seeds planted and an orchard established for study and comparison… twenty new species being added." The genus had seen extensive work by Charles Horton Peck, Murrill's former father figure in New York State.

Murrill's (1940h) only submission to Mycologia for the year was a minor paper reporting (with photo by Weber) "*Craterellus unicolor*" from pine-oak forests; much later transferred to *Clavariadelphus* (see Methven 1987).

THE VANISHING TYPES. Murrill's additions to the herbarium were laudable, but there became a problem of diminishing space. Lexemual Ray Hesler (1888–1977; Petersen 1978), mycologist at the University of Tennessee, visited Murrill in Florida more than once, post-World War II in all probability. On one occasion, Murrill reported a difficult situation in the Florida mycological herbarium. It seemed that coming to the herbarium one day, he found a lady employee with boxes and packets of Murrill collections spread out on a table. She was engaged in amalgamating dried specimens from several small boxes or packets labelled as synonymous, into larger boxes with more room in order to save space. Murrill was appalled, for type specimens were spared no mercy, so type and non-type specimens were being mixed as though a single collection. This was no way to run a well-curated herbarium.

Based on previous conversations with Hesler, especially their shared Cornell experience, and therefore aware of the meticulous curation of Tennessee's herbarium, Murrill asked Hesler if Tennessee would accept portions of Murrill types, just to keep them "pure." Although space for fungi at Tennessee was equally limited, Hesler readily agreed, and over the following months, Murrill types began arriving in Knoxville, eventually numbering over 200, where they reside to this day. No publicity was shed on these transactions, and they escaped the thorough search for Murrill types by Halling (1986).

THE SECOND WAR. The scanty biographic information about these years in Murrill's life do not touch on his opinion of the combatants in World War II. In fact, the U.S. maintained a bellicose neutrality; headlines carried news of the European (and much less, the Asiatic) combat, but Congress and the President stalwartly remained superficially neutral while loudly proclaimed their support for "the Allies." Franklin Roosevelt: "I have said this before, but I shall say it again and again and again; your boys are not going to be sent into any foreign wars." (Lend-lease Act 1941).

But among the belligerents in the fight, the hearts of most Americans were with "the Allies," England, France, Poland, and the rest of overrun Europe. One nation remained dubious, however. The Soviet Union, with its Communist government, had signed a non-aggression pact with Germany, and, at least on the surface, remained without obvious preparation for war.

"The Allies," however, were on the brink of defeat. Poland had been overrun from both west and east, France had largely been conquered (including the Museum National d'Histoire Naturelle in Paris), but after the calamitous withdrawal at Dunkirk, Britain resisted German incursion as best it could, with token air power. Hitler was enraged and loosed bombing air-raids over major British cities, starting in 1940 and extending into the following year (London Blitz 2024). Britain was stout of heart, but unequipped with all the weapons of war.

Neutral or not, Murrill, like many botanists, must have turned their thoughts to the "Blitz." Of the cities and military targets of the German Luftwaffe, what of the Royal Botanical Gardens at Kew and Edinburgh; the great universities of England—and cathedrals?

In July 1940, after Britain had sustained the loss of 11 destroyers to the German Navy over a 10-day period, newly elected British Prime Minister Winston Churchill requested help from President Franklin Roosevelt. Roosevelt responded by executive orders exchanging 50 destroyers for 99-year leases on British bases in the Caribbean region and Newfoundland. As a result, a major foreign policy debate erupted over whether the United States should aid Great Britain or maintain strict neutrality (Lend Lease Act 1941). Passed in March, 1941, the Lend-Lease Act established a system allowing the U.S. to lend or lease war supplies to any nation deemed "vital to the defense of the United States."

The ink was hardy dry on the Act when, in June, 1941, Germany invaded the Soviet Union in "Operation Barbarossa," violating their non-aggression pact. The USSR was in no shape to forcefully resist, but because it was now at war with Germany, it was brought under the Lend-lease Act, and the leaders of "The Big Three," Roosevelt, Churchill and Stalin met to plan strategy (German invasion of USSR 2024).

And yet, the U.S. remained "neutral," although the pressure grew to join forces with the Allies.

The final inflexion point came from an unexpected direction, the Imperial Armed Forces of Japan. On December 7, 1941, the Japanese Air Force bombed the American naval base at Pearl Harbor, Oahu, Hawai'i on a Sunday morning. Within 24 h, Roosevelt called on Congress to declare war on Japan. But Japan was only one of the Axis powers, and Germany was the first of them to formally declare war on the U.S. The die was cast—the U.S. was at war.

Of the attitude and actions of William Alphonso Murrill, free soul of Gainesville, Florida, we know almost nothing.

Mycoportal data again reveal most active fieldwork as winter; January and December of 1941. Some specific data points prove interesting. First, two month-long periods seem virtually clear. Except for March 30, there was a clear period from March 6 (Friday) to April 22 (Tuesday), and the venues in late April were all far away from Gainesville; August 5 (Wednesday) clear to Sept. 7 (Sunday). The cause(s) for these periods, of course, remain unknown. Second, several new locations were visited during the year: High Springs, Julington Creek, Keystone Heights, Lake Alfred, Madison, Orange Springs, Worthington Springs, almost all at a considerable distance from Gainesville. Third, in March there was a multi-day trip to the Gulf coast, with many specimens of pathogens harvested. Gasoline rationing did not take effect until spring, 1942, so travel remained relatively inexpensive during the 1941 collecting season.

CHICKENS AND ROOSTS. In addition to a full field schedule, Murrill allowed himself a rare stab at ironic humor.

One root of the story harked back to his early days at the Garden, where he tutored the young student, Lee Oras Overholts from Miami University, Ohio. Overholts (1911) followed Murrill's advice and spread his polypore collections over numerous genera and acknowledged Murrill in the resulting paper. Curtis Gates Lloyd was furious, and scolded Overholts for such unnecessary foolery. In finding fault with the product, though, Lloyd (1912), blamed the wrong culprit for the sin. Poor Bruce Fink, Overholts' professor at Miami, took the blame, and Lloyd and Murrill remained personal friends. (For more on this episode, see above).

Now, some three decades later, Murrill had the temerity to treat the boletes in a similar fashion. This time it was Walter Snell (1941; McLaughlin 1983) who strenuously objected in a MYCOLOGIA paper, calling the former editor out for using taxonomic characters usually reserved for species to divide the boletes into too many, unsupportable genera. Parenthetically, the Snell paper had been simmering since it had been read before a meeting of the National Academy of Science 2 years previous, giving it additional gravitas.

Murrill (1941) was incensed. His answering salvo was less than a half-page in his former journal (and given this unique paper space by friend, Fred Seaver). Short enough for economy, and certainly unusual for Murrill, the paper in full: "In my work with the polypores, when I tried to break up a large homogeneous group into convenient pieces there was a terrible howl about the lack of generic characters, which I must admit was partly justified. In the boletes I used the color of the spores to segregate genera, as was commonly done with other fleshy fungi. Now this treatment draws heavy fire because *no one had thought of it before*, and also because I did not split the eighty species into ten genera of exactly eight species each! Sorry, but there seems to be no pleasing all the people all the time." Fittingly, the title of the pocket-sized paper was "The devil and the deep."

Probably coincidental, Rolf Singer (1942a; Fig. 8.4), soon to spend a year in Florida on a Guggenheim grant, published a paper in LLOYDIA concerning his use of "small genera" to deal with the Agaricales. Invoking amyloidity of spores and hyphae, and especially presence or absence of clamp connections, he contended that even ancient type specimens could be more accurately placed within "small" genera. The paper was intended to point out particular type specimens that could be assigned to a "small genus" based on such anatomical and spore characters. Of the large number of taxa summarized, 52 were Murrill species, of which 10 required

Fig. 8.4 Rolf Singer. Dr. Rolf Singer, at about 35 years old. (Wikipedia)

new names or new combinations. Of course, it is not known when Singer's paper reached Murrill's eyes, but the subject must have been a prime candidate for conversation when the two met early the next year.

WAR-TIME BACK HOME. President Franklin Roosevelt proclaimed that a state of war existed against the Empirical Throne of Japan on December 8, 1941, the day after the Japanese attack. Less than a week later (December 11), Germany declared war against the United States and before the day was over, the reciprocal pronouncement was delivered by Congress. This time, the entire world was involved.

The Axis powers were well-prepared for the new state of affairs; the Allies less so. Roosevelt was a naval sort and the country's navy was better off than most, and 50 destroyers had already been loaned to Britain. Male conscription, "The Draft," had been predicted the preceding summer, and now was instituted. The country had to be mobilized at breathtaking speed. 1942 became a banner year for the dexterity of this Capitalist nation.

One facet of the war effort lends itself to a timeline—transportation. Two interrelated industries were involved: vehicle manufacture and the means to make it mobile. The first: production of new automobiles ceased (by proclamation) on January 1, 1942, less than a month after declaration of war, and purchase of new cars banned on February 22. Production of tanks, trucks, and personnel carriers (soon to be dubbed Jeeps) began as soon as industries could be identified and drawings for production produced. The fledgling shipbuilding and airplane industries were mobilized. Steel was needed everywhere. Hiring was fierce, employment approached full status. The Great Depression was overshadowed by the necessities of "Big Government" and wartime.

The American East coast—the western Atlantic Ocean—was the stalking ground for German submarines and U.S. oil tankers had become a prime prey. Even the Gulf of Mexico was unsafe. Yet this was the prime mode of transportation of crude oil from its wells in Texas and Oklahoma across the central states to the giant refineries in New Jersey, in the shadow of New York. Train and barge transport could only substitute a fraction of normal traffic. On April 9, 1942, gasoline deliveries were decreased to seventeen Eastern states, but to ensure fairness, rationing was required. On May 15, eight million motorists in those seventeen states registered for gas ration cards, and rationing began on the East coast on July 22 (Sarah Sundin 2022). On December 1, almost exactly a year after the Japanese attack on Pearl Harbor, country-wide gasoline rationing was instituted.

In June, 1942, construction began on a twin pipeline from Texas to Linden, New Jersey. Its dimensions tested the formulae for such things; the smaller was 24 inches diameter, the larger, 30 inches. Incredibly, the pipelines were opened only 1 year later, dubbed "the Big Inch," and transported 500,000 gallons of petroleum per day (NIST 2024).

Gas rationing, though, was really only a method through which to conserve rubber, not gas. Pre-war, raw rubber had been harvested largely in Southeast Asia, now under Japanese occupation. Substitutes for hevea rubber (e.g., guayule, chicle) were not proving practical, and gasoline was the proxy used to conserve rubber.

On Dec. 1, 1942, a nationwide "Victory speed limit" of 35 mph was instituted, since higher speeds removed tread more quickly. Lightweight "Victory Bicycles" had been introduced in March, but were issued only to war workers and certain professionals and only when public transportation wasn't available—and all other bicycle production, like automobiles, ceased.

WAR HEADLINES. Meanwhile, British forces fought Italians in northeastern Africa for control of Egypt and the Suez Canal. The headlines must have brought anxious memories. Even though Bresadola was dead, the Italian Tyrol remained, now under the rule of Mussolini. Paris was fully occupied by Germans, but its great historical gems had been spared destruction. Stockholm was still intact in neutral Sweden, and Berlin, always under threat from the air, remained largely undamaged. As a hybrid of the Atlantic and Pacific wars, "Free French" forces landed on Madagascar to thwart Japanese invasion of East Africa, and at the end of the year, Field Marshal Erwin Rommel and his "Panzer Division" of tanks rolled across the Mediterranean coast of Africa to secure it for the "Third Reich."

Mycoportal data (collections total 337): For Murrill, 1942 matched previous years for days in the field; virtually every week at least once (including Christmas Eve and Day), but June-July-August showed most frequency and richest taxon diversity. As usual, though, interesting ancillary data may also be harvested. First, although Murrill was accustomed to collecting basidiomycetes—particularly agarics and polypores, 1942 collections were dominated by plant pathogens, especially *Gymnosporangium* (e.g., *G. clavipes* and others, probably on hawthorn). *Puccinia, Tubercularia, Hypoxylon*, and others were also represented, with significant numbers of collections making their way to the Georgia Mycological Herbarium, New York Botanical Garden, USDA Bureau of Plant Industry Herbarium, and University of Michigan, where authoritative identifications could be obtained. In wide ratio, though, most collections remained at the University of Florida.

This "distortion" of Murrill's normal collecting could be linked to war-time conditions; they closely fit the mission of the Station, especially plant pathology, of which George Weber was head. Nicely, Murrill's (1942b; Murrill et al. 1942) interest in hawthorns might be reflected in their *Gymnosporangium* pathpgens. Less, collecting sites seem significantly lacking in destinations outside Gainesville. Perhaps gas rationing restricted mileage of Station personnel: even occasional excursions to Magnesia Springs or Neuman's Lake took a large fraction of a week's 3-gallon allotment of gasoline.

Murrill (1945a) related that he planted a "Victory Garden" and helped others do the same.

Although Murrill had already found a new outlet through which to organize and display his lists—the Experiment Station's mimeograph machine—1942 saw a great swell of mycological content. Summaries of boletes (Murrill 1942c), collybias (Murrill 1942d), russulas (Murrill 1942e), dark-spored agarics (Murrill 1942f), polypores (Murrill 1942g), amanitas (Murrill 1942h), Lactarii (Murrill 1942i), lepiotas (Murrill 1942j), and tricholomas (Murrill 1942k) appeared in close succession (Appendix 2). *Marasmius* (Murrll 1942l), treated for North American Flora by Pennington, now saw Murrill's attention, and finally, "lower Basidiomycetes," not Murrill's "long suit," appeared at the end of the year (Murrill 1942m). Why these reports appeared in their mimeographed state and not in recognized journals is not known, but Murrill's Florida life style may have been at odds with formal scientific presentation.

Both brief and fragmentary, Murrill (1942n) provided a key to the genera of resupinate polypores of Florida. Although eleven genera were keyed, only four were treated, two of them new. Seven more were covered in other publications. It was not Murrill's best or most thorough work.

In a small paper, Murrill (1942p) revealed: "Spiders listed below were collected by the author in and near Gainesville between the autumn of 1935 and that of 1936." If these dates are intercalated to the references concerning collecting of spiders as a clue to Murrill's first year of experiencing summer in Florida, that year was 1936.

MORE WAR HEADLINES. Russia (at that time better known as The Union of Soviet Socialist Republics—the USSR) had a habit of absorbing attackers from her west until the attacker's supply lines withered and/or winter set in—most likely both—and the attackers thought better of their incursion. So it was in 1942, as German forces advanced toward Leningrad (= St. Petersburg) in the north, Moscow in the center, and Stalingrad (= Volgograd) in the south. Although the indigenous peoples were at the point of starvation and frostbite, they were marginally better off than the German army, and after sieges of months or close to a year, the shivering Germans were forced to stumble back along the very supply routes around which they had employed "scorched earth" policies on their way in. Germany had over-flexed its dreams of "Lebensraum" by opening an "eastern front." For military rank and file, transfer to "the eastern front" was a mortal threat; especially capture by the Russians. Roosevelt's "lend-lease" program was a significant boost.

Meanwhile, late in 1942, the Allies (now including the U.S.) initiated "Operation Torch," the invasion of the north coast of West Africa, intending to sweep Romel's tanks away and, once in control, set out to invade Sicily on the way to Italy and France; a "southern front" for the Germans to think about.

"The Eastern War" for the United States was fought on the Pacific waters. The tatters of Pearl Harbor had been patched only poorly, but by "paste and patch," a fleet had been assembled. Japan correctly—secretly, of course—concluded that it had awakened a ferocious beast, nevertheless was psychologically trapped. The navel engagement at Midway was the result, and through a combination of code-breaking and stealth, the American "fleet" won out and the Japanese limped away. An overall plan was devised by the Americans to overrun and occupy, one by one, selected islands generally pointed toward the Japanese homeland. The plan became known as "Island Hopping." Meanwhile, back home, by executive order, Japanese-Americans were being interred in large, guarded camps across the U.S. for fear of undercover activities. The same would have been the fate of German-Americans, but they were indistinguishable in the overall population. It was not one of the nation's better moments. Eleanor Roosevelt, the First Lady, expressed regret and visited some of the camps.

At home, "Victory Saving Bonds" were sold to school-children, the "military-industrial complex" swelled, and "Rosy the riveter" became the "image du jour" for war-time posters.

Three gallons of gas a week could hardly support a tourist migration from the Northern states to the South, especially Florida, and the state's revenue was threatened as a result.

In reality, although the seasonal tourist invasion was lagging, Florida's economy was booming. Although Murrill had chosen Gainesville over other locations for his home base, in part, on its separation from the unexciting citrus landscape, this very industry was more than robust. For the first time, Florida passed California as the country's citrus capital, and a new-fangled process producing frozen, concentrated orange juice promised even higher demand, and therefore higher profits (Florida wartime economy 2024).

The tourists might stay home, but Florida's climate was nearly perfect for the military to establish training facilities for all the services. By the end of 1942, Florida had over 172 military installations for land-, sea- and air-based camps, including an astonishing 40 airfields. One base, Camp Blanding, occupied 180,000 acres and housed 55,000 soldiers at a time. The Great Depression was overcome with full employment. Many of the GIs would remember those days, and seek to return after the War (History.com).

ROLF SINGER (1906–1994). Destined to become legendary in the international mushroom fraternity, Rolf Singer spent 6 months in Gainesville in 1943, during which he worked with Murrill's specimens, particularly of boletes, shared some time in the field, and engaged in some contentious conversations. His pre-Murrill mycological experience requires some explanation, not only because he had seen more of the world before he met Murrill, but also because I have not seen an authoritative summary elsewhere. German by birth, with a Ph.D. from the University of Vienna, Singer's signature on a students' petition for peace put the Nazi regime on his trail. Folklore also points toward Martha's (aka Mimi, and Rolf's wife) probable Jewish family background. Regardless of the reasons, Mimi fled south through Switzerland to Italy, where she was joined by Rolf, who skied over much of the Alps to join her (Singer 1984; Horak pers. comm.). But Mussolini's attitude and partnership with Hitler made Italy only a waypoint, with the couple's next destination, Spain, where pro-Hitler Francisco Franco was propelling the country toward a civil war. Spain turned out to be no less problematic than Germany or Italy, and the Singers again fled, this time to Paris, France, which had not yet been overrun. There, Roger Heim, mycologist, negotiated a "petit bourse" for Rolf at the Paris Museum. The couple, by then, were becoming parents to a daughter. According to Mimi, N. L. Vavilov, Russian geneticist, also visiting the Museum in Paris in 1935, met Rolf and offered him a solid position in the Komarov Institute in Leningrad, Russia. There, the Singers lived in a "scientist's" house built under Peter the Great, and Rolf was not only free to work on agarics, but arranged at least two lengthy field trips to Russian Central Asia, accompanied by Larissa Vassileva, a student destined to be the leading agaricologist in Far Eastern Russia, headquartered in Vladivostok.

Leningrad was not only comfortable, but Rolf's work was honored [as of 2020, his desk was still set aside and pointed out to visitors (pers. obs.)]. Leningrad, though (see above) was inevitably a temporary venue as the Germans invaded the USSR and set siege to the city. Parenthetically, "Apollinary Semjonovich Bondarzev, the famous Russian polyporologist, received the order of Lenin for his role in organizing potato growing during the siege of Leningrad, but Vladimir Andreevich Tranzschel [Uredinales expert] died of hunger and cold…" (Singer 1984; Mycologia 36(3): 307. 1944).

In the preface to Bondarzev's (1971) book on Russian polypores, Prof. V. Savich wrote: [trans.] "The author [Bondarzev] introduces an original, natural system elaborated jointly with R.A. Singer."

Mimi was first to leave the USSR, arriving in the United States before Rolf, who made his way to the American West coast via Vladivostok. Perhaps through the good offices of Mimi, a temporary accommodation was made for Rolf at Harvard's Farlow Herbarium, where David Hunt Linder was Director. While there, Rolf received a Guggenheim grant for fieldwork in Florida, and it was during that trip that the Singers met William Alphonso Murrill. Parenthetically, after the War, the Singers moved to Argentina, from there to Chile, Brazil, and eventually to the Field Museum in Chicago. Through it all, Rolf Singer was prodigious in agaric research and simultaneously, the writing of the results in the monumental volumes entitled "The Agaricales in modern taxonomy" (the last revision, Singer 1986; Mueller and Wu 1997). Singer's bibliography is unique, for papers were faithfully penned in the language of the country in which the couple was living.

If Murrill had been blamed—by C. G. Lloyd—for employing excessive genera for the polypores (see above), Singer exceeded Murrill several times over for the agarics. His rationale, like those before him, was the search for "natural" groups based on ever-more-fastidious taxonomic characters. Singer's system, he testified, was in the lineage of Narcisse Theophile Patouillard, the Frenchman pharmacist visited at least twice by Murrill, and Victor Fayod (1860–1900; Wikiword 2024), a much less celebrated Swiss mycologist, but master of basidiomatical anatomy. To the gross taxonomic characters used by Persoon, Fries, and Murrill, were added those of Patouillard; hyphal systems and anatomy, structures not before measured, and by Fayod (1889; Fischer 1900; see also Singer and Smith 1946) and Singer to arrive at a vastly more complex but hopefully more natural system than previous. Murrill, of course, mostly employed the sensual characters, with spore measurements and shapes added to spore print color. After many years, when necessary, he borrowed Singer's observations (see above).

To sum up this "tempest in a teapot," the taxonomic characters emphasized by the two workers, Singer and Murrill, were almost like night and day, but Singer, at 36, had seen as many collections—at that time, Northern Hemisphere—of fungi as Murrill, at 74.

On herbarium labels dated as early as 1938, however, and more in 1939, Rolf Singer's name appears as "Com." (?). I suspect that this was Singer's way of indicating annotation. Mimi (Singer 1984) recalled only 1942 in Gainesville, but this was of half-year duration, and her dates are often inaccurate. But 1942 was mid-WW II, and 1938–1939 was Great Depression. Singer didn't get to the U.S. (i.e., Harvard) until the 1940s, so 1938–1939 is surely inaccurate. It could also indicate mycological name changes by Singer, but this should not have been indicated on the herbarium labels.

In 1943, early collecting (January) was almost daily, mostly in Gainesville, but some with less accurate location; "Florida." MycoPortal data show that Rolf Singer (see above) accompanied him. There are indications that Murrill was stalking particular genera during particular periods: i.e., April and September = *Agaricus*; July = boletes; April through mid-August, rich and varied collecting; from mid-August through early October almost clear (except Sept. 24).

In Mycologia 35(4) (Murrill 1943a); Singer's name first appears. In the description of *Armillaria floridana*, "Dr. Singer would probably refer the species to *Limacella*." The same sentiment was written under *Armillaria graveolens*. *Armillaria raphanica*: under "the latest classification species of the old genus *Armillaria* …would be placed by some in the genus *Lentinus*." (In fact, by the early 2000s, the species would be accepted as a *Lentinula*.) Most of the reported fungi were noteworthy by odor or color. Singer's 1942 paper commenting on 52 Murrill agaric names (see above) was surely in hand. A couple Lynchburg collections from the late 1920s were included.

In Mycologia 35(5) (Murrill 1943b): "Dr. Rolf Singer has kindly allowed me to use some of his notes on microscopic characters." This confirmed Mimi's comments about Murrill reading Singer's notes, but gave Murrill a passing grade for good manners. Under *Clitocybe peralbida*, a whole paragraph of Singer notes was transcribed, including structures hardly observed by Murrill.

That Murrill could transcribe Singer's notes on anatomical characters, involved the revelation to Murrill of what Singer thought Murrill's descriptions lacked. If nothing else, this served as a threat to the efficacy of Murrill's legion of new species, and for Singer, it was the awakening that most—perhaps all—of Murrill's type specimens required re-examination just to place them in Singer's "small genera" system. For Murrill, there was the possible deep realization that "the Naturalist's" observations were insufficient for a new era. How these contrary constructs were shared in the confines of the Florida herbarium is, of course, unknown, but neither man ceased his quest—for Murrill, new species for the Florida mycota; for Singer, the most coherent arrangement of the entire Agaricales. Whether their time together was jovial or volcanic can no longer be known, but the reports vary from agreeable (Singer 1984) to loud and furious (Kimbrough 2003). Undeniable, though, in 1942, Murrill was 73, Singer 36, almost exactly half Murrill's age. For both, patience was required.

Singer's Guggenheim grant was for fieldwork in Florida, and he arrived there at the end of July, 1942, and left almost exactly a year later. MycoPortal data shows that his first 6 months (August 1, 1942, through December) was spent in southeastern Florida. He seems to have arrived in Gainesville in time for field work on New Year's Day, 1943, and remained in or close to Gainesville until the end of July. Specimen label data show that he visited several of Murrill's favorite collecting sites, but Murrill's name does not appear on Singer's labels.

The mission of fieldwork did not necessarily include examination of herbarium specimens, and Singer fulfilled this mission, even though it meant harvesting new taxa from just under Murrill's nose, and/or confirming that Murrill had already named the organism in question, so Singer could simply augment descriptions. Both protocols required soft-speaking, of which both participants were adept. But Singer was a heavy smoker, both when laboring in the herbarium and the field, a habit antithetic to Murrill's life style. Their discussion on this topic, as always, remains unrecorded.

A serious exchange by Murrill and Singer of papers on Florida boletes occurred a few years later, and is covered in the next chapter of this narrative.

Martha "Mimi" Singer (1984), writing of Murrill's daily work routine, also implied that Murrill plagiarized Rolf notes on Murrill's type collections. She wrote: "Murrill usually worked during the early hours of the morning, between 3 and 6 AM, and during these hours, undisturbed, he would look not only at his own notes and specimens, but also at Rolf's notes on type specimens which later, in the identical wording, appeared in short notes, signed by Murrill. Rolf never minded this nor held it against his old colleague." As time passed, it became clear that Singer had, himself, copied Murrill's notes taken from the fresh specimens, with characters no longer available by the time Singer worked them over. Who was plagiarist and who was victim?

TAXONOMIC AND NATURAL AFFILIATIONS. Singer's taxonomic philosophy was pitted against that of American agaricologists, Alexander Smith, Calvin Kauffman, Lexemuel Hesler, and others. For Murrill it must have seemed like a philosophical maze, but he was not of a mind to alter his modus operandi. Therein lies a story.

By the time of Singer's Florida trip (1942–1943), Hesler had shifted his interests to include the fleshy fungi (= Agaricales). These dates, somewhat arbitrary, matched the period of Singer's time in Florida with Murrill. Because Hesler had already established a career as a plant pathologist, his switch to fleshy fungi required a learning curve similar to that by a layperson first introduced to mushroom hunting. After all, his Cornell days with Atkinson were far behind him. One circumstance, however, was pivotal: Hesler was already an academic with extensive contacts in the mycological fraternity. It was only natural that he turn to Alexander Hanchett Smith (1904–1986), at the University of Michigan, already a well-published agaricologist and mentor to graduate students in the trade, and soon, dual-authored papers began to appear (Smith and Hesler 1938). The first taxonomic group dealt with *Hygrophorus* (Smith and Hesler 1942), by then already a large genus split by some into a half-dozen smaller genera.

It was not long before some subtleties emerged: Hesler's hunting grounds were the southern Appalachian Mountains, just down the road from southwestern Virginia, and over the hills from Biltmore's school of forestry. To which of these geographical locales would the mushrooms of the Great Smoky Mountains National Park resemble? Probably not those of central Florida where Murrill lived, but quite possibly the Shenandoahs of Virginia, and farther north into the Catskills and Adirondacks, where Murrill (1916f; and Peck) had already sampled.

So the inquiry into distribution patterns was added to the taxonomic philosophy encapsulated into "lumpers" versus "splitters." In a paper dealing with "Hygrophorii," Smith and Hesler (1942) included several taxa described by Murrill (1916i), almost all from the American tropics. Murrill's types collected in Jamaica were unanimously not located by the later authors (but Murrill's type descriptions were transcribed in full) and Murrill's types from Florida were not studied. Comparison of descriptions, despite the publication dates, revealed the attitude of "modern" workers to Murrill's efforts.

On the subject of taxonomic "splitting," the long series of papers on *Hygrophorus* sensu lato took many years (Hesler and Smith 1963) for the divisibility of genera to be summarized." The point of greatest difference between various investigators hinges on what constitutes a discontinuity of generic magnitude. Although we held a conservative point of view in our early work, we have given serious consideration since then to the recognition of the genera as Singer (1951) has outlined them. In fact at one time we were almost ready to accept them. We finally returned to our original point of view because we felt that those erecting new genera were simply doing it on the basis of single outstanding characters of odd species. … On last analysis, however, the matter will probably continue to remain one of opinion, as it has in the past. Certainly each investigator may exercise his right of opinion, based on the material he has seen. One point, however, which we would like to make, is that it is to us more realistic to try to arrange the species of *Hygrophorus* in a natural classification within a large genus than to distribute them among a number of genera, and thus create the impression of a degree of discontinuity which we believe does not actually exist." A faith statement, hardly scientific. In the end, Murrill's (1916i) opinion was no better or worse than anyone else's.

In MYCOLOGIA 35(6) (Murrill 1943c), a small double-entendre was permitted. *Ithyphallus* was a stink horn genus, a fungus group universally accepted as phallic symbols, and *I. murrillianus* was described by Erdman West (1941) to honor Murrill. Murrill poked fun at the possible phallic symbol: "All the specimens of the above plant that have been brought to my attention have been determined by me as *I. rubicundus*. Mr. West's effort to do me honor is appreciated but I hardly think I deserve it."

WAR HEADLINES. Russian troops recaptured territory to the southwest of Moscow once the siege was broken. The first city of considerable size was Novgorod, which had been fiercely contested, and now was retaken by the Russian forces. Only a few days later, the Leningrad-Moscow rail link reopened, effectively signaling the broken siege of Leningrad.

The Allies established a beachhead on the Italian mainland at Salerno in September, 1943, and in the following March, set out for Cassino (of which the monastery atop the sheer escarpment was part). Allies dropped 1250 tons of bombs on Cassino, commencing their land offense on the monastery. The occupying Germans evacuated Cassino by mid-May. Within a few days, the Allies broke out of the new Anzio beachhead and made for Rome. Two weeks later, Hitler ordered withdrawal of German troops from Rome, and the Allies entered in early June (British History 2024).

Two days later (June 6), the vast Allied amphibian invasion crossed the English Channel and struck the beaches of Normandy, opening a serious western front against Germany, which now faced Russia on the east, Allies in Italy and invasion on the French coast in the west. In less than 2 weeks, the local coastal French city of Cherbourg was surrounded and the German military was trapped.

As though in retaliation, Germany launched the V1 rocket-bomb on London. It was unguided and so lived up to its purpose, terror through unpredictability.

In late March, Russian troops moved into Romania on the Black Sea for the first time, and 2 weeks later launched an onslaught on the Germans in Crimea. Within a month Sebastopol (aka Sevastopol) was liberated, and Crimea reverted to Russian control.

With significant lend-lease equipment and ammunition, Russian troops continued their western march, and in mid-July, entered Poland. Behind German lines, the Jews of Warsaw formed a garrison and resisted occupation, but within days the uprising was put down cruelly, with many civilians rounded up and exterminated.

With defeat after defeat evident, in the fall of 1944, some senior German generals set out to assassinate Hitler with a planted bomb. By chance, he survived, but insurrection had been revealed, and the senior military staff was sacked.

In late July, the Russians captured Lvov (aka Lviv), in northern Ukraine SSR, and announced a hand-picked Communist regime in Poland. As the eastern front was near collapse, the Russians entered Bulgaria. On the western front, Paris was liberated in late August with a tumultuous parade down the Champs Élysées. In following days, Brussels, Ghent, and Antwerp were liberated.

The German V2 rocket exploded in London, this time more powerful than the V1 and guided so it could be aimed. It was the latest brainchild of Werner von Braun, the German rocket engineer.

At year's end, the Germans broke through Allied lines in Belgium, and in atrocious winter weather, pushed toward the English Channel, with the hope of splitting the allied ground forces. It took the Allies several weeks to stop the advance and retake the occupied territory. The eastern and western fronts were closing in on Hitler in Berlin, and for the Allies there was distinct "light at the end of the tunnel." Secret meetings of Allied leaders were convened to reach agreement on the anticipated occupation of Berlin and the fate of Hitler. The Russians were obdurate; they had lost more life than all the rest of the Allies combined and had wrested more square kilometers of land as well. Russia was owed, they argued, the largest portion of Berlin (and its population) and should take possession of Hitler (Wikipedia WW II 2024).

In the Pacific war, the Philippines, Guam, Formosa (= Taiwan), and the Dutch East Indies (later = Indonesia) were in the hands of the Japanese. MacArthur ran the Pacific war from Australia. At last, Allied aircraft were threading their way through the Himalaya from India to deliver supplies to western China, and were putting pressure on the Japanese in Burma. Indonesia had been absorbed in 1942, but relief was anticipated as the Allies "island hopped" through the Pacific chains. Both the naval and air wars had been won by the Allies, but the Japanese had begun their suicide strikes, which inflicted serious damage, but not ultimate victory (National WW II Museum 2024).

While the Pacific War was horrendous and fought under tropical conditions, far more of the U.S. population had European roots, relatives and habits, and more GIs were in Europe than in Asia, so the European Theater was followed more closely.

WARTIME MYCOLOGY. Lest the timeline of the War appear irrelevant to the Murrill theme, at least some connection can be made through his discipline, mycology. At very least, two activities saw birth and development in service to war conditions.

First, a group of lowly molds well-known as food spoilers, the *Penicillium notatum* group, had been noticed to kill bacteria in contaminated cultures in the 1920s in England. The phenomenon came to be called antibiosis, but its biological dimensions—the depth and breadth of this behavior, remained an academic endeavor to be investigated. Now, the possibilities became exigent; how many bacterial taxa could be killed; what strength of this feature was necessary to be efficacious; how many different fungi had this property; how could production be scaled up to be practical? Laboratories sprung up across the Atlantic, with U.S. researchers taking the lead in making a minor discovery into a major commodity (Elder 1944; Waksman 1947). In record time, penicillin became demanded on the front lines around the world. Mycology suddenly took on cachet (Medical News Today 2024).

Second, for many years the deterioration of military gear, especially fabric, was recognized as a nasty problem generally, but especially problematic during war and doubly under hot, wet tropical climates. The problem was easy to observe, but went overlooked between the Wars. Finally, it was narrowed to the physiology of microorganisms, especially fungi. Laboratories were established to culture and test molds and to produce protocols to alleviate this biological scourge. No one solution was universal, but the establishment of research laboratories and culture collections served as a springboard for post-war mycological activities. Mycology took on a new look—fungal physiology and biochemistry (Emory Simmons pers. comm.).

The protracted visit by Rolf Singer, if it had any significant effect on Murrill's modus operandi, passed without perturbance. In addition to the papers discussed above, two others, one in Lloydia (Murrill 1944a), the other in the Florida Academy of Science Journal (Murrill 1944b) also appeared. Both were compendia of new mushroom species based on the usual sensual characters (with addition of spore statistics), many were monotypic (i.e., based on only the type collection; a practice frowned upon in the taxonomy community), and most collected some years previous. A lengthy concordance to "the Saccardo system" was appended to each.

Mycoportal data show another crowded field season for Murrill in 1944. Narrowly, the first half of the year shows him in the field several days most weeks with eclectic collecting. Rather few venues are listed as outside Gainesville. The "Sanibel Hurricane" could account for a lull in late October through early November. Kimbrough's (2003) report of Murrill strolling around the town, poking under hedges and trees, and prodding mushrooms from lawns would seem to ring true. Probably, the small monthly gratuity from the Experiment Station to Murrill presumably had been put in place by then and perhaps Murrill felt obliged to do his part to equalize the pact.

CELEBRATION. William Alphonso Murrill had seen "the life fantastic," and had chosen not to participate. He had seen the European capitals, but chose to live in a self-made cottage in Florida. His musical talent was appreciable, but he reserved it for "singin' conventions" and the tourist camp. He knew a naturalist's depth of the world around him, but shared his knowledge with young people and the public. Around the university he came to know the night-owls—custodians, graduate students—and the folks who got things done—secretaries, orderlies. In each of society's compartments he was willing to share himself, and each sector returned the favor with curious friendship. George Weber (1961) wrote it this way: "His many friends were solicitous of his happiness and were anxious to take him on collecting trips or bring their novelties to him for identification and an explanation. He was actively interested in the local garden clubs and organized a nature club that made weekly forays. The Girl Scouts work was pleasant. Church activities and Sunday School received much of his attentions, particularly in relation to his vast information of natural things, travels, experiences, and musical talent. He was kindly, patient, sincere, understanding, and worshipful to children and adults in his new environment. He lived alone and kept his own council regarding his past activities. He had no pension or retirement, and seemed to have no visible plan for the future." And somewhere he had a penchant for science, natural science, nature, mushrooms—collecting, drying, packeting, filing, writing, photographing—so many that mycologists everywhere became aware of Florida's untapped mycota, and came to

Gainesville to make his company. And through it all, his friends and associates wove a web and plotted an evening of celebration for this old man, ostensibly on his 75th birthday, October 13, 1944.

He could have said "No, thank you," but he didn't. Instead, he allowed himself to be drafted to show up for a birthday party, listened to praise, and enjoyed himself immensely.

Murrill had commented that he left Cornell with a thesis and a Sigma Xi key. Although he failed, even at that early date—1900—to mention a wife, now, in fall, 1944, it was members of Sigma Xi that did the groundwork of seeking out the wide circle of friends across the landscape to honor this man of many faces. The venue was the Bryan Lounge of the Student Union Building, a place well-known from many early mornings spent on the nearby couches.

While sitting "in a big chair" next to University President Hubbell, Murrill listened to George Weber, his Florida discoverer and colleague for two decades, as he sketched Murrill's life (so far), which must have been based on Murrill himself over years of conversation. Speakers extolled the old man and brought smiles on his behalf. In his autobiography (Murrill 1945a) were listed 26 names of notables, mostly mycologists, that had sent congratulations, although a scattering among them must have had difficulty finding the right words. Murrill's last autobiography chapter ended with several short passages from far-off greetings. As friends gathered around him at evening's end, several expressed the hope that he would write an autobiography to keep him fresh in their minds. He started within days, finished in mid-November, and the 164-page result saw the light early the next year.

As has been written about others, "It was his finest hour." He knew it, wrapped himself in it, and had a banner evening. He could walk just a bit jauntier as he teased small mushrooms, and glance back while he wrote herbarium labels.

Chapter 9
The Twilight Years

1945 REPORT. The country was at war almost the world over. In Europe, the lights were dimming for Germany. Italy had fallen and the Russians and Allies were waging a pincer operation with Berlin in the crosshairs. The Russians were winning, and fully intended to beat the Allies to seize Hitler and Germany's capital city. As the new year dawned, the U.S., under General George Patton, was mired in the "Battle of the Bulge" with Britain to the north and the Canadians and European soldiers to the south essentially stalemated.

TWO MOMENTS AWAY. For no other reason than memorialization, at least two GIs fighting in the neighborhood of Bastogne were to become mycologists post-war. Robert Lee "Gil" Gilbertson (Burdsall et al. 2012), Montana-borne, suffered a mangled leg and after repeated hospitalizations in Europe and stateside, was eventually returned to his unit until War's end. Howard Elson Bigelow (Barr 1988) parachuted behind German lines, was captured, shipped east, "rescued" by the Russians and held under austere conditions until released to U.S. forces. Both these GIs took advantage of the GI "Bill of Rights" and became leading professorial mycologists. Their life-stories were repeated by thousands of GIs and sailors.

FDR. Franklin Delano Roosevelt had been struck with poliomyelitis and was paralyzed below the waist. Both loved and detested by the voters, he had led the country through the Great Depression making government bigger, and closer to the individual citizen by establishing numerous commissions, agencies, departments, and czars—communicating partly through his patented "fireside chats" on radio—raising employment and wages and bringing essential services to millions through such new units as the Tennessee Valley Authority. The War exerted a mighty kick out of depression and into war-time economy.

FDR had found a "Shangri-La" in a small resort in Warm Springs, Georgia. He took to driving his special car through the countryside, stopping to visit the nearby farmers (Fig. 9.1). Especially, he enjoyed the special swimming pool, where dozens

R. H. Petersen, *William Alphonso Murrill: The Naturalist*, Memoirs of The New York Botanical Garden 124, https://doi.org/10.1007/978-3-031-93343-1_9

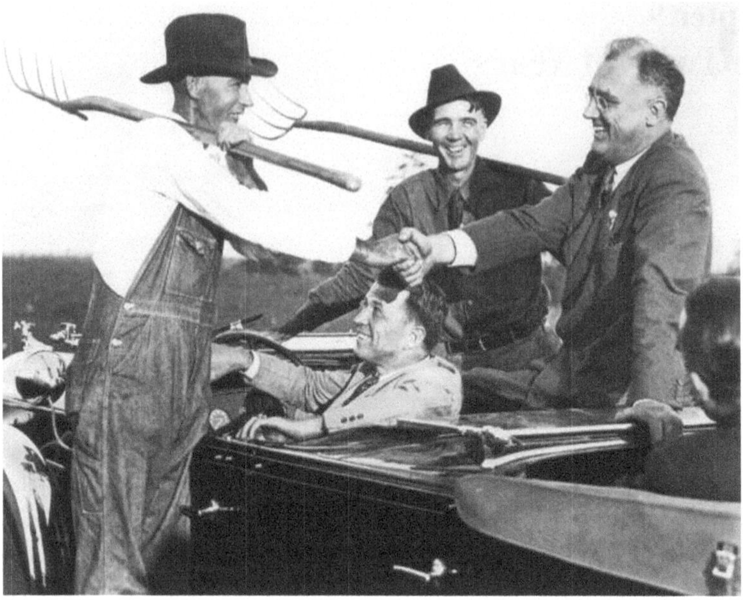

Fig. 9.1 Roosevelt in Georgia. Franklin Roosevelt greeting farmers from his car. (Getty Images)

of similarly stricken folks could frolic. On April 12, Franklin suffered a massive brain hemorrhage, and just that fast, America was robbed of the leader of the past generation. Harry Truman, haberdasher cum Midwestern senator, chosen as Vice President specifically for his low profile, was suddenly President—in the midst of a world war.

BACK TO BUSINESS. On May 8, 1945, the Russians and Allies were squeezing Berlin. Russian troops walked the streets—mostly rubble. The German high command tendered surrender. The war against Germany abruptly ended, and all over the Western world, the day was dubbed "V-E Day" (Final Battles 2024).

"THE BOMB." As history has emphasized, there existed a "science-fiction" research program, unknown to Truman (and the country), called "The Manhattan Project." Centered in several federal laboratories, it was testing the theory that if two atoms were fused, energy was produced. Isotopes, strange word to the public, had been split—"nuclear fission" as it was known—in small experiments, and now research was aimed at creation of enormous energy in massive explosions, essentially a bomb (American Heritage Foundation 2024).

At Roosevelt's death in April, the final touches were being organized. Truman required instant education. Two types of bombs were being assembled and readied for test. The demonstration took place in New Mexico, in the midst of open desert, and exceeded expectations. The two bombs could be assembled for use in war. Truman, sensitive to the massive loss of life that would result, nevertheless approved the program, calculating that the civilian lives lost through the bomb would be far less than through the invasion of the Japanese mainland.

The two bombs were loaded on separate ships and departed the West Coast for Saipan. Once there, plans were laid for their delivery over Japan. They were nick-named based on the shape of the bombs themselves: "Little Boy," and "Fat Man." Little Boy was dropped on Hiroshima on August 6, and wiped out the entire city and almost its entire population. Two days later, Fat Man, intended for Kokura but diverted due to military and meteorological complications, was detonated over Nagasaki, with even more havoc. The Japanese quickly—12 August—accepted the terms of surrender to the Allies, and the final papers were signed on September 2, aboard the battleship Missouri, in the presence of Douglas MacArthur, the Supreme Allied Commander of the Pacific Theater (Atomic Bombings 2024). The home front newspapers proclaimed "V-J Day."

In less than 6 months, the United States had lost its President and won two wars.

Almost simultaneous with the Hiroshima bombing, Russia attacked Japanese-occupied territory west of the Sea of Japan and north of Hokkaido. Once Japan surrendered to the Allies, Russia demanded reparations from Japan.

ANOTHER MOMENT AWAY. Just as in the European Theatre, some young men destined to become mycologists served in the Pacific theatre. Three have been identified: Leon Russell Kneebone (Koleno 2019), a paratrooper wounded in the fight for Leyte (Eastern Visayas, Philippines), Clark Thomas Rogerson (Samuels and Fogel 2003), a pharmacist in the Philippines, and Harry Delbert Thiers (Thiers and Halling 2003), a Warrant Officer on a hospital ship in Pacific waters.

A CHANGE OF COMMAND. For some years before Murrill's resignation from the Garden, he had worked on MYCOLOGIA with Fred Seaver, who served as one of the dozen or more associate editors. Upon Murrill's resignation and after some precarious months of impending demise of the journal, its publication was permitted to continue with Seaver as Editor-in-Chief. Now, in 1945, it was time for Seaver's retirement and a new editor was necessary.

The Garden's Board of Managers thrashed over a short list of candidates and chose Dr. Alexander H. Smith of the University of Michigan for the job—as with his Garden predecessors, an unpaid "volunteer." Unlike previous editors, Smith's time was taken up with teaching and mentoring graduate students—in short, a university professor—as well as mycological research and extensive fieldwork. His specialty was agarics, the same fungal group engaging the attention of Rolf Singer and Murrill. In return for the journal's work, the appointment provided a proverbial thumb by which to measure the pulse of American mycology. An announcement was published in MYCOLOGIA, so Murrill was soon officially aware of the change. His brainchild lived on.

The two had met and Smith had visited Murrill in Florida, although unlike Singer's interaction, Smith went unmentioned in print by Murrill. Murrill must have questioned whether he could continue to publish in MYCOLOGIA, but his production, albeit of short descriptions and a few disagreements with Singer, continued through the rest of his mycological career. Starting in 1944 (Murrill 1944c, d, e), frequency of short notes was prolific in 1945 (Murrill 1945c, h, i, j; Fig. 9.2), remained high in 1946 and 1947 (Murrill 1946b, c, d, e, 1947a, b, c), and slowed only over his remaining years (Murrill 1948a, 1949a, 1951a, b, 1952, 1953a, 1954, 1955). Interestingly, no more voluminous papers by Murrill appeared in MYCOLOGIA during those years: those seemed reserved for other outlets, some designed to produce income. Smith would continue his editorship of MYCOLOGIA into 1950, by which time Murrill had found other avenues for publication.

WRITING FOR MONEY. In the wake of his enjoyable birthday celebration in October, 1944, a succession of substantial booklets—for sale—appeared in 1945 and 1946 (and beyond), and like his autobiography, it would be easy to assume their publication dates as accurate. Reality, surprisingly, is that some of these were written well before the autobiography, but published only later. It is in this order that—when possible—they are reviewed here.

At least three such were taken down off the shelf, dusted off and published as though new. The first (Murrill 1945k), "Trees," and second, "Familiar Trees" (see below) were the products of a single reach to the shelf.

The preface signed July 1, 1940 (pre-War, wearying Depression), dusting of "Trees" was accomplished in 1945. The preface is worth quoting for what it reveals about Murrill as a practiced author. "Had I used the scores of photographs in my collection this book would be too expensive and also too large to carry about in one's pocket. Besides, there are many well-illustrated tree books already on library shelves and in the market." Rule number 1: don't tell the reader what he's missing: he'll be tempted to close the covers and look elsewhere.

Contents were straightforward as compared with several subsequent offerings. The first chapter recorded several topics of interest (or at least curiosity): tree worship, historic trees, Bible trees, useful tree products and tree ancestors (among others); quite possibly worth the price, $2 in 1945. A core of the book dealt with his own life experiences, related in the first person, and presaging the course set out in his autobiography, written in late 1944. His time at The New York Botanical Garden,

Fig. 9.2 Bearded Murrill. Bearded Murrill, perhaps about 1947, perhaps at his birthday party. (Courtesy Jim Kimbrough 2003)

for example, was taken up in just two paragraphs, clearly written to bolster his credentials to write the booklet, but these and other topics were not informative to the reader except as autobiographical.

Buried in the midst of descriptive material there was a section repeating a paper delivered on March 26, 1939, to the Southern Shade Tree Conference in Gainesville. Murrill reported that it had been published in THE FLORIDA TIMES-UNION. Featured was a key to Florida hollies, oaks, and spring wildflowers.

A significant list of fungi attacking trees in Florida was far over the head of most readers, both the trees and the fungi listed by Latin name, and the booklet ended with a bibliography of 40 books on trees, only one of which had been published contemporary to Murrill's original manuscript.

With a preface dated just four days later (July 4, 1940; Fig. 9.3), five years before his autobiography, and dedicated to Boy Scouts of America, "Familiar Trees" has been thought of as a guidebook for Scouts engaged in nature study. Up front, Murrill (1946f) laid out his life: "My youth was spent in Virginia, my best working years in New York, and my old age in Florida" (written when he was 71, published when he was 77).

Like others of its ilk, the text was in two parts. The first part was an allegory not unlike Cinderella. It was full of oblique references to Murrill's life, down to Edith (the "Ed" of Edna unmistakable). It was an amalgam of the Biltmore castle and forest, Murrill's leaving his former life for a position—high school teacher—in the big city, and on and on. It was only on p. 28 that the information became serious and the rest of the book was an assemblage of the features of the familiar trees of eastern North America.

With a preface signed on August 1, 1942, Murrill (1945b) finished work on "Historic Foundations of Botany…" Original writing was penned in the early days

Fig. 9.3 Informal photo of Murrill, clean shaven, probably before his birthday celebration, 1944. (Courtesy Jim Kimbrough 2003)

of World War II, and the country was whipsawed between the Great Depression and wartime. He wrote: "The author has met many leading botanists, both in New York and in his extensive travels in the Americas and Europe. It would be interesting to record his impressions of them but the story would be too long and the risk too great." Why the manuscript rested so long before printing is a mystery, deepened by its appearance in 1945, quite near the end of that same War. Dollars might have been the temptation.

Not specifically aimed at an amateur readership, Murrill could feel free to expand some of his major themes. As far as he was concerned, any discussion of botany had to start with geology, for plants were slaves to their environment, including rocks, and therefore, soils. Echoing his idol, Charles Darwin, Murrill laid out a model of succession. "The introduction of northern species occurred during successive periods of glaciation in the Pleistocene period, either gradually by ordinary means of distribution or carried by flood waters from the higher elevations to the new lowlands that were to be called 'Florida.' After that it became a question of adaption, struggle for existence, and survival of the fittest." He expanded further, using his experience in eastern North America, the Caribbean, South America and Europe.

The basics for his "Guide to the Flowers of Florida," written in 1944 and appearing in 1945 (see below) were laid down in "Historic Foundations." Extensive lists of eponymic generic and specific names appear in both. At the beginning of the chapter "Modern American Botany," he wrote: "The list of men who entered the botanical field at this time is a long one. I knew most of them." An advantageous position from which to write, but he did not allow himself many personal thoughts; simply short paragraphs of repetitive form in alphabetical order; an impressive disgorge of accumulated experiential knowledge.

Having tapped the vein of historic Florida botanical exploration since his arrival, now two decades past, Murrill expanded the idea. Large portions of the "Historic Foundations" publication are thumbnail biographies of American botanists, including mycology, plant pathology and green plants by taxonomic groups. Most are straightforward, but he allowed himself to expand his thoughts about people he knew (e.g., George Atkinson at Cornell, see above; Edward Angus Burt, also see above) including colleagues at the Garden (e.g., Nathaniel and Elizabeth Britton, see Appendix 1).

Some 15 years after the death of Curtis Gates Lloyd, his nemesis but friend, Murrill felt obliged to write the following: "Curtis Gates Lloyd, born at Florence, Ky., in 1859, built up in Cincinnati a good library and a collection of the larger fungi, many of which came from Florida. His best work was done on the puffballs. Having plenty of money and leisure, he spent much time abroad buying books and photographing specimens in various herbaria. Unfortunately, he suffered from an inferiority complex due to his lack of college training and fondly hoped that by attempting to pull others down he might elevate himself. Most of his photographic work is excellent and probably reliable but one cannot be sure of the personal observations, comments or descriptions. If he had received the proper training early in life, there is no telling what he might have accomplished, but he ardently desired to be considered a great mycologist." Lloyd's personality trait, whether measured accurately or not by Murrill, expressed itself repeatedly (Petersen 2023).

Murrill (1945l) finished the preface to "A Guide to Florida Plants" on April 15, 1944 (wartime in the Pacific and a few weeks before the Normandy beachhead). There he set out his intention: "This booklet is intended as a guide and help to teachers, students, tourists and those who simply make a hobby of some branch of nature-study. It is mostly supplementary to other literature. The field is broad enough and entertaining enough for any one lifetime."

Much in the booklet mined material already in "Historic Foundations…." In the first paragraph of the guidebook proper, otherwise devoted to the geological process producing the current flora, the following appears: "The herbaceous flowering plants were not present during the warmer geologic periods that developed luxuriant floras of trees and shrubs but appeared only in the colder parts of the Tertiary when plants had to dig in of necessity to escape freezing or drought. Herbs could grow under the summer sun and take a needed rest while winter raged." This mixture of technical and teleological verbiage would set the tone of the entire guide.

In an iteration of early botanists who visited Florida, there appears a semi-serious paragraph on the capitalization of species names which honor persons. "Not so long ago several kinds of specific names were capitalized but under the lead of the entomologists the practice was gradually abandoned. Now, even personal names are decapitalized, which I think is a mistake. It is unnecessary, undignified, devitalizing, and defeats the very purpose of the author in bestowing the name. If one received a letter with his name decapitalized he would be insulted, and justly so. Soon some simpleton will be proposing that the name of God be written in small letters."

Coverage of plant types—trees, shrubs, vines, herbs, etc.—is mostly treated without any means of identification except rummaging through the text. Those that bloom are divided by flower color, but then alphabetically. This is typical Murrill presentation; he already knows the taxonomic characters of the plants, and sees no reason to use them to arrive at a particular species. First comes the plant, then come its characteristics. When he finally comes to an organization of a group, it is an alphabetic list, not a key. The scattered keys begin to appear in mid-volume and then deal only with difficult and little-known plants. Welcome features include thumbnail synopses of historic botanists known for their work on various plant groups. For readers requiring more, Murrill often referred to the publications by John Kunkel Small, especially his monumental manual of the southeastern flora (Small 1933).

A closing grace note is an enumeration of some of Murrill's favorite collecting areas in the Gainesville neighborhood, bearing in mind that they were being described in 1944, some after 20 years of sampling. Especially for someone compiling a biography nearly a century later, these descriptions have been welcome.

"We were driving along a Florida highway, my friend and I, when we saw a large Indigo snake frantically endeavoring to escape our juggernaut. 'I believe I'll write something about the value of snakes and how much they need protection,' I naively remarked." Another booklet, "Reptiles Good and Bad," was born—for sale.

Over his Florida years, Murrill had written on subjects as wide as his adopted title, "The Naturalist." Many of these were other than botanical, but they were written for the public, mimeographed, and distributed for a pittance or free through the Experiment Station (see Appendix 2). With a preface dated April 12, 1945, though, he departed from his botanical theme for a booklet for sale, simultaneously displaying his knowledge and his conjecture on creatures not seen every day. The result was equally dubious (Murrill 1945m).

If reptiles tested Murrill's breadth, they were beyond the ken—or interest—of a reading public. But to a noted herpetologist, they were near and dear, and the booklet offered an opportunity to see what was being served up by a noted naturalist.

The reviewer's judgement was quick to be stated. "This is an outrageously priced pamphlet with 26 pages of text, and no illustrations. The text contains an unusual assortment of fact and fancy that is inextricably woven into a semi-conversational fabric." He seized on Murrill's habit for teleology: "A few of the gems are: ['The dinosaurs'] job was to change cold blood into warm, and this they did by racing over the earth's surface in the warm sunshine." Again: "If a big diamond-back [snake] bites you, better hurry.... The use of the pit [in the Pit Vipers] is unknown but it probably serves some purpose.... There is a literature list and a remarkable plea for restraint in the 'senseless killing of our innocent wild creatures.'" Final judgement was delivered: "Professional zoologists will find little of value and amateurs are likely to be badly confused. This, coupled with its expensive price [$2], forces one to withhold recommendation."

We know nothing of the effect on Murrill. He could well have smiled: the reviewer's words read quite like those by Lloyd on the number of generic names for the polypores, especially if the "Reptiles" booklet had been whipped up in the hopes of making a few dollars. One way or another, its review did not build an impenetrable wall—the booklets and guidebooks did not cease.

Only a month later (preface signed May 14, 1944) Murrill (1945n) finished work on "A guide to Florida animals." The intent was exactly the same as others in the series, namely: "This booklet is intended as a guide and help to teachers, students, tourists and those who simply make a hobby of some branch of nature-study." The layout followed that of the "Guide to Florida Plants," with mammals, birds, amphibians, fishes, reptiles, snakes, and shells each consuming a chapter. Insects (plus spiders) occupied half of the booklet. Murrill was candidly honest here and there, such as: "[under butterflies] … the Skippers are quite difficult as a group, even for experts … some of the Sooty-wings and Dusky-wings are not so easy. I had to send some of my catches to Washington [Smithsonian Institution] to make sure."

Added were stories and adages from "The Naturalist." Under flies: "I heard Prof. Apgar [see above] of New Jersey, make a statement once that I never forgot. He said that three green flies could eat up an ox before a lion could. It is estimated that the progeny of one house-fly, if undisturbed, would in the course of a single summer reach high into the billions. One of those double-or-nothing propositions! Mosquitoes are scarcely less prolific. The crane flies look like giant mosquitoes. About 300 species occur in the United States." It was Murrill at his best.

This booklet, moreover, also fell into the hands of a reviewer (Schneider 1946). "Out of the rich experience of many years' original observations in the field has come this fine handbook of Florida animals.... The author's keen appreciation of natural history has been combined with a delightful style of presentation in recounting the ramblings and reminiscences of a naturalist." What a difference a month makes!

Unlike all of the booklets, pamphlets, scientific papers and mimeographed material published in 1945, Murrill's (1945a) "Autobiography" was a product of acclamation. At least at his birthday celebration, there was a call for him to put his lifeline on paper for his friends to share. Composed in just 15 days from October 20 to November 4, 1944, when printed it occupied 164 pages. Only parts were written as factual—especially those dealing with his early years were written as though allegory, and important omissions can be detected when his life is assembled through other means. Little did any of the parties know that the document would be dissected nearly eight decades later in an attempt to understand Murrill's personality mirrored against his upbringing, experiences and—fair game—his omissions.

Bits of the autobiography have been scattered throughout this narrative. In one chapter he recounted his last days at the Garden, then made the statement: "His position at the Garden was held open for two years with the hope that he might come back, but the wild bird once free of the cage rarely returns." While his position as "Supervisor of Public Instruction" remained unfilled, I can find no evidence that this was in hopes for his return. In fact, a substantial administrative reorganization took place, perhaps levered by Murrill's departure, but it was the Board of Managers, through N. L. Britton that controlled the situation.

The theme of one chapter was a series of advice maxims born of long experience. The first dealt with money. It began with: "When money talks, what does it say? That depends on who is listening. To one it says 'Spend me freely and have a good time,' to another, 'Put me in a vault and worship me'; and to still another, 'Use me to enslave others and ruin their immortal lives.'" His recollection of his last years at the Garden furnished his life's choice, then (in the autobiography) he resented his own choice. He finished the thought this way: "If this advice, based on long experience and observation, should be generally followed, many losses, bitter tears, family quarrels and hopeless years would be averted; and otherwise futile turned to good account."

A chapter called "Retrospect" (Ch. XXVII) has some opening words: "In buoyant youth games and sports hold an important place; in the ambitious middle years achievement ranks high; and when old age comes creeping on one longs for appreciation." Considering his birthday party just days previous, those last words are worthy.

Toward the very end of the autobiography, there appears: "One of the main objects of matrimony is, or should be, the rearing of a family. Without children life is far from what it was intended to be. If it is impossible to have one's own, at least two should be adopted, preferably a boy and a girl. Think of the affection wasted on dogs, cats and other dumb brutes while little human beings go homeless." Yet, nowhere did he come to grips with his own life. His love story is probably described several times, always in the third person—"Three young Crusoes" being, perhaps the most transparent—but Edna is never mentioned in the autobiography, nor the birth and death of their son. Upon their move from Manhattan to the Bronx long before, he had written positively about his walks in the forest, "often with a faithful dog."

Entwined in this were Murrill's activities with the Girl Scouts and church groups in New York, abbreviated in his autobiography as: "All these varied activities secured a host of friends, both among adults and the young folks, which insured a full, rich life and also brought diversion from the stern scientific life in New York. Work with the Girl Scouts was revived [in Florida] and continued actively for many years. Entire days were spent with them in camping, swimming and nature-study." Obviously, one conclusion is cause-and-effect, these activities substituting for love and marriage, but the truth is, at this moment, unknown.

Murrill's graphic brush between his strict religious upbringing and liberal "modern" thinking took place at Sage Chapel at Cornell, and its ontogenetic result was included in his autobiography. "The humanities are still very much worthwhile but they are simply outclassed, at least temporarily. They, and also religion, must be adjusted to the discoveries and methods of science. Between true science and true religion there can be no real conflict. One is a matter of the head, the other of the heart. Man has always been religious and he needs the help of his emotions now more than ever [written in late 1944, advanced age and war time]. One might as well try to live happily without love as without religion."

The next two booklets have not been seen personally. The very rough contents, though, can be gathered from the reviews that followed each. Contents seem to loosely follow those for "Ferns," and "Flowers." The first is "Rocks" (Murrill 1946g); preface signed November 22, 1945. The review (Anonymous 1947a) reported: "This booklet, prepared for Girl Scouts doing nature study, should succeed well in its intended purpose of awakening interest in young people. In Part I, a considerable amount of geology and paleontology is woven into a story. In Part II, there is a more formal outline of historical geology. Part III is a description of common rocks, minerals, and precious stones."

Murrill could celebrate the new year by completion of his latest booklet, "Stars." With the preface signed January 1, 1946, the review of "Stars" (Murrill 1946h) appeared on the same page as the one on "Rocks" and obviously was prepared by the same reader. "This is a booklet prepared for Girl Scouts. The first part tells in story form some of the legends behind the naming of the constellations. Part II is a brief outline of the stellar universe; Part III gives brief descriptions of the 25 principal constellations" (Anonymous 1947b).

Three months later, Murrill (1946i) was still at it, this time—March 15, 1946—signing the preface of a book of 110 pages. Its title was misleading, for "Pore fungi," was only a small part of the text. In fact, of the three parts, the first was a 30-page allegorical fable of "the Punkies," featuring Robert, Dorothy, Elfie and a cast of fairies. The rest of the volume was occupied by an introduction to the fungi and chapters on the major groups of macrofungi. Several illustrations were line drawings by Murrill, not artistic nor large enough to be very informative (Fig. 9.4). Most of the halftones were composite photos of fruitbodies, which, while clear enough if

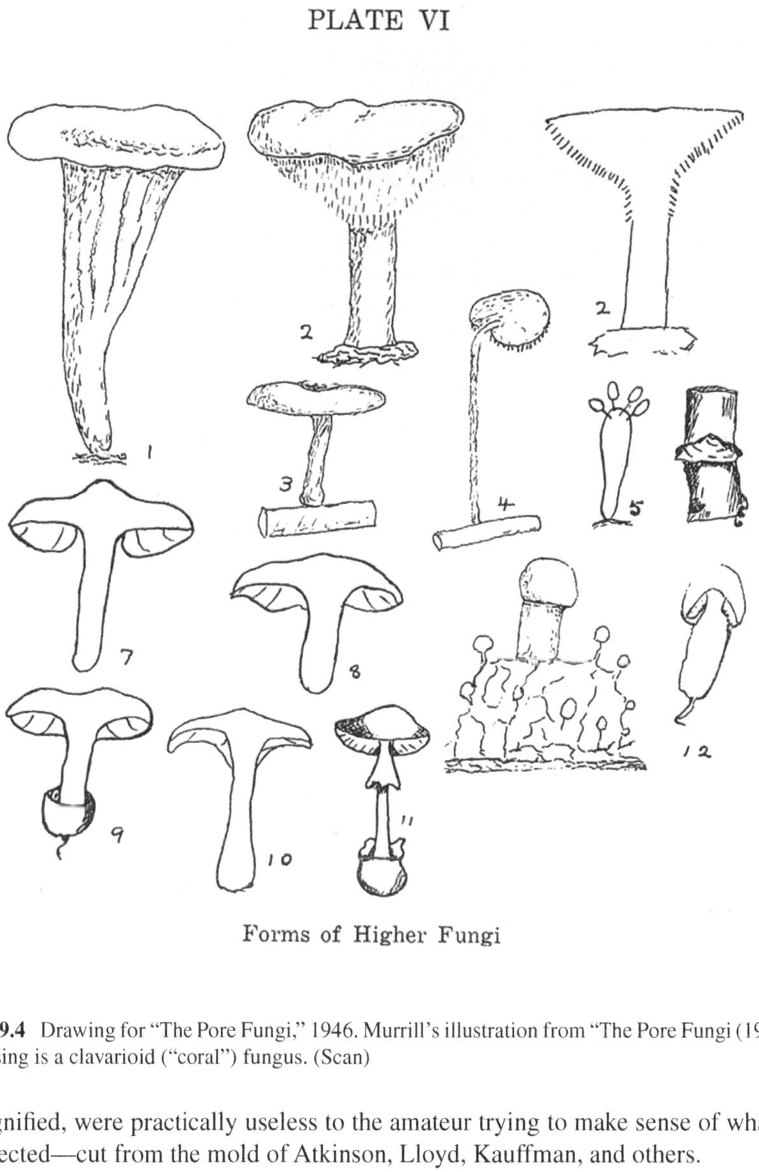

Fig. 9.4 Drawing for "The Pore Fungi," 1946. Murrill's illustration from "The Pore Fungi (1946)." Missing is a clavarioid ("coral") fungus. (Scan)

magnified, were practically useless to the amateur trying to make sense of what he collected—cut from the mold of Atkinson, Lloyd, Kauffman, and others.

In "Pore Fungi," two contemporary names stand out. Without fanfare, "Dedicated to Dr. Fred Jay Seaver" appears, without embellishment, perhaps a gesture of closer friendship than otherwise clued. Overleaf, in the preface, "In the poetic efforts I was aided by my sister Minnie, whose ability is well known." Most of these common insertions were sets of four-line, not necessarily rhyming, limerick-like verses placed to highlight a new chapter, all in the parable of the Punkies.

Fred Sparrow (1947) wrote the succinct review: "A paper-covered 'illustrated guide' composed of approximately one third punkies and two-thirds polypores and other fungi. Hardly a bargain at $3.00, even in these days of inflation."

Parenthetically, MycoPortal data shows that Murrill continued an active collecting schedule in 1946, especially in the first 3 months of the year. Sizeable openings appear between May 15 to June 28, October 11 to November 10, and the month of December, but no clues for alternate activities (including health-related) have been unearthed. The lifting of gas rationing is reflected in the number of collecting sites outside Gainesville, but not to excess. The vigor of this activity must be superimposed on a busy few months of writing (or compiling), traced below.

The time between the signing of the preface, April 5, 1946, and appearance in print of "Ferns," saw less delay, but the material in the booklet emphasized data taken (or remembered) from Virginia to New York (and beyond) gathered a half-century before (Murrill 1947d).

For the uninitiated reader, the setting for the story, especially in Florida, is far away, "at the base of the Catskills which ran down to the valley separating these mountains from the beginnings of the Appalachians." Most likely, the reference is to Yama Farms, a pleasant resort for Murrill from his last years at the Garden. For close friends and former colleagues, the references ring familiar. Of the two individuals to whom the booklet is dedicated (W.F. Clute and L.M. Underwood), both were pivotal in Murrill's longtime affair with ferns.

If love, pursuit, bullies, evolution, ferns, and Jamaica would appear difficult to knit into a plot, Murrill toiled on, perhaps somewhat awkwardly, dropping occasional clues that he was sketching the life of Lucien Underwood, Murrill's mentor in ferns and his friend and benefactor in New York City. "A Story" occupies 20-odd pages of the booklet, but boy wins girl, and they live happily ever after.

"Ferns are the most interesting and most beautiful of all plants." "The fern frond is Nature's pattern for a perfect leaf." "Although ferns were evidently created for beauty rather than for utility…." Words that begin the factual chapters. Extensive keys to groups, genera and species follow. Although written in Florida and aimed at its audience, repetitive references to Virginia stations and phenology create a decidedly northern effect. Unlike most other Murrill booklets, this one is illustrated. Photos lack contrast, but, although done with too coarse a pen, drawings are more informative.

Finally, on page 64, Florida (really Alachua County) ferns float to the top, but even then, "Northern species comprise about one-third of the fern flora [of Florida]." The best work, he admits, is the illustrated volume on southeastern ferns by Small (1938). Again, keys follow, with greatly abbreviated descriptions.

The volume ends with "Part III, Primitive Green Plants," in essence a short course in lower plant evolution, version Murrill (1947d). Two short chapters summarize the mosses and liverworts, with sketches of the significant workers in each and Murrill's time in the North. The Thallophyta, which in the table of contents include the fungi with the algae, finish the volume, with the fungi missing. It seems that much of Part III has been borrowed from previous Murrill works.

Again, less than a month passed, when, on May 1, 1946, Murrill (1946a) signed off on his next offering, "Flowers," this dedicated to the Girl Scouts.

Part I: a 30-page love-story of James and Elsie. Typical Murrill, laced with small references to places and events he wished to remember. The goal seems to be marriage and true love.

Part II is rich in floral information, but no way to get to the name of the flower (no keys except to Florida orchids, and Alachua mints).

For the biographer, Chapter VI, is brief but again stresses the power of Murrill's memory. "A chance visitor at my [Gainesville] house recently remarked as he was leaving 'Dr. Murrill, you have an ideal life; you should tell others about it.' Not an easy thing to do, but I will make an attempt."

Written in 1946, years after his migratory pattern commenced, he nonetheless extolled those days between Virginia and Florida. There appears the best description of that life and especially his Lynchburg cabins (living quarters plus studio). Included is a running essay of the changing flora through southern Georgia into Florida from the window of the train, a repeat of a prior publication.

Part III returns to botany, this time to the scope of the discipline and its subdisciplines. Although written, he attests, for Girl Scouts, nowhere is there an illustration or explanation of the structures of the flower which one might expect from the booklet's title.

Inside the back cover of "Pore Fungi" appears a list of "New pocket guides by Dr. W. A. Murrill" selling for $2. Listed are those on Stars, Rocks, [Familiar] Trees, Reptiles [Good and Bad], Autobiography, Florida Plants, Florida Animals, and Historic Foundations. "Illustrated guides, postpaid at $3 each" lists only "Pore Fungi," obviously with the intent for more. A small entry noted that Murrill's "Edible and poisonous mushrooms" and its chart were still on sale in New York for $8. That copies remained in stock might indicate that Murrill over-estimated their popular receipt.

In an attempt to follow the order of Murrill's "for sale" booklets, this narrative has gotten ahead of itself chronologically. A small hitch will hopefully restore order.

NEWS FROM MYCOLOGIA. Murrill's self-pride subsided gradually, and busy activities of 1946 took his time and energy. Alexander Smith, now Editor-in-Chief of MYCOLOGIA, teamed up with Rolf Singer (still at Harvard) to present a prodigious list of 198 generic names of agaric fungi to be discussed at the upcoming post-war International Botanical Congress in Stockholm in 1950 (Singer and Smith 1946). In most cases, the names were either furnished with a proposed lectotype species (and these choices, therefore, proposed for discussion at the Congress) or rejected. The War had postponed such congresses, but now the dam had had been breached and scores of nomenclatural proposals were being put forward for discussion by the nomenclatural authorities (Nicolson 1991; Petersen 2024).

Both authors knew Murrill, and were familiar with his many papers which followed the antiquated AMERICAN CODE OF BOTANICAL NOMENCLATURE. His name was hardly mentioned in their paper, but the generic names in his roster, usually pre-Friesian (and often pre-Linnaean), were considered illegitimate for this very reason. Murrill had faithfully concorded them with the more popular generic names used by Saccardo, who, likewise, was not mentioned, for he originated almost no generic names. The post-War mycological nomenclature battles were yet to erupt (Petersen 2024).

Their weighty paper, published in MYCOLOGIA (Smith as MYCOLOGIA Editor), surely was read by Murrill, but quite possibly with a slow head-shaking. After all, his modus operandi was immutable over many years, and would remain so. In his 1905 series on the polypores of North America (Murrill 1905a, b, c, d, e), it had taken 63 generic names to cover the subject. Now, in 1947, 53 of those names—of which about half were his own—were repeated to cover the polypores of Florida (Murrill 1947e). And the new summary was placed in LLOYDIA, as though thumbing his nose toward his friendly rivals.

Parenthetically, Rolf Singer's arrangement with the Farlow Herbarium was coming to an end, and within months he accepted an offer as Director at the Instituto de Botánico of the Fundación Miguel Lillo in Tucumán, Argentina [his last day at the Farlow was June 30, 1948 (Kyle Boyd, personal communication)]. For some years, much of his research would be published in the journal LILLOA, not widely distributed in the Northern Hemisphere.

A DUBIOUS GENEALOGY. In 1945, Murrill discovered a new outlet for descriptions of his Florida fungi. The Florida Academy of Science, known for its PROCEEDINGS, was in the process of inaugurating its own research journal, to be issued quarterly. Its audience might be limited at first as compared with the Elisha Mitchell Society's subscribers, or especially MYCOLOGIA, but it was respectable and willing. As author, he was William A. Murrill, University of Florida, more than enough cachet. But his judgement of the readership was not so respectful. The entire first page of his inaugural paper was a short course on mycological taxonomy and its societal gravitas. Examples ranged from *Penicillium* and Penicillin, "*Clitocybe*" *tabasecens* and root rot, to *Lepiota Morgani*, conjectured to produce "a drug similar to muscarin which is used in hysteria." But if this primer was thought necessary, the introduction ended with: "The nomenclature used is that of 'North American Flora,' but for the benefit of those following Saccardo a few of the names are changed at the end of the article to more familiar forms." Thus, both the baptized and unwashed readers were blessed. Sixty taxa were described, and the obligate concordance furnished (Murrill 1945b).

Buried in this cast was *Agaricus Blazei* [sic], humble at its start, but celebrated as time passed. Murrill's descriptions, by this time, were formulaic and the single collection of an *Agaricus* from Mr. Blaze's Gainesville lawn was no exception: an only collection to represent the name. The name and its description hibernated for many years; Stamets (2005) succinctly traced its checkered trail: "Decades later, specimens forwarded to Pennsylvania State University from Brazil via Japan were identified by P. Heinemann [*Agaricus* expert in Belgium] as best conforming to Murrill's species concept of *A. Blazei*" (Murrill 1945o; Heinemann 1993; conjectural words Stamets'). Almost a decade later, da Eira et al. (2002), doubtful of Heinemann's opinion: "The mushroom known for culinary and medicinal use, long

Fig. 9.5 Advertizing the powers of *Agaricus blazei*. Reduction of full-page advertisement for *Agaricus blazei*. (Fungi Magazine 2024)

thought to be *Agaricus blazei* Murrill, is, in fact, not so, but is *A. blazei* ss. Heinemann, here described as *A. braziliensis*." Since that time, both names (Fig. 9.5) have resonated with literature praising the attributes of commercially farmed strains (Fig. 9.6), probably more entities than presently formalized (Gyrofi 2007, 2008; Kumar and Kumari 2021; Mizuna 2002; Roza et al. 2016; Stamets 2000; Stijve et al. 2001, 2003).

Fig. 9.6 Mushrooms resembling *Agaricus blazei*. Putative *Agaricus blazei*. Note pigmented cap and "double annulus." (Google Images)

One of the separating characters employed by Murrill was the small size of the spores of *A. blazei*, but Freeman (1979a, b) reported them as significantly larger, casting additional doubt on the species. The *Agaricus* floristic monograph by Kerrigan (2016), surely the most authoritative on the subject, relegated Murrill's epithet as dubious, and posited that commercial strains under this name were probably *A. subrufescens*.

APPRECIATION. Murrill had already admitted that in old age, there was a yearning for appreciation. In the subsequent years, he put his encyclopedic Florida work on display: there still were agarics to be described and gathered into definable groups. He chose another "new" journal in which to summarize 25 years of Florida mycology.

In 1938, Calvin Kauffman had reviewed a new journal, Lloydia, a product of the Lloyd Library of Natural History in Cincinnati, Ohio. Although one might be tempted to attribute the journal to Curtis Gates Lloyd—he who played a role in Murrill's life—volume 1, number 1 of Lloydia appeared nearly a decade after C.G. Lloyd's death. The intent of the journal, succinctly announced on its cover, was biological science, broad enough to provide space for reports on all sorts of animals and plants (including fungi). Kauffman reported that volume 1, 1938, consisted of 254 pages.

Lloydia's Editor was Theodor Just of Notre Dame University, already editor of the American Midland Naturalist, a small but respected journal in its own rite. With such a wide coverage, Lloydia could have soon become a niche organ, but in terms of macrolevel natural history, it seemed to emphasize plants (writ large) and plant pharmaceuticals. For a decade or more, Murrill, Smith and Singer used it as an outlet for their systematic research.

Kauffman's review went on to briefly outline the history of the Lloyd Library, founded in 1864 by John Uri Lloyd, during Civil War time. Pharmaceutical literature—broadly speaking—was difficult to obtain and specimens of pharmaceutically interesting plants could be cached for future use. In his time, Curtis Gates Lloyd used the library to accumulate mycological specimens and literature and to produce his Mycological Notes.

Beginning in 1942, Murrill rationed LLOYDIA papers though 1947 to one per year, all describing new Florida fungi (a rough count arrives at the prodigious number of 273!). In 1948, though, he began a long series on Florida fungi, recapitulating his work (Murrill 1942o, 1944a, 1945p, 1946j, 1947e, 1948b, c, d, e, 1949b, c, d, 1951c, d, e). Interestingly, in every case, the LLOYDIA paper was preceded by previous mimeographed materials—in some cases by years (see APPENDIX 2). The numbers of species was significantly increased, of course, by his own donations, and the audience hopefully larger and more educated. As might be expected, as his Florida papers in LLOYDIA grew in number, Murrill ceased producing the informal mimeographed materials through the Experiment Station (Murrill 1947f).

MURRILL VERSUS SINGER. Perhaps the most significant of the "Florida" series was that on Florida boletes (Murrill 1948b), for it brought Murrill and Singer head to head.

To introduce the exchange, it is necessary to remember that Singer had spent a year in Florida, half of which was around Gainesville, often in the company of Murrill. Their conversations were reported generally in the preceding chapter, so the two knew one another. Upon returning to the Farlow Herbarium, Singer set to work on the task of summarizing the boletes of Florida, with the result coming in a series of four major papers (Singer 1945a, b, c, 1947). In the introduction to the second paper (Singer 1945b) Singer discussed his mission as follows: "For the first time an attempt will be made to treat taxonomically the Boletineae of a large region with due consideration of the anatomical and chemical characters which, according to the author's view, are of no less importance in the taxonomy of this group than are adequate macroscopical descriptions." This was obviously a poke at Murrill's abrogation of just such characters in his taxonomic descriptions. "Another factor determining the present arrangement was the author's wish to coordinate European and American nomenclature and classification." Singer's discussion of phenology and mycorrhizal associates was much more sophisticated than any by Murrill (1949e), although Singer had yet to sample the Central and South American mycotas. Singer detailed examples of *Pinus*- and *Quercus*-associated boletes, as well as bolete species characteristic of the summer or winter rainy seasons, but only very rarely both. Understandably, Murrill's name was hardly mentioned. In fact, Erdman West preceded Murrill in acknowledgments. Finally, Singer's taxonomic scheme (and naming) was miles more complex (and "names-rich") than that by Murrill.

That Singer's largest papers appeared in two journals, of which one—FARLOWIA—was probably unavailable for Murrill, may have delayed Murrill's receipt and reading. But finally, in 1948, he answered with one of the first of the "Florida" series (Murrill 1948b). As all of the series, it was preceded by a mimeographed hand-out some years previous. That "Florida boletes" could be summarized in 14 pages (compared to Singer's 261) gives some idea of the breadth and depth of Murrill's message. In his introduction, Murrill clearly stated that his paper was in answer to Singer's. "Although not entirely in agreement with Dr. Singer as regards certain minor matters, I admire his enthusiasm and trust he will continue to obtain firsthand information of our numerous interesting species as the seasons come and go. Perhaps 10 years from now we shall both see things somewhat differently."

Perhaps the sole glimpse of Murrill's recognition of change during his long career in mycology, he admitted: "The collapse of the American Code [of Nomenclature] necessitated changes in the nomenclature of the boletes resulting in confusion and complications. Therefore, I am considering here primarily the species, listing them for greater convenience in alphabetical order under Saccardo's genera (*Boletus*, *Boletinus* and *Strobilomyces*)." It took just 11 generic names to complete the task for Florida boletes.

UNFINISHED BUSINESS. Like most people's, Murrill's last decade takes on a regularity—perhaps even metronomically—about which tempo slows over time, and, therefore, about which less can be written or compiled. 1947 can be taken as an example.

MycoPortal, as usual, delivers data on Murrill's 1947 collecting. After a light schedule in January, February is without a single collecting date, March a single date, and April and May only two dates in each. To be sure, his booklet, "Ferns" (Murrill 1947d), written in 1946, appeared during 1947 (but date unknown). Three notes in Mycologia (Murrill 1947a, b, c) occupied less than two pages each, and a single mimeographed paper from the Experiment Station was not mycological (Murrill 1947f).

From mid-June through summer, collecting was frequent, especially July, when collections were gathered virtually daily. Ample rain is suggested—the introduction to one paper agrees. Almost as productive through fall, forays became less congested only in December, for another winter. Reasons for diminution of collecting are obviously not known—health, dry weather, competing activities—but probably not a state of torpor.

Two lengthy papers in Lloydia appeared just a year apart, both updates of numerous previous offerings. The first (Murrill 1947e) compressed his work on polypores, with Florida acting as filter. The second (Murrill 1946b) took up the boletes, a group thought to be only occasional or even rare in tropical climates. Both papers were packed with generic names by or resurrected by Murrill. And while neither was illustrated, keys to the species of large genera were furnished. No microscopy was included. It was almost as though his years prior to 1924 had ceased to exist and he was plowing new ground.

The bolete paper seemed to present a gateway to a long series of Lloydia papers, now sorting the Florida agarics according to genus; generic names had morphed only minimally, but the papers' titles used common derivative names rather than generic names—amanitas rather than *Venenarius*, lepiotas rather than *Lepiota* and *Chlorophyllum*, clitocybes rather than *Clitocybe*, *Monodelphus* and *Laccaria* (Murrill 1948d, e, 1949b, c, d, e, 1951e). The two non-agaric papers (Murrill 1951c, d) required more lengthy introductions for they represented fungal groups not often coming from Murrill.

SUCCESS STORIES FOR GIRLS. Written during 1949 (preface signed January 1, 1950), the latest—and, unknown at the time, last—booklet appeared, this one dedicated to the 4H Boys and Girls. Although most of the booklets had a fictional story preceding the scientific contents, this one dispensed with the science content altogether. It was simply 27 short stories—very short—tales of young girls, all mistreated, unloved, or neglected, but all of whom, through pure personality and/or skill, win success—in these tales, true love, loving marriage, and/or successful social position. Of the collection, Murrill (1950) wrote: "These simple stories are entirely original and purely fictitious. They are not meant to be unduly exciting but helpful, or at least stimulating." The general tenor of the stories resembles the Horatio Alger stories of the nineteenth century, which were popular during Murrill's early life.

This booklet, like others, was reviewed (Anonymous 1952). "There are 27 stories entirely original and purely fictitious. Many of them have a science basis. The author is a well-known scientist. He has many publications in the science field such as Stars, Rocks, Reptiles, Trees, Florida Plants, Florida Animals, Flowers, Familiar Trees, Ferns, and so on." Despite the inaccuracy of the list as scientific, Murrill would have been pleased, although a review was probably irrelevant.

In 1953, Murrill seemed to discover a new field for investigation—mushroom pharmacology. The note reported the preliminary results of tests of extracts from *Clitocybe subilludens*, described (as *Monadelphus subilludens*) by him in 1945. The introduction sought to clarify the generic nomenclature involved: described in *Monadelphus*, referred to as *Clitocybe*. Taxonomically, basidiomata resembled those of *Pleurotus olearius* "except in spore characters" (Murrill 1953b). As the species epithet indicated, *C. subilludens* (Fig. 9.7) shared characters with *Agaricus* (*Pleurotus*) *illudens*, a venerable, easily recognized fungus described by Schweinitz very early in American mycological history and separated from the well-known European *Agaricus* (*Pleurotus*) *olearius*, known to be toxic. Both species displayed a unique property—bioluminescence—reflected in its common English name, Jack O'Lantern mushroom (Forager/Chef 2024).

Fig. 9.7 Jack O'Lantern mushroom. Murrill's *Monadelphus subilludens* (now known as *Omphalotus subilludens*), source of "illudin" biochemicals. (Google Images)

Readers of Murrill's short paper were assumed to be familiar with his prior studies (unreported) in which mushrooms were fed to various animals ("… wild squirrels, cats, calves, students and myself …") and any apparent effects noted, almost exclusively edibility versus non-edibility (not necessarily toxicity). In this case, some other more laboratory-inclined personnel were enlisted, and when a gravid female white laboratory rat was fed bits of *C. subilludens*, instead of dying, she "voided her young and improved in health and appearance." Impressed, Murrill had his friend and colleague, George Weber, culture the fungus, so mycelium could be used for further experimentation.

Laboratory culture permitted biochemical extraction, and two such extracts seemed to produce efficacious results.

Unreported by Murrill, work was already underway elsewhere, notably his own professional home, The New York Botanical Garden, where biochemical research had been added to whole-plant taxonomy. Anchel et al. (1950) had found antibacterial antibiosis from Schweinitz's sister species and reported their results broadly. The biochemicals responsible for these results were tagged as illudins, including illudin S and illudin M (McMorris and Anchel 1965).

Parenthetically, over time, similar species were discovered elsewhere, and were grouped together in the genus *Omphalotus*, under which generic name some subsequent research has been performed. Significantly, illudins were found to exhibit anticancer agents (Kelner et al. 1987). The relationships among *Omphalotus* species were tested by Hughes and Petersen (1998) and Petersen and Hughes (1998). Even more recently, illudin extracts have been found efficacious against several tumor strains, and research along these lines continues to this day (Schubert et al. 2011), attempting to separate an antitumor agent from toxic agents.

Murrill, of course, did not live to see most of the illudin research, but, as in *Agaricus blazei*, he planted a seed which grew and blossomed years later.

A SWAN SONG. Many years ago, American baseball included the great pitcher "Satchell" Page, quoted as advising: "Never look back over your shoulder; they may be gaining on you." Murrill's last four publications, brief as they were, bring back issues almost as ancient as his own career. So it was with botanical nomenclature, Murrill's or International.

The most odious was his invariable use of two different generic names for a new species. Codes of nomenclature were clear that an author must intend his new taxon to belong in his intended taxonomic location. Nevertheless, Murrill (1952) titled a section of his paper, "*Tricholoma russuloides* sp. nov." Nothing could be clearer until some lines below: "A beautiful species, which in my nomenclature would be *Melanoleuca russuloides* (Murrill) comb. nov." Were a later author to agree that the species belonged in *Melanoleuca*, that name would be cited *Melanoleuca russuloides* (Murrill) Murrill, just as Murrill instructed. But another gambit might be to declare Murrill's species to be rejected altogether as violating the nomenclatural rule of being clear in placing a species name within the genus to which the author intends the species to belong. Even worse was Murrill's habit of describing species in "his nomenclature" but including a concordance of species epithets into the "nomenclature of Saccardo:" scores of new taxa were doubtful for they could bear one of two different names.

The same situation pertains for "*Amanita neglecta* Murrill" (1955): "In my nomenclature the species would be assigned to *Venenarius*." Likewise, *Amanita flavovolva* (Murrill 1953a) carried dual citizenship.

Surely, one of the more boisterous conversations between Murrill and Singer would have been over the accuracy of color notation. In reading scores of Murrill papers, I cannot find a single description in which colors are referred to by any other system than his own eyes. Early in Murrill's career, Robert Ridgway (1912) had published his color book, which became the standard for mushroom students rather quickly. Some years later, Maerz and Paul (1930) produced a color manual with far more shades of color than Ridgway, and popular enough to demand a second edition (Maerz and Paul 1950), reviewed by Singer and DiGilio (1952). Other manuals were available in Europe (Petersen 2016), but Murrill used none of them. For him, yellow was yellow and red was red and orange … well?

Many of Murrill's new species were based on single collections. On one hand, the type specimen was always clearly denoted with a herbarium collection number as voucher. But taxonomic dicta are clear; the more specimens by the originating author the better, for hopefully, the peculiarities caused by moisture, temperature and the like are more accurately reportable if several collections are deposited (and authoritatively identified), and better yet if they are from the topotype (the location where the type was collected). So "Known only from the type locality where it appeared at intervals for some time," (Murrill 1954, 1955) invites the disappointment that none of these ancillary appearances was sampled and preserved.

Finally, a note of praise. All very early codes of nomenclature completely ignored workers whose native language was not in Roman alphabet. All of the Middle East, Africa, and Asia was overlooked as using "barbarous tongues." Finally, when it was clear that the Japanese and others were beginning to describe new plants, the "Latin diagnosis" rule was added to the International Code. After all, even those far off workers must know the rudiments of Latin—the dead language. For Murrill, who had taught Latin to girls, it was only to choose which declension he wished. Once convinced, he applied the same formula again and again, but his names were validly published, not consigned to "nomina nuda."

THE PASSING OF MURRILL. On Murrill's personal condition in old age: "He lived alone and kept his own council regarding his past activities. He had no pension or retirement, and seemed to have no visible plan for the future."… "His physical needs were looked after by his Gainesville friends, and during his last few years he received some financial and medical support from the Florida Social Welfare Board" (Weber 1961). Overlooked was the small gratuity paid for his contribution of fungal specimens to the herbarium.

According to Murrill's death certificate, he entered Alachua General Hospital on November 27, 1957. He apparently passed out during his initial examination (Weber 1961). "William Alphonso Murrill died, after a month's hospitalization, on 25 December, 1957, in his 88th year, in Gainesville, Florida, where he spent the last quarter century of his distinguished career in an active retirement. To the very day of his final illness, he pursued a leisurely existence, enjoyed plants, people as he found them, and his own popularity" (Weber 1961).

"Fred Seaver, one of Murrill's co-workers at the Garden, shared that a polycystic kidney problem hospitalized [Murrill] in a small French town where he almost died and had no way to contact his wife or colleagues at the Garden" (Kimbrough 2003). Murrill, to my reading, neither mentioned the episode nor the disease. Moreover, Kimbrough's next sentence, "Murrill's wife traveled with him all throughout the US, but absolutely refused to accompany him abroad, having a deadly fear of water." The latter idea has been taken up above, and easily shown to be patently mistaken. Murrill's polycystic kidney disease bears the same scrutiny. Murrill's death certificate, however, states his cause of death as uremia, the result of diseased kidneys. Contributing conditions were cardiac deterioration and bronchial pneumonia, but these were detected only a couple weeks into his hospitalization. These facts were not repeated by Weber (1961), in spite of his years of collegiality with Murrill. Seaver, though, not only shared quarters with Murrill in the Museum building at the Garden, but remained in touch with him for the years after Murrill's departure. Such a malady could be supported by the totally circumstantial hiatuses to his perpetual collecting scattered haphazardly through the Florida years.

Further confusing information concerning Murrill's final days, Weber (1961) reported another condition, this one psychological. "He was hospitalized intermittently for over a year for correcting a nervous instability and physical exhaustion. His wife, with a feeling of neglect and with physical ailments resulting in inefficiency and moodiness, left his home and shortly thereafter served him with a divorce decree. Financial problems developed with which he was unable to cope; and as these events presented themselves, he became a man haunted, troubled mentally, alone, and discouraged. He left the Garden in an endeavor to escape the events that bore down on him, never to return, although several attempts were made to convince him that his responsibility to his science was most important. He returned immediately to the woods, flowers, birds and animals in the neighborhood of his birth near Lynchburg, Virginia, where he stayed with an aunt, and in the course of the succeeding months mildly regained composure, peace of mind, health and self-respect."

The possible intrusion of depression, in retrospect, might have been anticipated, although surely not labelled. For a boy praised by a mother more educated than most, winner of grammar school contests, day-dreamer of a world where living things spoke and guided, a "decision" to be a "naturalist" only lent an adolescent label to a young life filled with approbation and the resulting self-respect. This in apposition to young puppy love for girl schoolmates, seeking personal requite to satisfy feelings hardly definable, but visited time and again in later allegories, only to be deserted by the girl of his dreams. In his autobiography, he wrote: "One of the main objects of matrimony is, or should be, the rearing of a family. Without children life is far from what it was intended to be." Faced with the prospect of life-long loneliness amid memories of lost love and infant death, depression was only a moment away. Incredibly, his death certificate checks "never married," the ultimate rejection of Edna.

Equally curious, the headings on his death certificate calling for his father's name and his mother's maiden name were both labelled "unobtainable." Murrill was certainly not ashamed of his forebears, and I assume that because the certificate was completed after his death, such information was truly "unobtainable."

> During the last few years his applications became less and less regular, although his enthusiasm never diminished even up to the time of his last illness. Even then he reported of his own volition to the University Infirmary where he collapsed while the doctor was examining him. He did not want to die because there was too much interesting work yet to be done, yet he knew very well that the time had come. He said on his last birthday a few months [prior] that he expected to live to be a hundred. (Weber 1961)

A personal note from David Rose tells of Murrill's continued contact with the Garden even in his last year. Sometime after his leaving, Rosalie Weichert was put in charge of any business which came his way. Once Murrill was made aware of this connection, he faithfully remained in touch with Rosalie. In a note to her dated May 19, 1957, Murrill wrote: "Dear Rosalie—Thanks for the $3.00. Lots of rain here and many fleshy fungi appearing. See Life [Magazine] for May 13 with a very interesting article on 'magic' mushrooms. My *Psilocybe caerulescens* was collected by me right here in Gainesville. I think some northern species of *Panaeolus* have the same magic effect. Better try some. Sincerely, W. A. Murrill." These may be his last written words, and seem to reflect his insight into his own physical and mental shape.

Murrill's body was handled by the Williams Funeral Home, and buried (not cremated) on December 28. "His remains are interred in the Evergreen Cemetery, 401 SE 21st Avenue, Gainesville, Florida," owned and operated by the city of Gainesville (Weber 1961; Kimbrough 2003; Evergreen 2024). Kimbrough (2003): "After Murrill's death, Weber would go to the cemetery annually to see that the gravesite was clean and provide flowers around Christmas, the anniversary of [Murrill's] death." His gravestone is also simple, but when it was placed, and by whom, are both unknown now. A photo portrait of his gravestone (Fig. 9.8) includes a decoration of *Russula hixonii* Murrill. His epitaph: "Author-Naturalist-Editor" is apt. For some years he was remembered with a portrait and plaque near the Experiment Station Herbarium (Fig. 9.9).

Perhaps Murrill takes comfort as he reposes among the distant grandchildren of "Mother oak."

Fig. 9.8 Murrill gravestone with *Russula*. William Murrill's gravestone in Gainesville, Florida. Mushrooms putative *Russula hixonii*. (Courtesy Jim Kimbrough 2003)

Fig. 9.9 A perpetual honor. Portrait and plaque which decorated the Agricultural Experiment Station for some years. (Courtesy Jim Kimbrough 2003)

Chapter 10
A Postlude

From the time of the herbals, biologists (and clerics) have grappled with something summarized as a "natural system" of living things. Whether this term strove to reveal what God had on his mind during creation week, or the revelation of "survival of the fittest" not requiring a God, was (and remains) a personal decision, but one step has always dealt with grouping the living objects; what became known as "species" which are grouped into "genera," which can be gathered into "families," "orders" and "kingdoms." As put forward in the narrative above, mycology, including the fleshy fungi (and the subset, mushrooms), was not on the leading edge of this quest, the priests of the discipline exercised their individual mindsets over the questions which shrouded the problem: what is a species (see Chapter "Home and Away")? What is a genus (see Chapter "Poking and Proding")? What taxonomic breadth defined a genus—or subgenus—or section? How to define these terms, and employ them without prejudice.

Such questions, instead of becoming easier with time and study, become more imponderable. To be sure, Murrill's baptism in fungal taxonomy, his offerings in the polypores, especially in North American Flora, brought down the wrath of C. G. Lloyd and the silent wonderment of others. But like Saul after his conversion on the road to Damascus, Murrill's later conservative treatment of the agarics found a counter opinion from Rolf Singer.

The schism between Murrill's "lumper" tendency and Singer's "splitter" outlook only became more pronounced with their age and taxonomic experience. Singer lived 30 years beyond Murrill, and never relented on his use of smaller assemblages as genera (and therefore coining myriads of new generic names). Hesler and Smith (and Smith's students), shared Murrill's view, and felt obliged to comment directly (see below).

In the microcosm of this narrative, Singer's taxonomic philosophy, born of Patouillard and Fayod's European improvement of Fries's (and Persoon's) blueprint seemed pitted against that of American agaricologists: Lewis David de Schweinitz, William Henry Ravenel, Moses Ashley Curtis, Charles Horton Peck,

R. H. Petersen, *William Alphonso Murrill: The Naturalist*, Memoirs of The New York Botanical Garden 124, https://doi.org/10.1007/978-3-031-93343-1_10

Alexander Smith, Lexemual Hesler, William Murrill and others. For Murrill, Singer's "splitter" philosophy must have seemed like a philosophical maze, but Murrill was not of a mind to alter his own modus operandi.

By adding to the arsenal of useable taxonomic characters, "Post-Singer" years have produced selected endorsements— in fact, often expansion of his basic premise (see below).

Although directly mentioned rarely, mycology and fungi were considered temporary during Murrill's years at the Garden. First, his qualifications for a mycological curatorial and research position were poor—he had more cachet on ferns than fungi, but it is probable that Lucien Underwood endorsed Murrill for the position vacated by F.S. Earle. Second, Murrill mentioned mycological work as "temporary." A glance at his mimeographed materials in Florida reveals how widely he wished to cast his net. His age and sociological position had changed radically, but he still wished to deliver knowledge about stars and rocks as well as mosquitoes and snakes. Finally, at all ages, Murrill amasssed inventories — animate, geological, galactic — offering his inventories to anyone willing to enjoy them. At root, Murrill was an inventoryialist.

POST-MORTEM. Once the "type method" was accepted by the botany community (Europe and North America), a study of type specimens became necessary as a precursor to an understanding of subsequent collections and the taxonomic complexes to which they belonged. Singer (1942a, b, 1946a, etc.) understood this when he conversed and collected with Murrill in 1943. And so it was for L. R. Hesler (1958, 1959a, b, 1960, 1961, etc.). In preparation for more comprehensive taxonomic work, and especially for Hesler (1960), taxonomic characters overlooked by Murrill had to be added to Murrill's cast of taxonomic characters. Using *Russula* as an example, Hesler wrote: "In my effort better to understand the species of *Russula*, a study of certain North American types has been undertaken. Special attention has been given in these studies to the structure of the pileus cuticle and to the ornamentation of the spores. Studies of this kind have been reported by Singer, Burlingham, Schaeffer, Smith and others in the literature but in papers published to date illustrations of the cuticle and spores have been limited or lacking."

To be sure, Singer may have noted these characters in 1943, but his type studies and flagship volume (Singer 1943), although dense, were hardly illustrated, and written descriptions suffered from a paucity of terminology. It was said that Singer's keys were written exclusively for Singer.

Thus, although Murrill was a stickler for the type method, his type specimens required more fastidious morphological examination than even his colleagues could have wished.

THE PRESENT TENSE. In the closing decades of the twentieth century, the evidence of DNA as the carrier of genetic information had been established, and certain lengths of DNA were thought to be useful in sorting mushroom specimens by "species" (but see above). Even more recently, though, DNA has been coupled with macro- and micromorphological characters to more closely indicate the singularity or grouping of specimens. In some cases, even dried specimens have successfully yielded readable DNA, and Murrill's types have emerged for yet another round of examination. In *Russula*, especially difficult for DNA extraction, several papers have singled out Murrill's types once again (Adamcik and Buyck 2010, 2011, 2012; Buyck and Adamcik 2011a, b, 2013; Buyck et al. 2011). What emerge are often endorsements of Murrill's species as unique, but known only from their type location. Opportunity is presented to compare a larger suite of characters and therefore a firmer understanding of the mycota. Hopefully, Murrill would agree.

AN EXACERBATING FACTOR. Far from Murrill's control, however, have been factors operating against his accumulated herbarium. Physically, the necessity for physical space conservation has already led to Murrill types in Tennessee's (Hesler's) herbarium. Both at Tennessee and Gainesville, Murrill's collections spent years without air-handling considerations—warm and humid summers promote mold. The means used as insecticides have been varied, caustic and potentially carcinogenic. Murrill's penchant of describing new species based solely on single collections has not allowed subsequent workers opportunities to spread their expertise beyond the type, where DNA extraction is often restricted. In short, the twenty-first century has not been able to improve Murrill's specimens. He would be disappointed.

PROGRESSIVE DEVELOPMENT. In mid twentieth century, Hesler enlarged his interests from plant pathology to include the taxonomy of fleshy fungi (= Agaricales). This twist required a learning curve similar to that by a layperson first introduced to mushroom hunting. After all, his Cornell days with Atkinson were far behind him. One circumstance, however, was pivotal: Hesler was already an academic with extensive contacts in the mycological fraternity. It was only natural that he turned to Alexander Smith, at the University of Michigan, already a publishing agaricologist and mentor to graduate students in the trade, and soon, dual-authored papers began to appear. Murrill had had a similar experience with his early mentor, Charles Horton Peck, but no co-authorship has been found. Knowledge, skills, and the aptitude for applying them varied individually, and one significant import was Rolf Singer, and his interactions with the American fraternity is prominently mentioned above.

As noted in the narrative above, once in Florida, Murrill's published mushroom descriptions soon became formulaic and minimal, only adding the required Latin diagnosis (merely an abbreviation of the description) in his senior years. As time passed, other scattered agaric taxonomists retired and/or passed away, leaving Hesler in Tennessee and Smith in Michigan as co-workers. But Hesler's attention was geographic: the Smoky Mountains of eastern Tennessee, soon to be set aside as a national park, and the North Carolina Blue Ridge, through Coker's summer residence in Highlands in North Carolina, and Smith for Michigan and the upper Midwest. None of these was drawn to the tropics. In fact, their early work emphasized concordance with northern Europe. Only later did they begin to stray into monographic work as the logical product of their efforts.

But even though Gainesville was not considered tropical, Murrill was the sole exponent of collecting and naming agarics in Florida. Although inspection of Murrill's types has been essential in order to advertise inclusive geographic monographs, knowledge of Florida agarics has remained "skin deep." Murrill's "poking and prodding" of the Gainesville biome has not penetrated as deeply as those of other sections of the country. As a general naturalist, Murrill would have to be satisfied with his role in a process of revelation.

Appendices

Appendix 1: Abbreviated Biographies of Some Supporting Characters in This Narrative

George Francis Atkinson (1854–1918) (Fig. 1)

A superficial scan of the early years of George Francis Atkinson might elicit a stifled ho-hum: farmboy, born and raised in small town, rural schools. In fact, with a few particular substitutions—his antebellum birthyear being one—he might fit the template of William Alphonso Murrill. That he should find himself in a position of educating postgraduate students preparing for Ph.D.s must have been surprising even for him. Nonetheless, at 43, he was taking on a new young man of 28, who, like himself, had yet to select what nook of nature was to occupy his career; a three-page paper on a rare fern (one page was a rough drawing of the plant) in the Fern Bulletin was hardly enough on which to draw a conclusion. A few items, though, were interesting: the young man was recommended by two former Cornell graduates—they had, in fact, proselytized for a fellowship earmarked for him; he had flirted with Johns Hopkins and had accomplished the tasks set for him for admission; he had been teaching for several years (including at "seminaries" for girls!); and surely last, he came from below the Mason-Dixon line—from a slave state.

For the sake of accuracy, Atkinson was a winter baby, born in Raisinville, Michigan, a hamlet near Maybee, southeast of Ann Arbor. He was a lad during the Civil War, but this couldn't prohibit a juvenile enthusiasm for the army—the Union army. He remarked to a group of students that he had run away from home at 15, making his way as far as South Dakota (Thom 1956). The first brush with higher education was enrollment at Olivet College in Eaton County, Michigan, a small, liberal arts, abolitionist institution founded in 1845 to educate students regardless of gender, background, or financial need (Olivet College 2023; Humphrey 1961). The impetus for education isn't known, but the rudiments of Latin and mathematics must have been attained. For whatever reasons, Olivet was not enough, for after

R. H. Petersen, *William Alphonso Murrill: The Naturalist*, Memoirs of The New York Botanical Garden 124, https://doi.org/10.1007/978-3-031-93343-1

Fig. 1 George Francis Atkinson. George Francis Atkinson. Portrait from Lloyd's Mycological Notes, June, 1919, closely following Atkinson's death in autumn, 1918

some courses, he transferred to Cornell, in distant New York State, as a junior. In 1885 (age 31), he emerged from Cornell with a Bachelor of Philosophy (now considered somewhat less than a Bachelor of Arts), which would ultimately be his highest earned degree.

Armed with his formal degree and seeking employment, Atkinson embarked on a whirlwind revolving-door tour of Southern educational institutions. The first was the University of North Carolina (three academic years, 1885–1888). Next was the University of South Carolina (a single academic year, 1888–1889), then to a newly created position as Professor at the Alabama Agriculture and Mechanical Institution (now Auburn University; Fig. 2) for three academic years, 1889–1892. Possibly these experiences with undergraduates prompted him later to write several books intended as introductions of botany to fledgling students. Fresh from Alabama Polytech (where he had been awarded an honorary M.A.), in 1892, Atkinson accepted a position as Assistant Professor of Botany at his alma mater, Cornell, replacing William R. Dudley, who migrated to Stanford University in California. Atkinson was promoted the following year. In 1896, only four years after his return

Fig. 2 Early Auburn Campus. Early view of Georgia Agriculture and Mechanic College (now Auburn University), about 1885. Auburn University Archive

to Cornell, but upon the death of Albert Prentiss, Atkinson (1896a, b) succeeded him as Botany Department Head. As he stated in a retrospective summary of mycological collections made while in Alabama, Atkinson (1897c) recalled that Benjamin M. Duggar had been a mentor at Alabama Polytech, but had also moved on to the Cornell Botany Department to pursue a Ph.D., where he would become a good friend of Murrill.

Atkinson's (1886a) first recorded paper was published while at North Carolina and dealt with a larval pest of the tobacco industry, especially cigarette tobacco. Some extensive research undergirded the paper, including contributions by some other phytopathologists. His second (Atkinson 1886b), on a small insect damaging fruit trees in the South physically followed the first. Between the two, he set his early reputation on phytopathology, a reflection of the strength of the experience gained through the Agricultural Experiment Station run by Cornell. Dated within 1887, however, he burst out with nine more papers, all taking advantage of the fledgling Journal of the Elisha Mitchell Scientific Society, the eponym of the locally famous naturalist. Atkinson staked his claim as a naturalist, covering insects (Atkinson 1886a, b, 1887a), arachnids (Atkinson 1887b, c, d), earthworms (Atkinson 1887e) and birds (Atkinson 1887f). It is little wonder that Murrill could admire the breadth of his future professor, although based on this early record, botany was subordinated to zoology.

Parenthetically, 1888 had been a year of change for Cornell's botany program. Liberty Hyde Bailey, Asa Gray's last student—Gray died during the year—and a horticulturalist intent upon taking botanical knowledge from the lecture hall and academic laboratory to the practicing public engaged in agricultural pursuits—read farmers and horticulturalists—re-entered Cornell as the newly established Professor of Horticulture, stationed at Cornell's Agricultural Experiment Station in Geneva, New York. In fact, Atkinson acted as ambassador from the Cornell campus to the Station, while Bailey brought the agricultural message from the Station to the campus. In some ways, it was oil and water. While Bailey's vigor enlivened the personnel, it also roiled the routine of teaching and research.

As might be expected, excitement over the "New Botany" pedagogical innovations preached by Bailey led to some differences of opinion. This dichotomy played out in microcosm at Cornell, with the conservative Albert Prentiss as its botany head in the College of Liberal Arts followed by George Atkinson, and the effervescent Liberty Hyde Bailey at the Station, but soon to be Dean of the new State College of Agriculture with its version of a Botany Department (soon renamed). Charles Bessey, John Coulter, J. C. Arthur, Farlow, Gray, Dudley, Kellerman, and others were eager to push the discipline forward by promoting botanical education at all levels (Sundberg 2014; Petersen 2021).

Atkinson's time in the Southeast revealed that commercial crops were being attacked and lost to insects, and especially fungus damage, and that less research was being done in the region than he had seen in the North. In 1891 and 1892, for instance, he published almost exclusively on various maladies of crop plants, but sprinkled in were papers on wider subjects (Atkinson 1892a, b).

A bit out of mainstream as well were two reports on fungi collected on the Blue Ridge of North Carolina (Atkinson and Schrenk 1893; Atkinson 1893a), perhaps harbingers of Atkinson's future emphasis. A few details deserve attention. First, the introductory portion was a touristic advertisement of the botanical enchantment of the location. Second, the collections had been made in 1888 and 1889, but in August of both. Third, most of the identification of agarics was done by Albert P. Morgan in Ohio, one of the few specialists in the group. Morgan's recovery from poor health may have significantly delayed publication. Fourth, Hermann Schrenk was a student in Atkinson's lab, perhaps at Alabama Polytech. Fifth, Atkinson advised the reader that the nomenclature in the paper was intended to agree with that by Pier A. Saccardo, who was assembling the mycological taxonomic literature using his own brand of nomenclature.

Once back at Cornell, several papers appeared on ferns (Atkinson 1893b, c, d, 1894a, b). The papers were not basically taxonomic, but studies in developmental morphology. The research behind these papers resulted in a much more significant publication (Atkinson 1894a, with preface dated December, 1893) aimed specifically at students. Whether this book caught the eye of Murrill and furnished the reason for sending living material of *Asplenium ebenoides* to Cornell has not been recorded, but must be kept in mind. The book was also a tour de force for Atkinson's ability to render images into illustrations. The very small note on the fungus *Completoria complens* (Atkinson 1894b) was again a serious observation in developmental morphology.

For Atkinson, 1894 saw two major changes in research and publication life. First was an abrupt change in journals. Two residual papers appeared in the Elisha

MITCHELL SOCIETY JOURNAL, both looking back over his shoulder at fungus diseases in Alabama. All others (more than a dozen) were in BOTANICAL GAZETTE, John Merle Coulter's journal, and in the BULLETIN OF THE TORREY BOTANICAL CLUB, which had graduated from a local New York organ to a national botanical source. Coupled with these changes was a broadening of fungal groups to phycomycetes, ascomycetes and imperfect fungi, but still as plant pathogens. The small Cornell Botany Department, by then missing both Dudley and Prentiss, was acting like an appendage of the Experiment Station at the other end of Lake Cayuga.

Murrill's Ph.D. did not end their contact, which, while not constant, seemed cordial. Ithaca was only one day away from New York and housed a large and familiar herbarium. As time passed, Murrill was drafted into the nomenclatural cadre of Nathaniel Lord Britton, Director of The New York Botanical Garden, and the force behind NORTH AMERICAN FLORA, while Atkinson remained a follower of Saccardo and his SYLLOGE FUNGORUM.

Atkinson was at home in the field, and it was on a protracted collecting trip to the Pacific Northwest, in 1918, that he died of influenza, perhaps the "Spanish flu." By that time, Murrill had specialized in the polypores and author of several significant papers on the subject.

ELIZABETH KNIGHT BRITTON (1858–1934) (Fig. 3)

In these days, it should go without mention that this brief note deals with a special circumstance, namely a highly credentialled woman in science. That Elizabeth Gertrude Knight was female, an expert, practicing botanist, and married, in 2024 might actually be supportive in job hunting, although it might be more delicate to find a suitable position for the husband. Not so in turn-of-the-twentieth-century New York City, where occupations for women were quite proscribed. On the Lower East Side, girls were put to work with their sisters and mothers in their flats with "piece work." In higher social circles, etiquette was taught and "uppity" girls were screened through their formative years to understand their societal role(s).

Elizabeth was born in New York City in 1858, pre-Civil war but with clouds gathering on the horizon. Her family owned and operated a furniture factory in the vicinity of Havana, Cuba, thus contributing to the early deforestation of the island long before the Spanish-American War and American occupation. She spent much of her earliest years there and just these clues might suggest that she was used to servants, comfortable housing, sufficient food, and perhaps tutors. She also spoke fluent Spanish and developed a love of nature, prompted by her father (Howe 1934a, b). She was the eldest of five sisters, which might also have helped hone her sense of independence.

As she grew, she lived with her maternal grandmother in New York, where she attended a private school before enrolling at the Normal College of New York (Normal College 2023), an all-female institution presenting a three-year curriculum to train teachers, consistent with the norms of society. She graduated in 1875, age 17 (Elizabeth Gertrude Britton Wiki 2023) and remained there for a decade on faculty in the model school.

Fig. 3 Elizabeth Knight Britton. Early portrait of Elizabeth Knight Britton, botanist and wife of Nathaniel Britton. New York Botanical Garden Archive

During that decade, her fascination with plants, especially cryptogams must have grown, for in 1879, she was listed as a member of the Torrey Botanical Club. In the same year she travelled to Nova Scotia, a detail of which was reported later (Knight 1883a). Likewise, a trip to the Adirondack Mountains in 1881 also reported on observations of discordant albinism (Knight 1881). Against the report by Asa Gray, she exhibited specimens of the diminutive fern, *Schizaea pusilla,* from Nova Scotia, leading to its rediscovery in Newfoundland. In yet another paper, she reported on the sexual apparatus of *Eustichium norvegicum* [now *Bryoxiphium*], collected in Wisconsin (Knight 1883b). Elizabeth (as Britton 1886) wrote a paper on the flora of West Virginia, collections for which were surely made prior to the publication date. She was already well-travelled before her marriage. Interestingly, illustrations of women performing fieldwork in those days—including a photo of Elizabeth at work in the Adirondacks—show full length skirts and generous undergarments, and one wonders how Elizabeth observed the submerged leaves of an aquatic plant, especially a plant far from its natural range. Perhaps "intrepid" might be apt.

In 1883 her professional title was changed to Tutor in Natural Science, recognizing both her knowledge and her educational skills. Although her earliest papers did not reveal her specialty, by the early 1880s she was known as a bryologist (with a "minor" in ferns).

Back in New York, she and Nathaniel Britton were noted as a couple. She was a teacher, he an assistant to Prof. Newberry in geology at Columbia Mines. Their membership in the Torrey Botanical Club gave them occasions to be together as well as immersion in their avocations. Elizabeth later became the Editor of the Bulletin of the Torrey Botanical Club, perhaps the first female editor of a scientific journal. They were married in 1885, and coincidently elevated both their botanical achievements. Although not mentioned in the reports of the day, one must imagine a couple in which the bride was taller and proportioned sturdily, but one in which the individuals were seldom apart and devoted to one another. [There are specimen records of Elizabeth and Nathanial collecting together prior to their marriage, on Mackinac Island, Michigan.]

Nathaniel's position(s) was also shifting. In 1887, he was elevated to Instructor of Botany and Geology at Columbia Mines and soon took over Newberry's role, and coincident with the University's migration north in Manhattan, Britton became Head of the new Department of Botany. With Britton family resources, this provided a salary which could support the couple, important because Elizabeth resigned her job at Normal College when they married. Soon, she set about organizing the College's moss herbarium, gratis. In 1893, "with the aid of generous friends," the moss herbarium of August Jaeger of Switzerland was purchased for Columbia (Britton 1893). It contained several important exsiccati and nearly doubled Columbia's holdings.

While her curatorial duties were "off the books," her influence was felt nonetheless. Student Abel Joel Grout came to post-graduate studies in algae with Nathaniel, but he found a niche with Elizabeth and switched to mosses. He became one of the foremost bryologists of the next generation (Grout 1903–1910; Elizabeth Gertrude Britton Wiki 2023). The two were pivotal in forming the Sullivant Moss Society, with Grout as its first President. Later it became the American Bryological and Lichenological Society.

Some of Elizabeth's publications came in series. "Contributions to American bryology" began in 1889 in the BULLETIN OF THE NEW YORK BOTANICAL GARDEN. Later, in 1894, she began a series of eight illustrated papers under the title "How to study the mosses," published in THE OBSERVER, a magazine which turned out to be short-lived.

The purpose of the Brittons' 1888 trip to England was to study Rusby's collections from South America (Britton 1888), but the result grew into an unforeseen product—the tale of their stroll around the Royal Botanic Gardens at Kew and Elizabeth's innocent query of Nathaniel, "Why can't we have such a place in New York?" Although the story is expanded elsewhere in this narrative, suffice to relate that their musing was repeated to the Torrey Botanical Club, elicited enthusiasm and led to the founding of The New York Botanical Garden.

At the turn of the twentieth century, there was an awakening to the widespread destruction of the environment which expressed itself as extirpation of wildflowers. Clubs and societies were formed to preserve wild plants, and Elizabeth was at the forefront of the movement. The Wild Flower Preservation Society of America was the result, with such notables as Frederick Coville (USDA), President, and Charles L. Pollard (Washington), Secretary. The Board of Managers included Charles E. Bessey, Liberty Hyde Bailey, William Trelease, C. F. Millsbaugh, Alice Eastwood, and Elizabeth. Over the years, Elizabeth served two terms as Secretary (Howe 1934a).

Meanwhile, once Columbia's herbarium had been donated to the Garden, her curation of the growing moss collection went on unabated through fresh collections, donation or purchase of important collections and exchange with the world's

Fig. 4 Nathaniel Lord Britton. Nathaniel Lord Britton, Columbia student and botanist, Director-in-Chief of New York Botanical Garden. New York Botanical Garden Archive

herbaria. In 1912, she was named Honorary Curator of the Mosses, a title she held until her death (Howe 1934a; Barnhart 1935).

Nathaniel undertook a full program of floristic exploration of the Caribbean Islands over several years. He and Elizabeth were often the chief botanists, with her Spanish language skills making friends and easing difficult situations. It was written that of the 35 trips taken by Nathaniel, Elizabeth was on 34. One list of her destinations: Caribbean region: Bahamas, Barbados, Cuba, Jamaica, Puerto Rico, Trinidad and Tobago, Virgin Islands (U.S.); Atlantic region: Bermuda; North American region: Canada, United States (Elizabeth Britton JStor 2023).

Like numerous other Garden staff, the Brittons lived not far from the Garden itself, at 2965 Decatur Avenue, Bronx. Elizabeth died there in February, 1934 (Howe 1934b). Just four months later, Nathaniel succumbed in the same house.

Nathaniel Lord Britton (1859–1934) (Fig. 4)

His middle name was not a royal appellation, but recognition of his mother's family. In retrospect, his birth on January 15, 1859, was not as auspicious as Darwin's The Origin of Species in the same year, and the American Civil War came and went before Nathaniel was ten. Unlike so many other professional botanists, the Britton family was local to New York. The family home was in New Dorp, whose name was a reflection of the Dutch occupation of Staten Island. The island was closer to New Jersey than to Manhattan, and the Britton family owned considerable property, acquired since their immigration in the mid-1600s.

In his young years around home, Britton was befriended by a neighbor, Mr. John J. Crooke, graduate of Yale and accomplished collector of natural objects—shells, birds, minerals, plants (Howe 1934b; Britton 1911). Young Nathaniel absorbed much, including a penchant for learning the names of the plants, including wildflowers in his neighborhood, to the amusement and wonderment of his parents. Apparently, a young friend, Arthur Hollick, was a partner in the pursuit of natural science. As usual, we know nothing of the reverberations of the Civil War on local homelife and neighborhood. Slight of physique but well-endowed with intellect, Nathaniel seemed the stereotype of prodigy.

Mr. Crooke helped Nathaniel's admission into Columbia College School of Mines, at a somewhat precocious age. The school had been established only recently in 1864. It was the only institution with courses in natural history, albeit few, but Britton's application for admission was accepted, and four years later he was awarded a bachelor's degree as Engineer of Mines at age 20 (Merrill 1938). His friend, Arthur Hollick, was also a student there. Since the school was located in Manhattan (now considered downtown, but then uptown), Britton must have boarded nearby for it would have been an impossible daily commute from Staten Island across the harbor and north to uptown Manhattan. This is noteworthy only to testify to a degree of self-reliance at a young age and precarious physical vigor.

The Torrey Botanical Club (Rusby 1906), which Nathaniel joined as a student in 1877, played an important part in Britton's life, not only as an outlet for publishing his research discoveries and a gathering of like-minded souls including a sprinkling of influential individuals, but also by bringing him together with Elizabeth Gertrude Knight. The two became a couple. Elizabeth was not only an active member, but also a budding expert on mosses.

In previous narrations above, the "normal" peripatetic route of natural history academics was necessary: Murrill, Underwood, Atkinson and others moved through such migrations to attain their influential positions by the first half of the twentieth century. Britton was an exception; born in New York, educated in New York, married to a New Yorker, and first, second and ultimate jobs in New York.

Dr. John Strong Newberry (see below), Professor of Geology and Mineralogy at Columbia School of Mines must have appreciated Britton's enthusiasm and skills, for upon graduation in 1879, Nathaniel remained in place as Newberry's assistant. A couple of papers dealing with Staten Island plants appeared in the BULLETIN OF THE TORREY BOTANICAL CLUB with Hollick, Nathaniel's boyhood friend and fellow student, as senior author (Hollick and Britton 1879, 1880).

Trading on his degree and title as Assistant in Geology, Britton took on an appointment as Botanist and Assistant Geologist of the Geological Survey of New Jersey. He must have emphasized the botany portion, for in 1881 he published "A preliminary catalogue of the flora of New Jersey" (Britton 1881). Although only occasional lectures by Newberry represented botany in Columbia College (Merrill 1938)—not to be confused with Columbia School of Mines—the New Jersey flora paper seems to have served as a thesis with which Britton earned a Ph.D. from Columbia College in 1881 (Howe 1934b; Merrill 1938). It must be reminded that this flora was the turf of John Torrey and Asa Gray, botanists whose early floristic emphasis was regional with New York City as hub. A certain precociousness was evident on Britton's effort.

As gauche as it might seem, a few words must be inserted concerning Nathaniel's physical stature. Merrill (1938) characterized him as "slight in build and of frail physique…" but another popular measure (Gleason 1961) was more metric: "five foot three and a hundred and ten pounds… with a full, straggly beard." His stamina and single-mindedness, however, were unquestionable.

In the midst of this flurry of his accomplishments, he and Elizabeth married in 1885 (Merrill 1938). She would accompany Nathaniel on numerous trips to the New World tropics and Europe, in part to pursue her own bryological interests (Howe 1934b).

Perhaps recognizing a paucity of botanical instruction, in 1887, Britton was appointed Instructor of Botany and Geology in Columbia Mines (Merrill 1938), and to these subjects were added lectures in zoology. Higher appointments followed: Britton took over Newberry's role upon Newberry's retirement, became Adjunct Professor of Botany in 1890, and Professor the following year.

During his years at Columbia Mines, Britton honed his skills, published numerous papers and lectured to students—and when University President Seth Low negotiated the Morningside Heights property and Schermerhorn Hall was erected, the Brittons packed up and moved with it. Nathaniel became Head and Professor of Botany at Columbia University.

By this time, Elizabeth had posed the question prompted by Kew Botanic Gardens, and the idea was spreading like a brushfire in the Torrey Botanical Club. The New York Botanical Garden was becoming more than an apparition, and in 1896, Nathaniel became the first Director-in-Chief of the new endeavor, but continued to hold his Columbia position as Professor of Botany Emeritus.

"Britton's breadth of view is illustrated by the plan originated and consummated by him of bringing together in one place the scattered herbaria and botanical literature in New York, at the Garden, including the historical Torrey herbarium and the general herbarium of Columbia University, that of Barnard College, and the herbarium of the American Museum of Natural History. His interest was not limited to the higher plants as witnessed by his fortunately consummated plans of acquiring for the Garden the very important Ellis and Everhart collection of fungi, and the magnificent Mitten collection of mosses, both collections being exceedingly rich in historical material and containing many thousands of types. In the library field he planned and consummated the purchase of all the botanical works in the library of the Conservatoire et Jardin Botaniques de Geneve that were duplicated when the very important DeCandolle library was presented to the latter institution, thus vastly increasing the library resources of The New York Botanical Garden" (Merrill 1938).

Britton's personal wealth served as a gateway to the elites of the city, not the least of whom was Seth Low. In some of the lean years for the Garden, Britton knew which members of the Board of Managers might be receptive to a contribution or a subscription to strengthen the endowment and, thereby, increase income. The situation was always announced in the BULLETIN OF THE NEW YORK BOTANICAL GARDEN, a committee was formed, and subscriptions requested. The outcome as always the same: the endowment grew and Britton's ploy had been successful once more.

Merrill (1938) wrote at length about Britton's role at the Garden. "The challenge that Doctor Britton met was 250 acres of raw undeveloped land, no roads, bridges, fences, or buildings, an assured endowment of $250,000 of which the income only could be used, no laboratories, library, or herbarium, no staff, but with the hope of some financial support from the city and the potential hope that citizens of means would by gift or bequest, help support the infant institution. How well he met the challenge may be briefly indicated by the statement that in the thirty-three years of his directorate he saw the Garden develop from an idea to a well-developed tract of 400 acres, a commodious administrative building and museum, ample greenhouse and conservatory facilities, a great collection of books containing 43,500 bound volumes, one of the great botanical libraries of the world, a reference herbarium in excess of 1,700,000 specimens from all parts of the world, well equipped laboratories, an active and productive staff, a publishing institution sending its product to the ends of the world, its initial endowment of $250,000 increased to approximately $2,500,000, an auxiliary membership (annual, supporting, and life) of nearly 2000 individuals, and its annual city appropriation greatly increased. The end attained justified the faith of the moving spirit in the enterprise, but how many men, situated as Doctor Britton was in 1895, would have had the courage to accept the challenge, and had the faith that the desired end could be accomplished?

"Yet one wonders how many of [his colleagues] could have continued to be productive had they been faced with the multitudinal problems, large and small, that Doctor Britton met day after day, week after week, month after month, year in and year out, and still find time to produce technical paper after technical paper. It takes a rare combination of qualities, and above all persistence, abiding faith in one's work, and deep and unfaltering interest in the daily task to produce results under such circumstances, and these qualities were Doctor Britton's to an eminent degree. He worked unceasingly himself and he expected his associates to do likewise." (Merrill 1938).

A small tableau: "George Small, son of J. K. Small, a curator of museum exhibits, recalled that Britton expected his staff to put in 'full time' and to be on call daily. Members of the Garden staff took turns taking a full day of Sunday duty, during which they were supposed to make an inspection tour of the entire garden and check in with the guards; 'it was quite a chore,' Small remembered" (Mickulas 2007).

In Murrill, Britton found a softspoken naturalist, formally practicing under the title of mycologist, of entirely opposite background to Britton; rural, Southern, surely not wealthy, but agile of mind, and upon being given a position of influence (Assistant Director), an occasional gadfly on behalf of the staff, workers, and public. Moreover, when confronted by a wall of resistance, Murrill was not beyond a measure of chicanery as a challenge to Britton's "czarist" personality. The result was demotion of Murrill, and with it, ever growing non-research tasks (Gleason 1961), superficial triumph for Britton, and confirmation of Britton's mono-maniacal rule of the fiefdom of The New York Botanical Garden.

"On [Britton's] death, indicating his abiding faith in the institutions he helped to organize and to develop through their years of struggle, he bequeathed one-half of his estate in varying amounts, to The New York Botanical Garden, the Torrey Botanical Club, the New York Academy of Sciences, the Staten Island Institute of

Fig. 5 Addison Brown. Formal portrait of Addison C. Brown, New York jurist and one of founders of the New York Botanical Garden. (Wikipedia)

Arts and Sciences, and Columbia University. Doctor Britton's wife, Elizabeth Gertrude Britton, herself a botanist of note, died four months before him, and from this shock of separation Doctor Britton never fully recovered. There were no children." (Merrill 1938).

Addison C. Brown (1830–1913) (Fig. 5)

Addison Brown was a juror, lay authority on natural history and astronomy, and benefactor to causes pursuant to his chosen avocations. With additional detail, Brown was a Justice on the United States District Court for the Southern District of New York, presidentially appointed in June 1881; presidentially reappointed in October with Senate confirmation in two days. Upwards of 2000 decisions were written in 20 years on the bench. More: co-author with N. L. Britton of the three-volume Illustrated Flora of the Northern United States and Canada (in three editions); President of the Torrey Botanical Club for a dozen years; benefactor to the Club and to an infant New York Botanical Garden; sole author on numerous serious botanical items; similar credentials in astronomy.

With a lineage stretching back to early English immigration, Addison's father was a Massachusetts shoemaker. Value of education propelled Addison to Amherst College, the excellent, small liberal arts institution, at 18 years, but he transferred to Harvard the following year. Brown graduated in 1852, second in his class, and made the conscious choice that of the three occupations he foresaw—medicine, the ministry and law—he chose the least repulsive because he knew nothing about the law (Brown 1972). The result was an apprenticeship for a year and a year of Harvard law school, graduation in 1855, and a successful bar exam. He was qualified to practice law at age 25 (Addison Brown Prabook 2023).

Often, especially when he needed money, Addison was organist and choir master in various circumstances (Addison Brown Wiki 2024)

A series of individual practices and partnerships followed, accompanied, now and then, with investments in real estate, usually at the northern extremities of the city, where rural estates were being subdivided and developed. Surely, opportunities for financial growth were present, but not only was capital necessary, but also adroit sidesteps in a market entwined with the political bossism. The same could be said about a presidential nomination to the mid-level federal bench and swift senatorial confirmation—all the while indulging serious avocations, in Brown's case, botany and astronomy.

Addison Brown was one of very few instrumental in the birth of the Torrey Botanical Club and The New York Botanical Garden, separated by about a quarter century. His role in these as well as the American Museum of Natural History are taken up elsewhere in this narrative. In his will, a fund was bequeathed to the botanical garden to support a journal of colored illustrations of plants—it was fittingly titled ADDISONIA.

FRANKLIN SUMNER EARLE (1856–1929) (Fig. 6)

Franklin Sumner Earle, like so many other mycologists and phytopathologists, was a Midwesterner, born in Dwight, Illinois, a whistle-stop on the railroad from Chicago to Bloomington. His father, Parker Earle, about to decamp to the southern part of the state, already had deep experience in gardening and horticulture from his years in Massachusetts in association with Charles Mason Hovey, a seed and nursery man in Boston. Hovey also founded AMERICAN GARDENER'S MAGAZINE (later the MAGAZINE OF HORTICULTURE), for which he was editor for 34 years (Jercinovic 2007). Hovey's specialty was strawberries, and Parker Earle also marketed them after he moved to the deep, rich soil of Illinois. Parker and his wife, Melanie Tracy, produced three children, of whom Franklin was the first. The family business occupied the children as soon as they could work, so Franklin's life was steeped in both raising crops and marketing their produce.

Franklin was in his pre-teens when the Civil War erupted. Too young to serve in the Union military, patriotism for the anti-slavery cause was common. Having moved to deep southern Illinois (Union County), Parker was becoming noted for his crops, and, post-War, he developed a method to keep shipments of strawberries fresh on the rail trip to Chicago, where they brought a "princely" price. Franklin's biographers all mention his "sporadic" attendance in the University of Illinois during his 20s, but although he came under the tutelage of Professor Thomas J. Burrill, a noted naturalist and founder of the University herbarium, Franklin never received a degree (Jercinovic 2007).

Parker's reputation grew, much so that he acted as a horticultural judge at the 1876 Philadelphia Centennial Exposition. An even higher role, in 1884, Parker was selected as horticultural director of the World Industrial and Cotton Centennial Exposition in New Orleans. While in Louisiana he visited and fell in love with southern Mississippi, particularly the area around Ocean Springs, just east of

Fig. 6 Franklin Sumner Earle. Franklin Sumner Earle, plant pathologist at the New York Botanical Garden, then Director of the Agriculture Experiment Station of Cuba. (Wikipedia)

Biloxi—it was the deep antebellum South, plantation country, and surely tropical. It was the land of cotton and sugar cane. Parker vowed to move there.

While still living in southern Illinois, though, in 1885, Franklin, then 29, presented a paper on the "white rust" and other diseases of strawberry at the Mississippi Valley Horticultural Society meeting in New Orleans, reflecting his background with his father and with Burrill. The following year, he co-authored a paper with Burrill on the Parasitic Fungi of Illinois: Part II. Erysiphaceae, which embellished his cachet in phytopathology. Just as important to Franklin was his courtship and marriage to a neighbor's daughter, Susan Bedford Skehan. Their son, William Parker Earle was born in Illinois in 1897, but his daughters Melanie and Ruth were born in Ocean Springs. Soon after Ruth's birth in 1891, Susan died.

Around 1886 the Winter Park Land Improvement and Livestock Company was formed with Parker Earle as president. He and his two sons owned 97% of the stock. By the end of 1887, the company had invested in 15,000 acres of land in the vicinity of Ocean Springs. Parker and his sons, all with family, migrated from Illinois. In October, 1890, the Parker's farm was producing tomatoes, peaches and grapes over 80 acres. Their living seemed assured, but in October, 1893, the Parkers' investment was ruined as a major hurricane ripped across the coastal regions of southeastern Louisiana and southern Mississippi. Two thousand people were killed, and Parker Earle pulled up stakes and moved to the New Mexico territory.

Franklin, in the wake of Susan's death, became associated with the Mississippi Experiment Station and in 1891, became superintendent, serving until 1895. When Parker moved West, Franklin remained behind, now with three motherless children. The position as Director of the Mississippi Experiment Station burnished his cachet in administration and phytopathology, but in 1895, he became Assistant Pathologist for the USDA in Washington, in charge of the mycological herbarium. This brought him close to several additional mycological-minded workers (Petersen 2020a, 2021). The job added herbarium expertise to his credentials, a somewhat different experience than anything previous, but the following year he became the biologist and horticulturalist at the Alabama Agriculture Experiment Station. This position spread his name throughout Alabama agricultural circles.

Sources differ as to the dates of his appointment in the Alabama Polytechnic Institute (API; now known as Auburn University), in Auburn, Alabama. First, his name appeared as Professor of Biology and Horticulture (API catalog, 1897–1898) then as Professor of Botany (Catalogs 1899–1900, 1900–1901). In the API catalog of 1900, under Biology and Horticulture, the following explanatory material appeared: "The work in this department will be largely confined to the study of the lower forms of plant life and to vegetable physiology, or the study of plant life as living beings.... The instruction will be by text-book, supplemented by lectures and by practical laboratory work. The students will be required to make careful microscopical drawings of objects studied. They will also collect material in the fields and determine it so far as their instruction will admit."

During the summer of 1897, Earle and a new colleague, Charles Fuller Baker, collected widely in the area around Auburn (Jercinovic 2007). API bestowed an honorary M.S. on Earle in 1901, just as he left to become Assistant Curator of Fungi at The New York Botanical Garden, still a very young institution. This position put him under the Director, Nathaniel Lord Britton (about whom above). It was Earle's sixth job in a decade, and he was a mature 45 years old.

The position sounded similar to his time with the USDA, cataloging specimens collected chiefly by others, and the fungal specimens held at the Garden were also dominated by fungi as parasites causing diseases of plants. In addition, however, were important historical materials, especially the Ellis herbarium. Moreover, Director Britton encouraged his curators to actively collect specimens in "interesting" areas to make the Garden herbarium more cosmopolitan. Already, philanthropic endowment funds could support some excursions and exploration.

Earle's prior experience was chiefly with tropical or subtropical climates, and he had become a special expert on sugar cane and its agricultural complications. The actual plant was of ancient origin, now assumed to have been domesticated in India and New Guinea (Warner 1962). Catering to Northern Hemisphere seemingly insatiable appetite for sweet, sugar cane was introduced to Pacific, Caribbean, and South American colonies. In the United States, the coast along the Gulf of Mexico was suitable, setting the stage for Earle's activities in agriculture.

Irrespective of sugar itself, the tropical climate in which it thrived was intriguing to Garden Director Britton, and he envisioned monographic research of the floras of regions of American tropics, and for sociopolitical reasons, Caribbean islands seemed good targets for such exploration and collecting, not only of herbarium material, but also living examples for glasshouses erected specially for public education (Britton 1903a, b). For Franklin Earle, though, his expertise—and interest—lay in living, not preserved, plants. After an initial introductory visit, in 1904 Earle accepted the position as Director of the Estación Agronómica at Santiago de las Vegas, Cuba. Soon, he added his own plantation at Herradura, which he maintained until his death. He acted as a business consultant, president of the Cuba Fruit

Exchange and later even as a USDA investigator, but sugar, its growth, health, and industry remained his major occupation (Earle 1928). Throughout, he maintained a good relationship with The Garden, visiting periodically and occasionally depositing specimens.

William Ashbrook Kellerman (1850–1908) (Fig. 7)

In war-time 1812, two distilleries were built near the crossroads settlement of Ashville, Pickaway Co., Ohio. Owned by the Staige brothers, Richard Staige owned most of the land of the present-day town. After many years he sold his distillery to Mahlon Ashbrook, who was in business with his brother Absolom. The Ashbrooks owned a large store run by sister Iva, "Ivy," and her husband Daniel Kellerman. The Ashbrook's enterprises failed in mid-century and Ashville lulled until the railroad and its depot came to the town. One son of Ivy and Daniel was born in 1850 and his middle name was an heirloom of his mother's family.

Of the early life of William Ashbrook Kellerman very little has survived, but the Civil War was fought during his formative years. He progressed through schooling, probably through the public school(s) of the area. There is a hint that he taught in local schools for a few years, armed with only a public school diploma. How he came to entertain ideas of college is unknown but the funds for matriculation at Cornell University were found, perhaps through one of Cornell's competitive scholarships. He probably started in 1870, only two years after the school's opening, for he graduated with a B.A. in 1874 as one of the first three classes of that institution. Only two years before Kellerman's arrival, Albert Nelson Prentiss helped found the university as chair of the new Botany Department, and Kellerman was likely to have had classes with him (Petersen 2021). Likewise, how he secured a teaching position at the State Normal School in Oshkosh, Wisconsin, is unreported, but he remained there for five years. Married in 1876, Kellerman determined that he needed European exposure (and concomitant sophistication), so William and Stella (née Dennis) embarked on a two year-residence (1879–1881) in Göttingen, Germany, gateway to the Harz Mountains and residence of Christiaan Hendrik Persoon (1755–1837) years before, thereafter in Switzerland, earning a Ph.D. from the Polytechnik in Zürich (Willard 1909). Two years in Europe must have been expensive and the Kellermans' source of support is unknown. Moreover, the birth of a son, Karl Frederic (1879–1934) must have come dangerously close to their embarkation. Upon return, Kellerman taught at the State Agricultural School in Lexington, Kentucky (later the University of Kentucky) for a single academic year.

In the early spring of 1881, a life-changing event transpired. The fledgling Ohio Agricultural and Mechanical College (begun in 1870) inaugurated a new department, Horticulture and Botany, to be led by an assistant professor with the munificent salary of $1500 per annum. A final roster of six candidates was considered, with Andrew P. Morgan, Joseph C. Arthur, William Rand Lazenby and William A. Kellerman among them. Kellerman had hardly returned from Europe with his Ph.D. diploma in his pocket. Of the candidates, Morgan had the thinnest credentials as only a self-taught botanist and public school teacher in Dayton, Ohio, and no collegiate experience, but he was chosen and started working. Unceremoniously, in

Fig. 7 William Ashbrook Kellerman. William Ashbrook Kellerman, official portrait cover of Lloyd's "Mycological Notes," August, 1910, after Kellerman"s death in Guatamala, in 1908

three months he withdrew (or was dismissed), the circumstances now unknown but with a commendation by the Board of Trustees (Ellett 1970). It is probable that because the department was inaugurated to make horticulture a major presence, Morgan, the self-taught botanist, could not cope in this discipline (Meyer 1983). In addition, a likely cause may have been poor emotional health of Morgan, but in the meantime Kellerman had moved to Kentucky and Arthur to Cornell's Experiment Station thence to Purdue. William Rand Lazenby was appointed as the replacement for Morgan, with the title of Professor of Horticulture and Botany effective in the fall of 1881.

After a year in Kentucky, 1882, Kellerman began his next seven years in Kansas, three on faculty at the State College and four as Kansas State Botanist, including teaching and research, the latter mostly floristic and sometimes co-authored by Stella. His first textbook on botany appeared in 1883 (Kellerman 1883), barely a year after his arrival in Kansas. Kansas State College had been founded in 1863, taking advantage of the Morrill Act of 1862, which mandated land grants to the states to establish colleges in which the curriculum would include agriculture.

Kellerman's potential first identifiable major financial contribution to mycology *per se* came with the start of the JOURNAL OF MYCOLOGY in 1885—co-editors were Job Bicknell Ellis and Benjamin Matlock Everhart. Dr. Everhart was already furnishing financial support for Ellis's research and gathering of immense exsiccati and he was surely a contributor to the start of Kellerman's new journal. The journal idea was not totally new. John Merle Coulter had conjured up the BOTANICAL BULLETIN in 1875, quickly changing its name to BOTANICAL GAZETTE. Both botanical research and its community had reached a point where communication—both scientific and social—was valued.

The first four volumes appeared on schedule, but scattered rumors indicated financial difficulties. Indeed, starting with volume five, the USDA took over, but even that arrangement met with difficulties, and in 1902, Kellerman, by then at Ohio State University (Stuckey 1970) took back the JOURNAL OF MYCOLOGY with volume eight and maintained it until his death in 1908.

The circumstances of Kellerman's death left several spheres to be mourned—the JOURNAL OF MYCOLOGY, his teaching and interactions with students, his second journal, OHIO MYCOLOGICAL BULLETIN, his place in the botanical community, and, of course, his marriage. On one front, an answer came from New York City. William Alphonso Murrill, also with a Cornell Ph.D., was the mycologist on staff at The New York Botanical Garden, itself not yet at maturity.

The JOURNAL OF MYCOLOGY testified to a large synapse in Kellerman's psyche. Again and again he seemed enthusiastic to pass on whatever he learned to those who had not been exposed. It was true of his teaching: Willard (1909) wrote of Kellerman's time in Kansas: "Professor Kellerman had a most magnetic personality, and probably not one teacher in a hundred possesses the power to rouse the interest and enthusiasm of his students as he did. This was not by any artificial effort, but because of the overflowing of his own enthusiasm together with a rare gift of presentation."

Kellerman was not a man to be denied and it took him 10 years—1891—to land a faculty position at that same Ohio State College that had passed him by, but now he was Head of the Botany Department. Kellerman's (1898) "A Text-book of Elementary Botany, Including a Spring Flora" was written in Ohio.

Kellerman had been at Ohio State barely a year when the Chicago Columbian Exposition was conceived in 1892. Among the planned exhibits was a major building devoted to forestry. Kellerman (1892) put together a prize-winning exhibit of the timber trees of Ohio as a showpiece for the state's bounty. Over 21 million visitors viewed the fair during its six-month life in 1893. It was yet another example of Kellerman's intent to communicate scientific knowledge to a broader audience.

In 1903, Kellerman again saw the need for the lay public, including the farmers in central Ohio, to be educated on mycological and phytopathological subjects. He conjured up the OHIO MYCOLOGICAL BULLETIN, really a series of pamphlets featuring elementary lessons on fungi—their diversity, life cycles, structures, exponents and literature—all in very few pages of cheap paper. For Kellerman, the OHIO MYCOLOGICAL BULLETIN was edited in tandem to the JOURNAL OF MYCOLOGY which he had revived the previous year. Subscriptions (as dues in the "Ohio Mycological Club") he set at 10 cents a year, but advertising was welcome, including recently published mushroom books. The series ended with no. 87 at his death in 1908. In the last years, outside workers submitted short papers and photos of mushrooms. Although also titled as "University Bulletin, series 7 no. 12. Botanical series no. 13," its editing, formatting, addressing and mailing depended on Kellerman (and perhaps Stella). The series ended when his death evaporated the editorial input and financial backing.

In 1905, Kellerman "discovered" Guatemala (Kellerman 1906; Lowden 1970). This "vacation" trip with several friends, including two couples, revealed a tropical biota, and he endeavored to return with students in order to collect plants and fungi, the latter especially parasitic, and to experience the Guatemalan people and their milieu. Subsequent trips became annual. The departure date was set just before or on Christmas day. The primary itinerary was always by train from Ohio to New Orleans, the port of embarkation and reentry, thence by steamer along the coast of Yucatan and arrival at Belize (5 days), capital of British Honduras. One additional day over to Gulf of Honduras, led to the eastern entry to Guatemala, Puerto Barrios. Rail and muleback rose from sea level to 5000 ft. elev. and Guatemala City. From this capital, rail was available to the Pacific coast, thence south to San José north to Champerico. From every small town along this entire land route collecting trips were made on foot and by muleback. Lowden (1970) presented detailed itineraries. Higher plants, cryptogams and fungi were all collected in large numbers. Baggage included camping gear, plant presses and food. Each specimen required logged data and preservation, and work often extended into the night.

Between the trips of 1906 and 1907, Kellerman proposed a formal program and curriculum, "Tropical School of Botany." Such a concept, of course, had to pass muster at several levels of administration, but his proposal materialized with the receipt of a letter from the president of the University stating: "The Board of Trustees granted your request for a leave of absence next winter and the organization of the school of botany." The new School was approved with the proviso that the University would not be liable for any financial investment whatsoever. His leaves were on-salary, and for the 1908 trip, $250 was appropriated for making collections for the Botany Department. His salary could cover the trip expenses for himself, or the normal expenses back in Columbus, but not both. Again, his own funds were required to make up the shortfall. Plans were laid to carefully select a few students to enroll when Kellerman departed for the 1908 trip, during which he was overtaken by malarial fever and died in Guatemala in March, 1908. His final journey and death were described by Taylor (1910).

(Anonymous 1908b): "When the January number of the Journal of Mycology was issued, no one had the remotest idea that the succeeding number would record the death of its founder, editor and publisher. Like a thunderbolt from a clear sky came the cablegram from Zacapa, Guatemala, announcing the death of William Ashbrook Kellerman, March the eighth." Willard (1909): "He was fearless and outspoken in expressing and maintaining his views. In his character and ideals of conduct he was upon a plane that was impregnable to criticism. He was an inspiration to all with whom he came in contact, and his death was a loss to botanists the world over, and an irreparable one to his friends." (Griggs et al. 1908): "Until handicapped by a difficulty in hearing which arose in the last few years, he was one of the most

Fig. 8 Pierre Lorillard. Pierre Lorillard IV, whose Bronx estate formed a large portion of the New York Botanical Garden property. (Wikipedia)

active participants in the discussions of the [Ohio Naturalists] club, always adding interest and profit to its meetings. Many of his papers, since published, were read for the first time in its sessions. … Never did a meeting pass at which he was present without his reporting some new and interesting discoveries he had made during the month." (Lloyd 1910): "There are few men so enthusiastic in their work that they are willing to make personal sacrifices in order to carry it on."

Pierre Lorillard IV (1833–1901) (Fig. 8)

That Pierre Lorillard IV was best known as a thoroughbred horse owner, for his house ("The Breakers") in Newport, Rhode Island, which he sold to the Vanderbilts, the huge family estate in the Bronx (see Fig. 6,3), or his son's introduction of "the Tuxedo" attire at their 13,000 acre summer home in southern New York State at what became known as Tuxedo Park—and not his tobacco fortune, inherited through his paternal forebearers since the early 1800s—tells a tale of a philanthropist with "old money" (Pierre Lorillard Wiki 2022) and not one of the newly rich "Robber Barons." Unlike several of the wealthy elites who were members of the organizing committee of the Garden, Lorillard was not directly involved in the establishment of the institution (Pierre Lorillard Wiki 2022).

The Lorillard family tobacco empire can be traced to at least the mid-1700s and Pierre Abraham Lorillard (1742–1776), a French-Huguenot immigrant who opened the first family shop on Park Row in 1760. The first Lorillard tobacco mill opened in the Bronx in 1808, and its bricked replacement still exists on the NYBG grounds—"The Snuff Mill" (currently under the name, The Lillian and Amy Goldman Stone Mill). The family exhibited their products at the Columbian Exposition in Chicago in 1893 (Sinon 2022). Its major production facilities were erected around 1900, in

Fig. 9 Seth Low. Seth Low, Mayor of Brooklyn, later of New York, Trustee of Columbia University, later President, member of Board of Managers of the New York Botanical Garden. (Wikipedia)

Jersey City, NJ, close to railheads for ease of shipment of incoming and outgoing materials.

Shortly after the Lorillard Bronx parcel of land was sold and then dedicated to the new New York Botanical Garden, the P. Lorillard Company merged with the American Tobacco Company, but the following year, the U.S. Supreme Court dismantled the merger as anti-competitive, and Lorillard was again independent (NJCU 2022).

Seth Low (1850–1916) (Fig. 9)

If the attitude of the academic community of 1890 resembled that of 2023, faculty would have been suspicious when the former Mayor of Brooklyn was elected President of the assembly of colleges under the Columbia name. After all, his only prior acquaintance with such an institution was his own student time at Columbia College, ending twenty years prior. To be sure, he was a Trustee, but that must be merely ceremonial. To all this he would have assented, but his practical experience in hurly-burly Brooklyn politics could serve him well—in spite of the difference in the constituency—but Manhattan politics required a delicate hand at the helm.

The Low family had come to America three generations before young Seth Low, but it was his namesake, Grandfather Seth (1782–1853), who, with his son Abiel, moved to the town of Brooklyn and started a company importing tea and silk from China. The A. A. Low & Brothers import business grew and prospered, so the next generation of Lows not only bought and sold goods, but owned their own clipper ships as well. Even the Opium Wars did not unduly hinder the business. Seth (II) attended the reputable boys' boarding school, Brooklyn Collegiate and Polytechnic Institute, known as Poly Prep, followed by Columbia College in Manhattan. The trip from Brooklyn to downtown Manhattan must have exposed teenage Seth to the ghetto conditions of the Lower East Side, a far cry from his family's comfortability. He graduated from Columbia in 1870 at age 20. A short trip abroad was followed by absorption into the family business (Seth Low Wiki 2023). None of his growing years seemed destined to mold him into the reformer he became.

Brooklyn grew around him, from the large town of 1830, to small city by 1850, to city of a half-million in 1870. Seth married Anne, the daughter of U.S. Supreme Court Justice Benjamin L. Curtis, a worthy match for the son of the wealthy Low family, but again, not indicative of a political life. But counter to his background, Seth was involved with the "Progressive Movement," a loose organization urging a larger role by government in the lives of the destitute. His participation brought him into a circle of local movers and shakers, and his name became generally known in local reform circles.

In the 1880s, though, governance of Brooklyn was controlled by Hugh ("Boss") McLaughlin, a Democrat whose machine was corrupt, but which "bought" allegiance by doling out jobs, food and living supplies to the poor. In the upcoming race for Mayor, the Republicans struggled for a candidate to face McLaughlin, and with only two weeks left for campaigning, they settled on Seth Low, a distinctly compromise candidate (Seth Low Wiki 2023). To everyone's surprise, Low won: out of some 80,000 votes cast, his margin of victory was 5,000. Suddenly, in 1881, at 31, the sideline reformer was Mayor of Brooklyn, albeit with only a two-year term.

Buoyed by the Progressive leaning of the election, Low set out to set new rules for Brooklyn (Swett 1960). The result was a "mixed bag." In the "pro" column: Low was able to wrest control of city (not yet borough) governance from state control to "home rule" (Fredman 1972); textbooks in public schools became universal, not the dole of machine regulars; teachers were subjected to an examination before being hired; new schools were established to serve an additional 10,000 students (but see below); Civil Service was introduced for city employees, based on merit; beer gardens were allowed to serve on Sundays if they remained "orderly." Doubtful: public schools were racially and religiously integrated. In the "con" column: property taxes were raised from \$2.33/\$100 to \$2.59/\$100 over Seth's two years, but the added revenue was used (in part) to build new schools (Seth Low Wiki 2023).

Low inherited a pre-existent skirmish. The Brooklyn Bridge opened in 1883, after a protracted 13 years of design and construction, all of it under the Manhattan political machine of Boss William Tweed and his sycophants in Tammany Hall. An emolument as Brooklyn mayor, Low served on the New York Bridge Company, and led an unsuccessful campaign to oust the Roebling family from the project. The brush with Tweed power was to be resumed some years later.

A second term followed, won by an even slimmer margin than the first. Low cast his lot with the "mugwump" faction of the Republican Party (Mugwumps Wiki 2023) to support Democrat Grover Cleveland, breaking with the rank and file Republicans. As his term passed, he refused to run again, the Democratic candidate won, and with him, machine politics and McLaughlin returned. In 1885, Low was the "golden boy" of reform, progressive politics, but politically unemployed (Seth Low Wiki 2023).

In 1888, the Low import-export company was dissolved. Among other factors, development of larger, ocean-going steam-driven vessels had made clipper ships obsolete. Seth's portion of the dissolution made him quite rich when added to the inheritance from his father and his income over the years.

Surely based on his prospective philanthropy rather than his recent political period, Low was elected a Trustee of Columbia College, and upon the death of Frederick Augustus Porter Barnard (1809–1889; Columbia University Archives, 2023), Low's fellow trustees tapped him for President. Although two terms as Brooklyn Mayor were not exactly a straight line to the job, his skills as a reformer bode well for an accumulation of colleges overdue for leadership. Low's inauguration took place early in 1891.

According to Sloan (1980) and Keating (1973), as early as 1885, Low had already made pronouncements about the role and scope of a university. He had declared at an alumni dinner that "It would be well if the Alumni understood that Columbia is a University and is on the way to be the first University in the United States."

Not to detract from the Seth Low/Columbia's narrative, 1889 saw the gathering of some women in a rented brownstone on Madison Avenue, not far from Columbia's scattered colleges, which stalwartly had refused to consider women as students in spite of the remonstrations of President Barnard. Upon his death, the women formalized their program as Barnard College. One of Barnard's first faculty members was Effie Almira Southworth, instructor in botany (Petersen and Knudsen 2022). It would take a decade of negotiations before Columbia and Barnard colleges joined programs so that men and women could take classes together. When Columbia moved uptown, Barnard College did so also, settling across the street on Broadway (Barnard College wiki 2023).

The Columbia of 1890 was perceived as "a rich old institution, much behind the times, which neither deserved nor desired any assistance from the city or the city's residents"—despite the fact that President Barnard had initiated organizational and curricular changes to make Columbia more inviting and inclusive. It was the epitome of "town versus gown." By the early 1880s, he had even convinced the trustees and faculty to add electives to what had been a fully prescribed curriculum (Topper 2011). Barnard was the quintessential American educator of the nineteenth century, even in the face of changes being imported from European institutions altering the lecture-memorization-recitation methodology.

Even before Low's inauguration in December, 1890, he and the trustees laid the foundation for a generous sabbatical and pension plan that probably softened faculty attitude toward the new President, and sweetened the deal for many prospective faculty. Predictably, the sabbatical and pension plans were duly noted in the newspapers: Columbia began to get publicity, albeit on items of doubtful enthusiasm by "the common man."

As part of his inaugural address, Low spoke prophetic words about the future of higher education. "There is but one New York on all this continent, and, for the purposes of technical and professional training, her location in New York supplements the work of Columbia with advantages not elsewhere to be had." Low continued: "The real world is not to be found in books," but in actual contact and experience with "men and women of living flesh and blood; and the great city [of New York] can supply the human quality which the broad-minded man must not suffer himself to lack" (Topper 2011). This from a two-term mayor of Brooklyn.

In 1895 President Low articulated even deeper insight: "If a university strives to absorb that which is characteristic of the city in which it does its work, I am persuaded that every city university will have a flavor of its own that will draw to it persistently the men who want that thing" (Sloan 1980). Low himself worked to place his university in the center of a larger configuration of urban educational institutions that represented the opportunities available and the limitations inherent in New York City as a cultural entity.

One of Low's first acts upon taking office was to call a three-day consultation of the faculty on the subject of university reorganization. Over time, the faculties had been organized along department lines; the graduate Faculty of Political Science had been strengthened; a new graduate Faculty of Philosophy had been formed; a graduate Faculty of Pure Science had been established. The College of Physicians and Surgeons was being fully integrated into Columbia, the School of Law had been reorganized, and an arrangement of collaboration had been completed with the New York Teachers College. Steps were also under way to establish reciprocal relations with Union, Jewish, and General Theological seminaries. Columbia's seeming dominance was not the same as exclusivity. Low's proposed reorganization was "*e pluribus unum*" on an academic scale. Low even explored an attempt to make Columbia the only major university in the city by proposing that New York University consolidate with Columbia, actually a scheme for Columbia to absorb NYU. NYU President Henry McCracken and his trustees declined the offer, a situation bound to change within a few years (Sloan 1980).

Shortly thereafter, Low wasted little time in fulfilling his inaugural promise to develop Columbia's professional education in medicine. Columbia's medical school was the College of Physicians and Surgeons (P&S) which, at the time, was a semi-independent wing of the college. In February 1891, Low and the trustees fully incorporated the College of P&S and thus gave Columbia another mark of university status. Thanks to the timely generosity of the Vanderbilts and Sloans during the late-1880s, the College of P&S had recently been expanded and remodeled: it was a shining example of modern medical technology when Low and the Columbia trustees brought it into the fold (Topper 2011).

As the absorption of P&S was materializing, the Trustees and Low were also plotting to do the same with the College of Law. Its curriculum was classical—books and memory—headed by Theodore W. Dwight (1822–1892) for 30 years, and considered countrywide as the paragon of legal education, except, of course, by the Harvard Law College and its legal orbit. Harvard had mutated its curriculum to be case-based (Topper 2011; Turow 1977). Whereas Dwight's methodology taught Columbia students from theory toward practice, at Harvard the law was taught from practice—from the study of particular cases—toward theory.

Such was the situation when President Low informed Professor Dwight that, first, Dwight would henceforth report to him and not to the Trustees, and second, that Harvard's case method would become the new model of instruction at Columbia.

This seemingly dictatorial act was exacerbated by the recruitment of a law professor from Harvard to the Columbia faculty the previous year. In any case, Dwight submitted his resignation shortly thereafter, setting off a media frenzy: what was reported as a likely renovation at Columbia College, threatened to destroy the Law School of Columbia. This supposed project came to be known at Columbia as the "Harvardizing" of the Law School (Topper 2011), spurring some departures of students but also applications of a like number.

Once inaugurated, Low set out to marry donations from the super-wealthy to publicity for the university (Topper 2011). A new publication, the COLUMBIA COLLEGE BULLETIN, published details of sizeable philanthropic gifts along with news of the various colleges, and featured "an enumeration of the special advantages offered [to contributors] by Columbia College." In 1892, Low proposed a tremendous challenge to the trustees and other wealthy New Yorkers and ignited the largest outpouring of philanthropy in Columbia's history. As a Columbia alumnus, and having been a trustee for nearly ten years before becoming President, Low also recognized that Columbia's scattered sites around "49th and Madison" could no longer accommodate growth.

Low suggested that it was time to move the entire collection of colleges to a new, more spacious location in the city near the recently closed New York Hospital's Bloomingdale Asylum for the Mentally Ill (1821–1889; Bloomingdale Insane Asylum Wiki 2023), at 116th Street and Broadway—the site that would be called "Morningside Heights." Low's plan struck a chord with his wealthy friends and associates. In just one year, Low raised more money for Columbia than the total sum of its gifts since its founding. Cornelius Vanderbilt (II), William Colford Schermerhorn (Chairman of the Board of Trustees; William Colford Schermerhorn Wiki 2023), and J. Pierpont Morgan (who would himself become a Columbia trustee in 1903) contributed $100,000 each. Low's father, A. A. Low, contributed $15,000 to the cause, and [Seth] Low himself pitched in $5,000 (Topper 2011).

In May, 1895, Low made the front page of the New York Times with his gift of $1 million for a new university library in memory of his father (Fig. 10). Once again, the philanthropic dominos cascaded. Schermerhorn accepted Low's gesture as a kind of friendly challenge and gave $300,000 to Columbia for a new Natural Sciences building. Low's million-dollar gift brought Columbia nationwide acclaim in daily newspapers and in mass-circulation magazines. The library was only one part—albeit the centerpiece—of Columbia's new campus at Morningside Heights.

For Seth Low and the evolving university, the establishment and growth of its natural science faculty was less than earth-shaking, given the size and intricacy of the whole. The Columbia Herbarium, for instance, had a long history, and had been moved about from one old building to another, until Nathanial Lord Britton considered it well-located in his report of 1887 (Britton 1887), only to see it moved to Morningside Heights a few years later, and eventually become part of The New York Botanical Garden.

Another program evolved in the middle of the decade. Low announced establishment of the Low Professorship in History. In fact, the endowed chair lured William M. Sloane away from Princeton. Coupled with the faculty position were twelve

Fig. 10 Low Library. Seth Low Library of Columbia University, built with a large donation from Seth Low in late 1890s in memory of his grandfather. Note woman in typical turn-of-the-twentieth-century dress. Columbia University Archive

merit scholarships for applicants from "any school in Brooklyn, either public or private," an equal number for Brooklyn boys to Columbia and Brooklyn girls to Barnard (Topper 2011).

In the spirit of the era, 1895 to 1898 marked Teddy Roosevelt's time as New York Police Commissioner, and his relentless pursuit of Boss Tweed and Tammany Hall. Teddy and Seth were good friends. Another example of philanthropy; Carnegie Hall opened in 1891 for a season of concerts.

It was in 1895 that Low chipped in $25,000 toward the founding of The New York Botanical Garden. For him, the donation was pocket change, but just as with Columbia, and the Museum of Natural History, it joined similar tokens from other super-wealthy friends and acquaintances (Britton 1915).

Following his strong concern for the well-being of those at the other end of the social ladder, Low also served as President of the University Settlement Society in Manhattan's Lower East Side. At the Settlement, college students—primarily from Columbia—provided social services to the poor; it was an early version of service-learning. Low firmly believed that Columbia students "should have the opportunity to study the social conditions as they existed in fact and outside of mere theoretical speculation." The Settlement also served as an experiment station for Columbia economists and sociologists, whose statistical analyses of city poverty helped give to Columbia a worldwide reputation in the collection and use of social-sciences statistics (Topper 2011). Low took to heart the work of his friend, Jacob Riis (1890), whose famous publication, How the Other Half Lives, brought to light the miserable condition of the city's poor and called the citizens of New York to lend a helping hand. It was in these ties that the "Ashcan artists," mostly illustrators for Philadelphia newspapers, moved one by one to New York and there, painted scenes of everyday life on the streets.

Persistent "gown versus town" made it difficult to persuade middle class New Yorkers to suppress their anti-intellectual prejudices and embrace the idea of college life. The prescribed classical-curriculum, with its emphasis on lecture, recitation, and theory, had little appeal for pragmatic Americans of the 1890s. In fact, its rigidity even drew criticism from prominent "captains of industry" like Andrew Carnegie and Leland Stanford. Carnegie declared that the duty of university leaders "was to make our young men fit to be American citizens and not to waste their time trying to make poor imitations of Greeks and Romans." Stanford frequently complained about the lack of "definite technical knowledge" that college graduates brought with them to their job interviews at his railroad company. His wish in 1891, when he established the future Stanford University, was that it would offer "a practical, not a theoretical education" (Topper 2011).

In 1897, long after his Brooklyn terms ended, Low was characterized as the "first Czar mayor of America," this based on Low's concept of centering power in the office which he occupied. The flaws, of course, were that the power could be used corruptly or unwisely and could easily create fiefdoms. In spite of such leanings, though, Low was put up for mayor by the Republicans in 1897. He finished second. There was reason to think his heart was not in running, but when the election of 1900 peeked over the horizon, they again sought Low, this time as a fusion candidate. He won, but had to give up his position at Columbia in order to serve.

The greater city of New York, however, had been in existence from only 1898, and Seth Low had been part of the commission which drafted the document of borough merger. Formation of the "greater city" was the amalgamation of its five constituent populations. When Low won, he became the Mayor of New York City, which was composed of five boroughs: Manhattan, Brooklyn, Queens, Bronx and Staten Island. His term was only two years (Fredman 1972): he ran again in 1903, and lost.

From his position of Columbia President and city Mayor, Low had been able to favor programs of his choice, but once his term as Mayor ended, he effectively passed from any significant role in growth of Columbia and the founding of The New York Botanical Garden. His public service roles, however, persevered: he lived on to play a part in the post-War (WW I) peace missions, and other diplomatic roles. His philanthropy was strong and varied, extending even to his last will and testament.

Although not within the span of this narrative, Columbia suffered from another social weakness: deterrence of admission to Jewish young men. I can find no mention of this situation under Seth Low—it may have been part of the City's folklore and so not the subject of news reports—but by the 1920s it was generally known and understood (The Eye 2023; The Current (magazine) Wiki 2023). An alternate institution, Seth Low Junior College, was opened in Brooklyn and Jewish males were encouraged to enroll there, not in Manhattan, especially if a program in pre-med or pre-law were intended.

Fig. 11 John Strong Newberry. John Strong Newberry, founding faculty member of Columbia School of Mines and mentor to N. L. Britton and A. Hollick of the New York Botanical Garden. (Wikipedia)

John Strong Newberry (1822–1892) (Fig. 11)

Notable for his role in educating Nathaniel Lord Britton at the Columbia School of Mines, John Strong Newberry was not a mainstream botanist, but a geologist and especially, a paleontologist. It is said that in his early years, he was fascinated by the fossils that turned up in the coal which passed through his father's business in Ohio (John Strong Newberry Wiki 2023). A visit with James Hall, the New York State geologist and stratigrapher also bolstered Newberry's interests. In those early years, Newberry collected plant specimens and gathered them into a herbarium, and wrote a catalog of the plants in the vicinity of Cuyahoga Falls (JStor Global Plants 2023). As was usual for a number of the players in this narrative, natural history, including geology, was taught in the electives of the medical curriculum. Newberry received his M.D. from Cleveland Medical College in 1848. Again, as expected of educated American men, Newberry spent two years in Europe with his wife, bolstering his medical and paleontological chops. Like others, he forsook medicine in 1855 (age 33) and began his participation as geologist (and sometimes as botanist) to several expeditions to the Western regions. The first was to Oregon and California to scout possible routes for the Pacific Railroad. Two years later he joined the Ives Expedition to the Colorado River, then to the San Juan River. His experiences in the field and his ability to arrange the large cache of specimens won him a high reputation (Fairchild 1893).

The Civil War interrupted his exploratory adventures, but he joined the Union Sanitary Commission, and was billeted to Kentucky where he excelled as an administrator of hospitals. At War's end, he became associated with the Smithsonian Institution and was honored by selection as one of the original fifty members of the new National Academy of Sciences (Sloan 1980).

In 1866 (age 44), Newberry assumed the Chair of Geology and Paleontology at the still young Columbia School of Mines. Over time, other positions of responsibility and honor were regularly added. In 1868 he became President of the Lyceum of Natural History, just two years after a disastrous fire had destroyed the Lyceum's meeting quarters and all its belongings except its library. As Pumpelly (1893) wrote: "In 1867 he was President of [AAAS]. From 1868 until his death he was President of the Lyceum of Natural History. He was one of the judges at the [Philadelphia] Centennial Exhibition in 1876. From 1880 to 1890 he was President of the Torrey Botanical Club. He was appointed to the Ohio State Geological Survey in 1869, a job which commanded only part time. In 1884 he was appointed one of the paleontologists of the United States Geological Survey. In 1887 he was elected an Associate Fellow of [the] Academy [of Arts and Sciences]. In 1888 the Murchison Medal was awarded to him by the Geological Society of London, and the following year he was elected First Vice-President of the Geological Society of America." White (1902), in an obituary, concurred.

After the fire of 1866, the Lyceum floundered without a permanent home. For several years it met in Mott Memorial Hall, then changed its meeting place to the New York Academy of Medicine, and finally in 1883 moved to Columbia's old Hamilton Hall, where Newberry also had his office. The library was shifted from place to place before finally being transferred to Columbia, where the college took care of the books in exchange for their being made accessible to Columbia students (Sloan 1980). This arrangement set the stage for future agreements, including that with The New York Botanical Garden.

The New York Academy's peripatetic existence fed the desire among its scientists for a permanent and prestigious scientific center. At the same time, the lack of a building of its own brought it into an ever-closer relationship with Columbia College. Both the desire for a visible center for the city's scientific activities and growing interinstitutional connections would become increasingly significant (Sloan 1980). Membership in the Academy, however, grew steadily over those years of Newberry's guidance (Fairchild 1893).

In the midst of Newberry's mammoth contributions to New York City's science in the nineteenth century, perhaps his most lasting legacy rested in his students, a number of whom went on to leadership roles in New York natural history.

Louis Pope Gratacap (1851–1917) was trained in mineralogy and conchology and after graduation in 1879, joined the staff at the American Museum of Natural History. Charles Arthur Hollick (1857–1933), Nathaniel Britton's neighbor and friend, pursued graduate studies in botany under Newberry, serving as his personal assistant from 1881 to 1891, and then took over the teaching of paleobotany at Columbia after Newberry's death. Hollick became a member of the scientific staff of The New York Botanical Garden. Bashford Dean (1867–1928) became an instructor at Mines after his studies under Newberry and later worked in ichthyology as curator at the American Museum of Natural History. Dean was also active in the New York Microscopical Society and the Academy of Sciences. Nathaniel Lord Britton studied botany under Newberry, served as his assistant from 1879 to 1887, and in the 1890s, became the first head of Columbia's Botany Department, secretary of the Faculty of Pure Science, a key figure in the Academy of Sciences, and the Torrey Botanical Club. In 1896 he became the first director-in-chief of The New York Botanical Garden (Sloan 1980).

Nathaniel Lord Britton spoke of Newberry in glowing terms. After all, Newberry had seen the world and had an immense knowledge of natural history. When four pedestals were erected in front of the museum building at the Garden, Britton saw to it that one of the busts was that of Newberry, even though Newberry did not live to see The New York Botanical Garden or the twentieth century.

Lucien Marcus Underwood (1853–1907) (Fig. 12)

In upstate New York, Woodstock would become famous for a music festival, not the birthplace of Lucien Marcus Underwood, but his childhood was typical of several other biologists. His upbringing was on the family farm, where the children worked as soon as they could retrieve eggs from the roost. Surely, his flair for making collections—flowers, grasses, newspaper clippings—and once in school, rocks, insects and other objects, pointed to a personality type. He was not an exceptional student until the textbooks were complex enough to challenge his intellect. The needs of farm chores did not leave time for extensive school attendance, but his parents must have recognized the promise of their young son, and at eleven, he enrolled at nearby Cazenovia Seminary, a school of some note (Petersen 2021). Lucien was not a resident student, but walked the three miles to and from the Seminary daily (Curtis 1908).

This dual life continued for several years, but during the school year 1870–1871, two divergent tides converged. His teacher at Cazenovia suggested that Lucien was college material, perhaps helped by his winning both the mathematics and scholarship prizes, but simultaneously, circumstances on the farm needed his complete attention. For two years the farm was his occupation, but in 1873, at age 20, he re-entered Cazenovia for the spring semester. Again he excelled academically, and fastened his attention on attending college (Curtis 1908). The new college in nearby Syracuse became his target, and he entered in the fall of 1873, in the science course.

Here, though, came another of those odd occurrences which separated Underwood from others of his (or our) age. He soon lost respect for the science course for it did not demand enough scholarship to satisfy him. So much so that he began investigating alternative curricula, finally deciding to add the Latin course to his science course, even though it meant a great deal of preparation and sophistication in Latin language. And finally, he also began gathering a herbarium, especially of local fern species. He graduated with a B.A. in 1877.

Saving money from summer jobs, Lucien was able to visit New York City on his way to Philadelphia for the Centennial Exposition in 1876. As expressed by one biographer: "… thus giving him his first view of the outside world." (Curtis 1908: 4).

Like many other budding scientists (Petersen 2019, 2021), Underwood decided to begin a career in teaching. His first teaching job was as principal of the rural Morrisville Union School, but the students and conditions were so appalling that he lasted only one year. In that year, though, he also studied advanced textbooks, and obtained a master's degree from Syracuse University in spring, 1878. He returned to the scene of his earlier education, now to teach at the Cazenovia Seminary for the

Fig. 12 Lucien Marcus Underwood. Lucien Marcus Underwood, academic botanist and nomenclaturalist at the turn of the twentieth century. (Google Images)

academic year 1878–1879 and found time to publish his first botanical paper, this on the ferns of Herkimer County (Underwood 1878). (For a complete index of Underwood's publications, see Barnhart 1908.)

Underwood's expertise in geology resulted in a thesis, "The geological formations crossed by the Syracuse and Chenango Valley railroad," and a Ph.D. from Syracuse in 1879. The degree was the gateway to teach at a higher level than Cazenovia, but as it was for many others, the next years were peripatetic.

His first appointment was as Professor of Natural Sciences at Hedding College, in Abington, Illinois. His teaching load was suffocating, but he managed to start a manual of the ferns of North America. In 1880, he began an appointment in geology and botany at Illinois Wesleyan University, in Bloomington, where he stayed for three years. His manual of ferns appeared (Underwood 1881), but the printer stock for a second edition was lost in a warehouse fire. During this period he managed a courtship with Marie Antoinette Spurr, and they were married in 1881 (Lucien Marcus Underwood Wiki 2022). In time, the fern manual was revised through a sixth edition in 1900 (Curtis 1908).

During his time in Bloomington, Underwood became seriously interested in the Hepaticae, the "liverworts." As he had done for the ferns, he gathered material for a manual which was published soon (Underwood 1884). He developed a close relationship with the nearby Illinois State Laboratory of Natural History, and over the next couple years, he published entomological papers as well as botanical, demonstrating his wide knowledge of natural history. His infatuation with liverworts, though, dominated his research time, including the gathering of an exsiccati, Hepaticae Americanae, together with Orator F. Cook. A chapter of an anticipated manual of the Hepaticae was published (Underwood 1893), but additional parts did not appear. His fieldwork took him to most parts of Illinois and he established friendships with several other workers in natural history, several in Indiana. Before long, in 1885, the Indiana Academy of Sciences was founded, with Underwood playing an active role. At about this time, the Underwoods welcomed a daughter, Helen, to the family.

In 1883, his old alma mater called him back to Syracuse University as instructor in geology, zoology, and botany, and in three years he became professor. This included, though, added teaching—analytical chemistry, mathematics, and human physiology (Howe 1908). His research output continued, if anything, at a faster pace. Work on hepatics dominated.

Notable was Underwood's pedagogical philosophy. Laboratory work was introduced when possible, with the necessity of lab materials and equipment. Enrollment grew rapidly and eventually, repetitious sessions were necessary; more laboratory participation, and more demonstration material. He was a devotee of the "new botany."

Even "vacation" time was well-spent. In 1884, he visited Asa Gray in Cambridge, and became involved in the annual winter meetings of the AAAS (Anonymous 1884). In 1884, at the Philadelphia meeting, he met several botanists that he had known only through correspondence. At the 1885 meeting in Ann Arbor, Michigan, he roomed in the same house as J. C. Arthur, Charles Barnes and John Merle Coulter, all three involved in editing Botanical Gazette, and soon developing botany at the University of Chicago (Petersen 2023).

The summer of 1887 was spent in Georgia, Tennessee, and Virginia, collecting in the southern Appalachian Mountains for the Smithsonian Institution. To recollect the trove of earlier collections by Moses Ashley Curtis and Asa Gray was surely a goal, but Underwood's first priority was the ferns and liverworts. The summer of 1888, however, found him in southern California (Curtis 1908), probably prompted by work on the cryptogam collections of earlier California workers (Underwood 1888).

During this period, Underwood's attention to ferns and hepatics was eroded by the fungi: A Century of Illustrative Fungi (Underwood and Cook 1889), with the addition of Generic Dynopses of the Basidiomycetes and Myxomycetes. The addendum displayed a familiarity with the higher fungi and slime molds not previously evident, and surely required a period of preparation. His co-author, Orator Fuller Cook (1867–1949), was a student at Syracuse who graduated with a Bachelor's degree in 1890, while Underwood was away. He replaced Underwood as instructor of botany. They co-authored several papers and an exsiccati over the next decade or more. Cook was to publish several papers on botanical nomenclature among many others. Another such student was Carlton Clarence Curtis (1864–1945), who, after a B.A., M.A. and Ph.D. from Syracuse University, took education from Cambridge University and the University of Leipzig. When Underwood arrived at Columbia University in 1896, Curtis was at the Brooklyn Polytechnic Institute, but moved to Columbia as instructor of botany in 1899. He was to become Underwood's major biographer (Curtis 1908).

Granted a leave of absence from Syracuse and a fellowship from Harvard, Underwood set out to study the earlier hepatic collections of Sullivant and Taylor, but before his Harvard time was finished, he received a request from the USDA to report on the "orange disease" in Florida (Underwood 1891). But even this interruption was interrupted, for he travelled to and within Cuba, always collecting and learning. On his return trip to Harvard, he collected in some locales in Georgia.

During his short time remaining at Harvard, Underwood was tendered a call and accepted a position at De Pauw University in Greencastle, Indiana. The school was in the midst of changing its name from Indiana Asbury University, named for Bishop Francis Asbury of the Methodist Church (DePauw University Wiki 2022). To a twenty-first century eye, Underwood's move would seem improbable, but Curtis (1908) wrote: "This was the first time in his career that he had the opportunity to direct his attention to botany alone, and it is noteworthy that he accepted this position at a lower salary than he was receiving and also at the same time declined a more remunerative position in another institution in order to specialize more closely. There now followed a period of work under the most congenial surroundings and during these four years he published numerous papers on the lower groups of plants." [The DePauw University herbarium was purchased by NYBG in the 1980s.]

In 1892, Underwood attended the AAAS meeting in Rochester, New York. The botanists had become increasingly dissatisfied with the European "*Lois*" of botanical nomenclature, and a small group, including Underwood, had formulated what became known as "The Rochester Code" (Petersen chronology 2022). Two important items departed from the *Lois*: (1) Otto Kuntze's recent "Revision Genera Plantarum" (Kuntze 1891) traced Linnaeus's use of binomial nomenclature to 1735 (Zanoni 1980), but the Americans wanted to use Linnaeus's "Species Plantarum" of 1753, and there already was an undertow that nomenclature of cryptogams might require even later dates; and (2) the Americans based their nomenclature on the "Type Method," in which all ranks of plants were based on "types," ranging from type species of genera and type specimens of species. A third "Canon" stated that the earliest published name took "priority" over all subsequent names. This rule of priority ran counter to the European view (Underwood 1892a).

Not only was Underwood one of the committee which drew up the Rochester Code, but his participation before and during the meeting resulted in his election to act as the American representative to take the Rochester Code to the upcoming International Botanical Congress in Genoa, Italy, the following year, 1892 (Curtis 1908).

Like his previous journey through New York City to Philadelphia, the voyage to and from Europe and travel across the continent opened new horizons for Underwood, now a mature 42 years old. To the days in Genoa, he added visits to the great herbaria in the capitals and the offices of the great botanists. Curtis (1908) added: "The interest thus aroused led him repeatedly to visit England and the Continent, in all making eight trips for the purpose of comparison and study at various botanical centers." In each venue, there was exchange of knowledge and assessment of personalities (Underwood 1892a, b, c). Underwood could add an international exposure to his national reputation (Howe 1908).

Parenthetically, in his papers resulting from his travels, no mention of wife, Marie, or daughter, Helen, was made, nor was there any note of an Underwood family by his major biographer, C. C. Curtis (1908).

His continental exposure also renewed his interest of a summary of the cryptogams of North America (Howe 1908), and he took pen in hand to write Dr. Nathaniel Lord Britton at Columbia University, who had been advocating just such a national summation of higher plants. Before long, a standing board of editors was created for the "Systematic Botany of North America." Soon, the Board passed responsibility back to Britton and Underwood, to be published by the not-yet operational New York Botanical Garden by the pair. The project changed its name to NORTH AMERICAN FLORA, and added the fungi to its menu.

At the 1893 AAAS meeting in Madison, Wisconsin, Underwood was selected as one of ten botanists commissioned to found the Botanical Society of America, and he served as President in 1899–1900. Simultaneously, he was serving as the Vice-President of Section G of the AAAS, Biological Sciences, and offered a paper on hepatics as his closing address (Underwood 1894). Underwood found himself deep in the politics of science.

But on the home front, the educational ambitions of De Pauw were larger than its pocketbook, and the botany department was very soon scheduled to close, surely to the dismay of Underwood. He quickly accepted a position at the Agricultural and Mechanical College of Alabama in far-off Auburn in the far-off state of Alabama in the heart of the impecunious Confederacy, to start in January, 1896 (Auburn University Wiki 2022; Underwood and Earle 1897). One wonders that once more, Marie and Helen had to pack the household belongings, and again try to accommodate Lucien's single-minded pursuit of research and its necessary teaching.

The small Alabama college had been swept up—as East Alabama Male College—under the land-grant program in 1872, and became co-educational just four years before Underwood's arrival. The College's catalogues from 1896 through 1901 advertised a curriculum in Biology and Horticulture, with Professor F. S. Earle as faculty. Without data on Earle's date of arrival, he was resident when Underwood joined him. They must have gotten on well, for they collaborated on some significant publications, interestingly, on fungi. Although Earle had published previously in mycology (q.v.), fungi were only a minor interest for Underwood. They did fieldwork together, collecting in several parts of the state. Three papers resulted (Underwood and Earle 1896a, b, 1897), all chiefly phytopathological. In 1897, a comprehensive text on previous Alabama mycology was furnished (Underwood and Earle 1897), obviously mining a very large database of mycological literature. As sole author, some smaller papers were published (Underwood 1896a, b, c, 1897a, b, 1898a, b, c, 1899a).

In those years, The New York Botanical Garden was becoming an official reality, and Nathaniel Lord Britton juggled his position as Botany Department Head at Columbia University to become Director at the Garden (but retained a title at Columbia in order to promote collaboration between the two institutions). Surely with his urging, a letter of inquiry was written to Underwood in Alabama. Underwood accepted, and—with family—pulled up stakes for the migration to the metropolis (Curtis 1908). In 18 years, this was his ninth faculty job. "Up to this period his life had indeed been a varied one, but it is not to be inferred that this was due to any uncertainty of purpose or lack of perception. Three times he had accepted less remunerative positions in order to confine his work more closely to the lower forms of plant life and now for the first time he had the opportunity of realizing the ambition of his life." (Curtis 1908). When he joined Columbia in 1896, it was located at 49th street in Manhattan, but within a few months, it was to move north to Morningside Heights, perhaps involving yet another migration for the Underwoods (Lucien Marcus Underwood Wiki 2022). On the new campus, the brand new Schermerhorn Hall greeted faculty of Natural Sciences, and the Seth Low library was rising nearby.

The data for Underwood's (1899b) magnum opus, MOULDS, MILDEWS AND MUSHROOMS must have been gathered over a long time period, for the details of the text were very complete, keys were furnished for every fungus group, occasional figures were furnished, the bibliography was luxuriant, and the print rather small (Petersen 2020a, b). Its size (19 × 12.5 cm) conformed to Underwood's reference to a "pocket guide" (Howe 1908). Murrill's arrival in Columbia's neighborhood was imminent, and Murrill also wrote of his membership in the Torrey Botanical Club, so the two men had every opportunity to become acquainted personally, having exchanged previous correspondence on Virginia ferns.

In his new appointment, responsibility kept pace with rewards for Underwood. As it had for the Brittons, the Torrey Botanical Club provided a group into which Underwood fit comfortably. Soon, in 1898 until 1902, he became Editor for the Club's publications, by then including the BULLETIN, MEMOIRS and TORREYA, and the significant Index to American Botanical Literature (Barnhart 1908).

Under his Associate Editorship, NORTH AMERICAN FLORA began accepting monographic manuscripts. He became one of the Board of Scientific Directors of the Garden, and in 1901, was elected Chairman. In his first report for the Scientific Directors (Underwood 1902) he reported the appointment of two new staff members: Prof. Franklin Sumner Earle in Plant Pathology, and Dr. Charles Arthur Hollick in paleobotany (Howe 1933). Just two years later, he would write: "[The Board] regret to report the loss of Professor F. S. Earle from the staff of the Garden, but can announce that very satisfactory reciprocal relations have been entered into with him in his present position as Director of the Agronomic Station in Cuba which will bear directly on the work of tropical exploration in progress by the Garden. Dr. William A. Murrill has been appointed Professor as Earle's successor and has entered upon his work with enthusiasm, and his publications of original work are already attracting attention."

In the midst of the crisis of 1907 (Crisis of 1907, 2023), Lucien apparently lost most or all of his savings, suffered a fit of insanity, resulting in the injury of his wife and his suicide at their Connecticut farm.

Appendix 2: Murrill Mimoegraphed "Bulletins"

Murrill, W. A. 1937. Flora of Alachua County, Florida (preliminary checklist). Probably mimeographed. 36 leaves (rest of citation missing).
Murrill, W. A. 1938. Bryophytes of Alachua Co. Mimeogr. Contrib. Herb. Univ. Fla. Agr. Exp. Stat., 9 leaves.
Murrill, W. A. 1938. Mushrooms of Alachua County (a preliminary checklist). Mimeo. Contrib. Univ. Fla. Agr. Exp. Stat., 25 leaves.
Murrill, W. A. 1938. Fungi of Alachua County [cover title]. Basidiomycetes of Alachua County (Polypores, puffballs, etc.) [first page title] Mimeo. Contrib. Univ. Fla. Agr. Exp. Stat., 16 leaves.
Murrill, W. A. 1939. The Polygonaceae of Alachua Co., Florida. Mimeo. Contrib. Herb. Univ. Fla. Agr. Exp. Stat., 3 leaves.
Murrill, W. A. 1939. The Lamiaceae. Mimeo. Contrib. Herb. Univ. Fla. Agr. Exp. Stat., 7 leaves.
Murrill, W. A. 1939. Trees of Alachua Co. Mimeo. Contrib. Herb. Univ. Fla. Agr. Exp. Stat., 10 leaves.
Murrill, W. A. 1939. Roses and their relatives in Alachua Co., Florida. Mimeo. Contrib. Herb. Univ. Fla. Agr. Exp. Stat., 7 leaves.
Murrill, W. A. 1939. Alachua goldenrods. Mimeo. Contrib. Herb. Univ. Fla. Agr. Exp. Stat., 1 leaf.
Murrill, W. A. 1939. Alachua orchids. Mimeo. Contrib. Herb. Univ. Fla. Agr. Exp. Stat., 9 leaves.
Murrill, W. A. 1939. The Amanitaceae of Alachua Co., Florida. Mimeo. Contrib. Herb. Univ. Fla. Agr. Exp. Stat., 4 leaves.
Murrill, W. A. 1939. The Ericales of Alachua, Florida. Mimeo. Contrib. Herb. Univ. Fla. Agr. Exp. Stat., 6 leaves.
Murrill, W. A. 1939. The Euphorbiaceae of Alachua Co., Florida. Mimeo. Contrib. Herb. Univ. Fla. Agr. Exp. Stat., 3 leaves.
Murrill, W. A. 1939. Our blazing stars (Laciniacea). Mimeo. Contrib. Herb. Univ. Fla. Agr. Exp. Stat., 1 leaf.
Murrill, W. A. 1939. Vines of Alachua County: list of Alachua vines, tropical American geology, origin of Alachua vines, common names of vines. Mimeo. Contrib. Herb. Univ. Fla. Agr. Exp. Stat., 7 leaves.
Murrill, W. A. 1939. Additional Alachua Basidiomycetes. Mimeo. Contrib. Univ. Fla. Agric. Exp. Stat., 7 leaves.
Murrill, W. A. 1940. Bartram's *Ixia*. Mimeo. Contrib. Herb. Univ. Fla. Agr. Exp. Stat., 2 leaves.
Murrill, W. A. 1940. Insectivorous plants of Alachua Co., Florida. Mimeo. Contrib. Herb. Univ. Fla. Agr. Exp. Stat., 3 leaves.
Murrill, W. A. 1940. Alachua hollies. Mimeo. Contrib. Herb. Univ. Fla. Agr. Exp. Stat., 3 leaves.
Murrill, W. A. 1940. Florida hawthorns (*Craetaegus*). Mimeo. Contrib. Herb. Univ. Fla. Agr. Exp. Stat., 16 leaves.

Murrill, W. A. 1940. Spring wild flowers (woody plants) of Alachua County, Florida. Mimeo. Contrib. Univ. Fla. Agr. Exp. Stat., 6 leaves.
Murrill, W. A. 1942. Florida boletes (Boletaceae). Mimeo. Contrib. Univ. Fla. Agr. Exp. Stat., 6 leaves.
Murrill, W. A. 1942. Florida Collybias (*Collybia*, *Gymnopus*). Mimeo. Contrib. Herb. Univ. Fla. Agr. Exp. Stat., 6 leaves.
Murrill, W. A. 1942. Florida *Marasmius* (*Marasmius*, *Marasmiellus*). Mimeo. Contrib. Herb. Univ. Fla. Agr. Exp. Stat., 6 leaves.
Murrill, W. A. 1942. Florida Russulas (*Russula*). Mimeo. Contrib. Herb. Univ. Fla. Agr. Exp. Stat., 15 leaves.
Murrill, W. A. 1942. Florida dark-spored agarics. Mimeo. Contrib. Herb. Univ. Fla. Agr. Exp. Stat., 12 leaves.
Murrill, W. A. 1942. Report of progress. Mimeo. Contrib. Herb. Univ. Fla. Agr. Exp. Stat., 3 leaves.
Murrill, W. A. 1942. Florida polypores (Polyporaceae). Mimeo. Contrib. Herb. Univ. Fla. Agr. Exp. Stat., 45 leaves.
Murrill, W. A. 1942. Florida Amanitas (*Amanita*, *Venenarius*). Mimeo. Contrib. Herb. Univ. Fla. Agr. Exp. Stat., 7 leaves.
Murrill, W. A. 1942. Florida Lactarias (*Lactarius* Pers.) Mimeo. Contrib. Herb. Univ. Fla. Agr. Exp. Stat., 9 leaves.
Murrill, W. A. 1942. Florida Lepiotas. Mimeo. Contrib. Herb. Univ. Fla. Agr. Exp. Stat., 7 leaves.
Murrill, W. A. 1942. Florida Tricholomas. Mimeo. Contrib. Herb. Univ. Fla. Agr. Exp. Stat., 6 leaves.
Murrill, W. A. 1942 (Dec.). Lower Hymenomycetes of Florida. Mimeo. Contrib. Herb. Univ. Fla. Agr. Exp. Stat., 6 leaves.
Murrill, W. A. 1943 (Dec.). Florida Thelephoraceae. Mimeo. Contrib. Herb. Univ. Fla. Agr. Exp. Stat., 10 leaves.
Murrill, W. A. 1943 (Dec.). Florida hydnums. Mimeo. Contrib. Herb. Univ. Fla. Agr. Exp. Stat., 6 leaves.
Murrill, W. A. 1943 (Dec.). Florida clavarias. Mimeo. Contrib. Herb. Univ. Fla. Agr. Exp. Stat., 2 leaves.
Murrill, W. A. 1943 (Nov.). Florida white-spored agarics. Mimeo. Contrib. Herb. Univ. Fla. Agr. Exp. Stat., 64 leaves.
Murrill, W. A. 1943. Florida rosy-spored agarics (Pluteanae). Mimeo. Contrib. Herb. Univ. Fla. Agr. Exp. Stat., 12 leaves.
Murrill, W. A. 1943. Florida rusty-spored agarics (Pholiotanae). Mimeo. Contrib. Herb. Univ. Fla. Agr. Exp. Stat., 21 leaves.
Murrill, W. A. 1943. Florida boletes (Boletaceae) Mimeo. Contrib. Herb. Univ. Fla. Agr. Exp. Stat., 18 leaves.
Murrill, W. A. 1943. Florida Gasteromycetes. Mimeo. Contrib. Herb. Univ. Fla. Agr. Exp. Stat., 10 leaves.
Murrill, W. A. 1943. Lower hymenomycetes of Florida (Auriculariaceae, Dacrymycetaceae, Tremellaceae, Hypochnaceae). Mimeo. Contrib. Herb. Univ. Fla. Agr. Exp. Stat., 7 leaves.

Murrill, W. A. 1943. Florida white-spored agarics (Agariaceae pars.) Mimeo. Contrib. Herb. Univ. Fla. Agr. Exp. Stat., 28 leaves.

Murrill, W. A. 1943. Florida Thelephoraceae. Mimeo. Contrib. Herb. Univ. Fla. Agr. Exp. Stat., 15 leaves.

Murrill, W. A. 1943. Florida hydnums (Hydnaceae). Mimeo. Contrib. Herb. Univ. Fla. Agr. Exp. Stat., 8 leaves.

Murrill, W. A. 1943. Florida clavarias (Clavariaceae). Mimeo. Contrib. Herb. Univ. Fla. Agr. Exp. Stat., 3 leaves.

Murrill, W. A. 1946[?]. Introduced woody plants of the Gainesville region (Preliminary list). Mimeo. Contrib. Herb. Univ. Fla. Agr. Exp. Stat., 12 leaves.

Murrill, W. A. s.d. Florida dark-spored agarics. Mimeo. Contrib. Herb. Univ. Fla. Agr. Exp. Stat., 10 leaves.

(at this point, Murrill seems to desert the mimeograph mode and instead publish in the Journal of the Florida Academy of Science and Lloydia)

References

Adamcik S, Buyck B (2010) Re-statement of *Russula levyana* Murrill as a good and distinct American species of *Russula* section *Xerompelina*. Cryptog Mycol 31(2):119–135

Adamcik S, Buyck B (2011) Type studies in American *Russula* (Russulales, Basidiomycotya): species of subsection *Decolorantinae* described by H. C. Beardslee, G. S. Burlingham and W. A. Murrill. Cryptog Mycol 32(3):323–350

Adamcik S, Buyck B (2012) Type studies in American *Russula* (Russulales, Basidiomycota) in and out of subsection *Roseinae*. Nova Hedwigia 94(3–4):413–428

Addison Brown Prabook (2023). https://prabook.com/web/addison.brown/1068179

Addison Brown Wiki (2024). https://en.wikipedia.org/wiki/Addison_Brown

American Heritage Foundation (2024). https://ahf.nuclearmuseum.org/ahf/history/science-behind-atom-bomb/#:~:text=The%20U.S.%20developed%20two%20types,device%20with%20a%20plutonium%20core

American Stock Exchange (2023). https://www.nyse.com/publicdocs/American_Stock_Exchange_Historical_Timeline.pdf

Anchel MA, Hervey MA, Robbins WJ (1950) Antibiotic substances from Basidiomycetes. VII. *Clitocybe illudens*. Proc Natl Acad Sci 36:300–305

Anonymous (1884) The Botanical Club of the A. A. A. S. Bot Gaz 9(10/11):156–160

Anonymous (1885) The Botanical Club of the A.A.A. S. Bot Gaz 10(9/10):336–341

Anonymous (1892a) Proceedings of the Botanical Club of the forty-first meeting of the A.A.A.S., Rochester, NY, August 18–24, 1892. Bull Torrey Bot Club 19:281–298

Anonymous (1892b) Proceedings of the Botanical Club of the A.A.A.S. Bot Gaz 17:285–290

Anonymous (1904) Professor Earle's new position. J N Y Bot Gard 5(54):107–109

Anonymous (1908a) Discussion of the species question. Am Nat 42(496):272–281

Anonymous (1908b) William Ashbrook Kellerman. J Mycol 14(2):49–63

Anonymous (1920) Descriptive guide to the grounds, buildings, and collections. Bull N Y Bot Gard 10(38):89–212

Anonymous (1947a) (Review) Rocks, by William Alphonso Murrill. 62 pp. Q Rev Biol 22(2):178. [lists the price as $1]

Anonymous (1947b) (Review) Stars, by William Alphonso Murrill. Q Rev Biol 22(2):178. [lists price as $1]

Anonymous (1952) Review: William A. Murrill. Success stories for girls. Gainesville, Florida (University of Florida). 1950. 63 p. Sci Educ 36(1):673

Anonymous [Committee of the Botanical Club of AAAS] (1893–1894) List of Pteridophyta and Spermatophyta growing without cultivation in northeastern North America. Mem Torrey Bot Club 5:5–359

R. H. Petersen, *William Alphonso Murrill: The Naturalist*, Memoirs of The New York Botanical Garden 124, https://doi.org/10.1007/978-3-031-93343-1

Armitage KC (2009) The nature study movement: the forgotten popularizer of America's conservation ethic. Univ Press of Kansas. 291 p
Arthur JC (1908) The physiologic aspect of the species question. Am Nat 42(496):243–248
Arthur JC, Barnhart JH, Britton NL, Clements F, Cook OF, Coville FV, Earle FS, Evans AW, Hazen TE, Hollick A, Howe MA, Knowleton FH, Moore GT, Rusby HH, Shear CL, Underwood LM, White D, Wright WF (1907) American Code of Botanical Nomenclature. Bull Torrey Bot Club 34:167–178
Atkinson GF (1886a) The cigarette beetle (*Lasioderma serricorne* Fab.). J Elisha Mitchell Sci Soc 3:68–73
Atkinson GF (1886b) Notes on the orchard *Scolytus* (*Scolytus rugulosus* Ratz.). J Elisha Mitchell Sci Soc 3:74–75
Atkinson GF (1887a) Singular adaptation in nest making by an ant, *Cremastogaster lineolata* Say. Am Nat 23:770–771. [Reprinted in J Elisha Mitchell Sci Soc 4: 88–89]
Atkinson GF (1887b) A new trap-door spider. Am Nat 22:583–594. [Reprinted in J Elisha Mitchell Sci Soc 4: I6–26]
Atkinson GF (1887c) A family of young trap-door spiders. Entomol Am 2:86–92. [Reprinted in J Elisha Mitchell Sci Soc 4: 26–33]
Atkinson GF (1887d) Descriptions of some new trap-door spiders, their nests and food habits. Entomol Am 2:109–137. [Reprinted in J Elisha Mitchell Sci Soc 4: 33–52]
Atkinson GF (1887e) A remarkable case of phosphorescence in an earthworm. Am Nat 23:773–774. [Reprinted in J Elisha Mitchell Sci Soc 4: 89–9I]
Atkinson GF (1887f) Preliminary catalog of the birds of North Carolina, with notes on some of the species. J Elisha Mitchell Sci Soc 4:44–87
Atkinson GF (1892a) Botany at experiment stations. Science n. s 20:328–330
Atkinson GF (1892b) Cryptogamic botany and plant pathology. N Y Agric Exp Stn (Cornell) Bull 49:306–316
Atkinson GF (1892c) An automatic device for rolling culture tubes of nutrient agar-agar. Bot Gaz 17:154–156
Atkinson GF (1893a) Additional note on the fungi of Blowing Rock, N. C. J Elisha Mitchell Sci Soc 10:78
Atkinson GF (1893b) Unequal segmentation and its significance in the primary division of the embryo of ferns. Bull Torrey Bot Club 20:405–408
Atkinson GF (1893c) The extent of the annulus and the function of the different parts of the sporangium of ferns in the dispersion of spores. Bull Torrey Bot Club 20:435–437
Atkinson GF (1893d) Symbiosis in the roots of the Ophioglossaceae. Bull Torrey Bot Club 20:356
Atkinson GF (1894a) The study of the biology of ferns by the Collodion method. Macmillan & Co., New York. 134 pp
Atkinson GF (1894b) *Completoria complens* Lohde. Bot Gaz 19:467–468
Atkinson GF (1896a) Albert Nelson Prentiss. Science 4(93):523–524
Atkinson GF (1896b) Albert Nelson Prentiss. Bot Gaz 21:283–289
Atkinson GF (1897a) The preparation of material for general class use. Bot Gaz 23(1):372–376
Atkinson GF (1897b) Experimental morphology. Science 46:3–23
Atkinson GF (1897c) Studies and illustrations – mushrooms: I. Cornell Univ Agric Exp Stn Bull 138:337–366
Atkinson GF (1897d) Some fungi from Alabama. Bull Cornell Univ (Sci) 3(1):1–50
Atkinson GF (1898) Method of teaching botany in the secondary schools. Asa Gray Bull 6:102–106
Atkinson GF (1899) Studies and illustrations of mushrooms II. N Y Agric Exp Stn (Cornell) Bull 168:491–516
Atkinson GF (1900a) Studies of American fungi. Mushrooms, edible, poisonous, etc. Andrus & Church, Ithaca. vi + 275 pp
Atkinson GF (1900b) Some wood-destroying fungi. Spec Rep Geol Surv Louisiana 9(Sept. 1899):331–338

Atkinson GF (1901a) Studies of American fungi. Mushrooms, edible, poisonous, etc. Andrus & Church, Ithaca. 322 pp
Atkinson GF (1901b) Studies on some shade tree and timber destroying fungi. Cornell Univ Agric Exp Stn Bot Div 193:200–235
Atkinson GF (1901c) Tree-destroying fungi (*Polyporus*). Am Gard 22:267–268
Atkinson GF (1902) Studies of some tree-destroying fungi. Trans Mass Hort Soc 90:109–130
Atkinson GF (1903) Studies of American fungi. Mushrooms, edible, poisonous, etc, 2nd edn. Henry Holt & Co, New York. 323 pp
Atkinson GF (1908) A new poisonous mushroom. Bot Gaz 46:461–463
Atkinson GF (1909a) Preliminary notes on some new species of Agaricaceae and *Clavaria*. Ann Mycol 7:365–376
Atkinson GF (1909b) A remarkable *Amanita*. Bot Gaz 48:283–293
Atkinson GF (1909c) A new edible species of *Amanita*. Science n. Ser 29(754):944
Atkinson GF (1910) Motions relating to the rules for nomenclature of the fungi proposed for action at the third International Botanical Congress in Brussels, 14–22 May 1910. Ithaca, NY
Atkinson GF, Schrenk H (1893) Some fungi of Blowing Rock, N. C. J Elisha Mitchell Sci Soc 9:95–107
Atomic Bombings (2024). https://en.wikipedia.org/wiki/Atomic_bombings_of_Hiroshima_and_Nagasaki
Auburn University Wiki (2022). https://en.wikipedia.org/wiki/Auburn_University
Barnard College Wiki (2023). https://en.wikipedia.org/wiki/Barnard_College
Barnes CR (1899a) [Review] Ganong WF 1899 The teaching botanist. Bot Gaz 28(4):276–277
Barnes CR (1899b) [Review] Coulter JM 1899 Plant relations, a first book of botany. Bot Gaz 28(4):278–279
Barnhart JH (1908) The published works of Lucien Marcus Underwood. Bull Torrey Bot Club 35(1):17–38
Barnhart JH (1935) The published work of Elizabeth Gertrude Britton. Bull Torrey Bot Club 62(1):1–17
Barr ME (1988) Howard E. Bigelow, 1923–1987. Mycologia 80(5):595–598
Beardslee HC (1918) The Russulas of North Carolina. J Elisha Mitchell Sci Soc 33(4):147–198
Bessey CE (1880) Botany for high schools and colleges. American sciences series. Henry Holt Books, Boston. 611 pp
Bessey CE (1893) Botany at the Madison meetings. Am Nat 27(321):821–828
Bessey CE (1908) The taxonomic aspect of the species question. Am Nat 42(296):218–224
Bessey CE (1914) A notable botanical career. Science n.s. 40(1019):48
Bloomingdale Insane Asylum Wiki (2023). https://en.wikipedia.org/wiki/Bloomindale_Insane Asylum
Blue Ridge Parkway (2023). https://www.blueridgeparkway.org/parkway-map/
Bondartsev AS (1971) The Polyporaceae of the European USSR and Caucasia. Translated from the Russian. USDA & NSF. Printed in Jerusalem, Israel. 896 pp
Briggs BG (1991) One hundred years of plant taxonomy, 1889–1999. Ann Missouri Bot Gard 78(1):19–32
British History (2024). https://www.historic-uk.com/HistoryUK/HistoryofBritain/World-War-2-Timeline-1944/
Britton NL (1881) A preliminary catalogue of the flora of New Jersey. Geological Survey of New Jersey, New Brunswick. xi + 223 pp
Britton EG (1886) Botanical notes in the great valley of Virginia and in the southern Alleghanies. Bull Torrey Bot Club 13:69–76
Britton NL (1887) Columbia College herbaria. Bot Gaz 12:9–11
Britton EG (1888) An enumeration of the plants collected by Dr. H. H. Rusby in South America. 1885–1886.—III. Pteridophyta. Bull Torrey Bot Club 15:247–253
Britton EG (1893) The Jaeger moss herbarium. Bull Torrey Bot Club 20:335–336
Britton NL (1903a) New paths at the southern end of the Garden. J N Y Bot Gard 4:128–129

Britton NL (1903b) A new appropriation for construction. J N Y Bot Gard 4:149–150
Britton NL (1906) Additional members of the Garden staff. J N Y Bot Gard 7(1):4–5
Britton NL (1908a) Professor Underwood's relation to the work of the New York Botanical Garden. Bull Torrey Bot Club 35(1):39–40
Britton NL (1908b) The taxonomic aspect of the species question. Am Nat 42(296):225–242
Britton NL (1911) John J. Crooke. J N Y Bot Gard 12(139):160–161
Britton NL (1913) Report of the Director-in-Chief. Bull N Y Bot Gard 8(29):171–185. [March 18]
Britton NL (1915) History of the New York Botanical Garden. Publ. Priv. 16 pp
Britton NL (1916) The New York Botanical Garden and Columbia University. Columbia Univ Q 18:352–359
Britton NL (1919) Dr. Henry Allan Gleason appointed first assistant. J N Y Bot Gard 20(230):39–40
Britton NL (1924) Report of the Secretary and Director-in-Chief. Bull N Y Bot Gard 12(44):181–193
Britton NL (1925) Resignation of Doctor Murrill. J N Y Bot Gard 26(301):13
Britton NL (1928) Doctor W. Gilman Thompson. J N Y Bot Gard 29:13–14
Bronx Park Wiki (2023). https://en.wikipedia.org/wiki/Bronx_Park
Brown A (1908) The Elgin Botanic Garden, Its Later History and Relations to Columbia College, the New Hampshire Grants and the Treaty with Vermont in 1790. New Era, Lancaster, PA. iv + 57 pp
Brown A (1972) Judge Addison Brown: autobiographical notes for his children. Carr Publishing, Inc, Boyce. 193 pp
Brown M (2022) Discovering William A. Murrill. Chestnut J 36(3):25–26
Burdsall HH, Blackwell M, Nakasone KK (2012) Robert Lee Gilbertson, 1925–2011. Mycologia 104(6):1521–1524
Burlingham GS (1908) A study of the Lactariae of the United States. Bull Torrey Bot Club 14:1–109
Burt EA (1922) The North American species of *Clavaria* with illustrations of the type specimens. Ann Missouri Bot Gard 8(1):1–78
Buyck B, Adamcik S (2011a) Type studies in *Russula* subgenus *Heterophylla* from the eastern United States. Cryptog Mycol 33(2):151–169
Buyck B, Adamcik S (2011b) Type studies of *Russula* species described by W. A. Murrill. 1. *R. roseisabellina*, *R. sericella*, and *R. obscuriformis*. Mycotaxon 115:131–144
Buyck B, Adamcik S (2013) The *Russula xerompalina* complex (Russulales, Agaricomycota) in North America. Scripta Bot Belg 51:117–131
Buyck B, Bessette A, Adamcik S (2011) *Russula hixonii* Murrill, a rare and intriguing southern species of uncertain systematic position, rediscovered in Georgia, USA. Cryptog Mycol 32(4):403–412
Cambridge Code (1935) Briquet J 1935 International rules of botanical nomenclature. G. Fischer Verlag, Jena. 152 pp
Chamberlain CL (1899) [Review] Fischer A 1899 Fixirung, Färbung und Bau des Protoplasmas. Kritische Untersuchungen über Technik und Theorie in der neueren Zellforschung. Gustav Fischer, Jena. 362 pp
Cheeseman TF (1907) Notes on botanical nomenclature; with remarks on the rules adopted by the International Botanical Congress of Vienna. Trans New Zealand Inst 40:447–465
Cifuentes J, Petersen RH, Hughes KW (2003) Campanophyllum, a new genus for an old species name. Mycol Prog 2:285–295
Clements FE (1908) An ecologic view of the species conception. Am Nat 42(496):253–264
Coker WC (1917) The amanitas of the eastern United States. J Elisha Mitchell Sci Soc 33(1–2):1–88. pls. 1–69
Coker WC (1918) The lactarias of North Carolina. J. Eisha Mitchell Sci. Soc. 34(1/2):1–62
Coker WC (1923) The Clavarias of the United States and Canada. Univ. North Carolina Press, Chapel Hill. 209 pp
Coker WC, Beardslee HC (1921) The collybias of North Carolina. J Elisha Mitchell Sci Soc 37(1/2):83–107

Columbia University Archives (2023). http://library.columbia.edu/libraries/cuarchives/presidents/barnard_frederick.html
Cook OF (1898a) Stability in generic nomenclature. Science n.s. 8(189):186–190
Cook OF (1898b) The method of types. Science n.s. 8(198):513–516
Cook OF (1899) Four categories of species. Am Nat 33:287–297
Cook OF (1900) The method of types in botanical nomenclature. Science n.s. 12(300):475–481
Cooke WB (1954) On Overholts' conservatism. Mycologia 46(5):683–688
Couch JN, Matthews VD (1954) William Chambers Coker. Mycologia 46(3):372–383. [with portrait]
Coville FV (1895) A reply to Dr. Robinson's criticism of the "List of Pteridophyta and Spermatophyta of northeastern America". Bot Gaz 20(4):162–167
Cowles HC (1908) An ecological aspect of the conception of species. Am Nat 42(496):265–271
Crisis of 1907 (2023). https://www.stlouisfed.org/-/media/project/frbstl/stlouisfed/education/lessons/pdf/the-panic-of-1907-jp-morgan-and-the-money-trust.pdf
Curtis CC (1908) A biographical sketch of Lucien Marcus Underwood. Bull Torrey Bot Club 35(1):1–12
da Eira AF, Dudukh MY, Stamets P, Wasser SP, de Amazonas MAL (2002) Is a widely cultivated culinary-medicinal Royal Sun Agaricus (the Himematsutake Mushroom) indeed *Agaricus blazei* Murrill? Int J Med Mushrooms 4(4):278–320
Daly DC (2016) Systematics and systematists at The New York Botanical Garden. Brittonia 68(3):230–237
Dedication (2024). https://babel.hathitrust.org/cgi/pt?id=mdp.39015076549289&seq=23
Degener O (1971) Fred Jay Seaver, 1871–1970. Bull Torrey Bot Club 98(3):172–174
Denson LA (1926) The drought of 1925. J Elisha Mitchell Sci Soc 42(1/2):109–112
DePauw University Wiki (2022). https://en.wikipedia.org/wiki/DePauw_University
Dewey LH (1892) A botanical congress and nomenclature. Bot Gaz 17(6):199
Donk MA (1949) New and revised *nomina generica conservanda* proposed for Basidiomycetes (Fungi). Bull Bot Gard Buitenzorg Ser 3(18):83–168
Earle FS (1903) Report on a trip to Jamaica. J N Y Bot Gard 4(37):1–10
Earle FS (1909) The genera of the North American gill fungi. Bull N Y Bot Gard 5(18 February):373–462. [In table of contents, listed as "issued separately, January 21, 1909"]
Earle FS (1928) Sugar cane and its culture. Wiley, New York
Eaton A (1820) Botanical exercises, including directions, rules and descriptions, calculated to aid pupils in the analysis of plants; with a labelling catalogue, for the assistance of teachers. Websters & Skinners, Albany, 171 pp
Eichengreen B (1992) The origins and nature of the Great Slump revisited. Econ Hist Rev 45(2):213–239
Elder AL (1944) Fungi for penicillin production. Mycologia 36(3):307
Elizabeth Gertrude Britton Wiki (2023). https://en.wikipedia.org/wiki/Elizabeth_Gertrude_Britton
Ellett CW (1970) A history of the Department of Plant Pathology, Ohio State University. Manuscript, Library, Ohio State University. 37 pp. http://kb.osu.edu/dspace/bitstream/handle/1811/75737/1//ARV_OSU_Centennl_at_Histories_PlantPathology.pdf
Engler A, Prantl K (1897) Die Natürlichen Pflanzenfamilien. 4 vols. Leipzig
Evergreen (2024). thiswondrousplace.org
Fairchild HLR (1893) A memoir of Professor John Strong Newberry. Trans N Y Acad Sci 12:152–168
Farlow WG (1910) A consideration of the *Species Plantarum* of Linnaeus as a basis for the starting point of the nomenclature of Cryptogams. Publ. Priv, Cambridge, MA. 10 pp
Farr ML (1973) An annotated list of Spegazzini's fungus taxa. Biblioth Mycol 35:1–1661. (in two volumes)
Fayod V (1889) Prodrome d'une histoire naturelle des Agaricinés. Ann Sci Nat Bot Sér 7 9:181–411
Federal Reserve Summary (2024). https://www.federalreservehistory.org/essays/great-depression

Fernald ML (1901a) Some recent publications and the nomenclatural principles they represent. Bot Gaz 31(3):183–197

Fernald ML (1901b) The instability of the Rochester nomenclature. Bot Gaz 31(5):359–367

Final Battles (2024). https://www.youtube.com/watch?v=AI1N6ioL4Ho

Fischer E (1900) Victor Fayod 1860–1900. Verhandl Schweiz Naturf Gesell 83:XXXII–XXXVI

Fitzpatrick HM (1932) The Mycological Society of America. Mycologia 24(6):515–516

Fitzpatrick HM (1934) The Mycological Society of America. Mycologia 26(1):108–112

Fitzpatrick HM (1944) The first twelve years of the Mycological Society of America. Mycologia 36(1):1–17

Florida and the Depression (2024). https://fcit.usf.edu/florida/lessons/depress/depress1.htm

Florida Depression Report (2024). https://dos.fl.gov/florida-facts/florida-history/a-brief-history/the-great-depression-in-florida/

Florida Farm Bureau History (2024). https://floridafarmbureau.org/history/#:~:text=The%20Great%20Depression%20of%20the,shippers%2C%20who%20often%20controlled%20harvesting

Florida Wartime Economy (2024). http://npshistory.com/brochures/casa/world-war-ii.pdf

Fontinele A (2023) A brief history of taxes in the U.S.: https://www.investopedia.com/articles/tax/10/history-taxes.asp

Forager/Chef (2024). https://foragerchef.com/know-thy-enemies-jack-olantern-mushrooms/

Fordham University (2023). https://www.fordham.edu/about/fordhams-history/our-story/

Fredman LE (1972) Seth Low: theorist of municipal reform. J Am Stud 6(1):19–39

Freeman AE (1979a) *Agaricus* in North America: type studies. Mycotaxon 8(1):1–49

Freeman AE (1979b) *Agaricus* in the southeastern United States. Mycotaxon 8(1):50–118

Fries EM (1821) Systema Mycologicum, vol 1. Sumtibus Ernesti Mauritti, Gryphiswaldae

German Invasion of USSR (2024). https://encyclopedia.ushmm.org/content/en/article/invasion-of-the-soviet-union-june-1941

Gibson WH (1895) Our edible toadstools and mushrooms and how to distinguish them. Harper & Bros, New York. 337 pp

Gleason HA (1920) Report of the first assistant. Bull N Y Bot Gard 10(39):230–234

Gleason HA (1961) Thumbnail Sketches of Botanists. 137 pp. Unpublished. Archives, New York Botanical Garden

Gray A (1879) The Botanical Text-book (sixth edition). Part I. Structural botany; or organography on the basis of morphology. To which is added the principles of taxonomy and phytography, and a glossary of botanical terms. Ivison, Blakeman, Taylor, & Co, New York. 442 pp. [for nomenclature, see 346–352]

Greene EL (1887) Botanical nomenclature in North America. J Bot 26:326–329

Greene EL (1894) Manual of the Botany of the Region of San Francisco Bay. Cubery & Co., San Francisco. 328 pp

Griggs RF, Landacre PL, Hambleton JC (1908) On the death of William Ashbrook Kellerman. Ohio Naturalist 8(2):361–362

Grout AJ (1903–1910) Mosses with Hand-lens and Microscope: a non-technical handbook of the more common mosses of the northeastern United States. Published by Author, New York. 5 parts. 416 pp

Gyorfi GA (2007) Study on the mushroom species *Agaricus blazei* (Murrill). Int J Hort Sci 13(4):45–48

Gyorfi GA (2008) Growing experiments with a medicinal mushroom *Agaricus blazei* (Murrill). Int J Hort Sci 14(4):7–10

Halling RE (1986) An annotated index to species and infraspecific taxa of Agaricales and Boletales described by William Alphonso Murrill. Mem N Y Bot Gard 40:1–120 + portrait

Harris H, Harris WE, Wiebe D (2006) Yama Farms: a most unusual Catskills Resort. Cragsmoor Historical Society, Cragsmoor. 93 pp

Hathi Trust. https://babel.hathitrust.org/cgi/pt?id=coo.31924000412084&view=1up&seq=7

Heinemann P (1993) Agarici austroamericani. VIII. Agaricaceae des regions intertropicales d'Amerique du Sud. Bull Jard Bot Belg 62:355–384

Hesler LR (1958) Southeastern Agaricales II. J Tennessee Acad Sci 33(3):186–191. [*Tricholoma* types]
Hesler LR (1959a) Southeastern Agaricales III. J Tennessee Acad Sci 34(3):162–166. [*Collybia* types]
Hesler LR (1959b) Southeastern Agaricales IV. J Tennessee Acad Sci 34(3):167–171. [*Marasmius* types]
Hesler LR (1960) A study of *Russula* types. Mem Torrey Bot Club 21:1–59
Hesler LR (1961) A study of *Russula* types II. Mycologia 53(6):605–625
Hesler LR, Smith AH (1963) North American Species of *Hygrophorus*. Univ Tennessee Press, Knoxville. 416 pp
History.com. https://www.history.com/topics/world-war-ii/world-war-ii-history
Hoffman C (1970) The depression of the nineties. An economic history. Greenwood Publ. Co. 326 pp
Hollick CA, Britton NL (1879) The Flora of Richmond County. Staten Island, New York. 36 pp
Hollick CA, Britton NL (1880) Flora of Richmond County, N. Y. Additions and new localities, 1879. Bull Torrey Bot Club 7:11–12
Holmes CC (1908) Lucien Marcus Underwood. Bryologist 11(3):41–44
Hoste I (2022) Conflicting ideas on author citation: François Crépin and *Lois de la Nomenclature botanique*. Huntia 19(1):1–25
Howe MA (1908) Lucien Marcus Underwood: a memorial tribute. Bull Torrey Bot Club 35(1):13–16
Howe MA (1933) Arthur Hollick, February 6, 1867–March 11, 1933. Bull Torrey Bot Club 60(8):537–553
Howe MA (1934a) Elizabeth Gertrude Britton. Wild Flower 11:26
Howe MA (1934b) Nathaniel Lord Britton. J N Y Bot Gard 35:169–180
Hughes KW, Petersen RH (1998) Relationships among *Omphalotus* species (Paxillaceae) based on restriction sites in the ribosomal ITS1-5.8S-ITS2 region. Plant Syst Evol 211:231–237
Humphrey HB (1961) Makers of North American Botany. Ronald Press, New York. 265 pp
Index Herbariorum (2023) Index Herbariorum: a guide to the location and contents of the world's public herbaria. Part I. The herbaria of the world. Regn. Veget. 120, 8th Ed
Jercinovic E (2007) Josephine Skehan and the mountains near Gray. New Mex Bot 39:1–6
John Strong Newberry Wiki (2023). https://en.wikipedia.org/wiki/John_Strong_Newberry
JStor Global Plants (2023). https://plants.jstor.org/stable/10.5555/al.ap.person.bm000006029
Kauffman CH (1918) The Agaricaceae of Michigan. Mich Biol & Geol Surv Publ 26, Biol Surv 5, Lansing, Michigan. vol I, 924 pages. Vol II, 172 plates
Keating JM (1973) Seth Low and the Development of Columbia University, 1889–1901. Dissertation, Teachers' College
Kellerman WA (1883) The elements of botany, embracing organography, histology, vegetable physiology, systematic botany and economic botany. J. E. Potter & Co, Philadelphia. 360 pp
Kellerman WA (1892) The forest trees of Ohio; for the World's Columbian Exposition. Ohio World's Fair Commission Bull 5:1–7
Kellerman WA (1898) A text-book of elementary botany, including a Spring Flora. Eldredge & Brother, Philadelphia, p 300
Kellerman WA (ed) (1906) Notes from mycological literature, XX. J Mycol 12(4):164–183
Kelner MJ, McMorris TC, Beck WT, Zamora JM, Taetle R (1987) Preclinical evaluation of illudins as anticancer agents. Cancer Res 47(12):3186–3189
Kennedy JM (1968) Philanthropy and Science in New York City: The American Museum of Natural History, 1868–1968. Dissertation, Ph.D., Yale University. 277 pp. http://www.proquest.com/docview/302390353?pq-origsite=gscholar&fromopenview=true
Kephart H (1913) Our Southern highlanders. Outing Publ. Co., New York. 395 pp. Hathi Trust: https://babel.hathitrust.org/cgi/pt?id=njp.32101035687209&view=1up&seq=13
Kerrigan RW (2016) *Agaricus* of North America. Mem N Y Bot Gard 114:1–572
Kimbrough JW (1972) Keys to the Fleshy Basidiomycetes of Florida by Wm. A. Murrill. Publ. Priv., Gainesville. 199 pp

Kimbrough J (2003) The twilight years of William Alphonso Murrill. Mushroom 21 22:47. (with portrait)

Knight EG (1881) Albinism. Bull Torrey Bot Club 8:125

Knight EG (1883a) Submerged leaves in *Limnanthemum*. Bull Torrey Bot Club 10:34

Knight EG (1883b) On the fruit of *Eustichium norvegicum* Br. Eur Bull Torrey Bot Club 10:99–100

Koch R (1876) Untersuchungen über Bakterien: V. Die Ätiologie der Milzbrand-Krankheit, begründet auf die Entwicklungsgeschichte des *Bacillus anthracis*. https://edoc.rki.de/handle/176904/5139

Koch R (1893) Ueber den augenblicklichen Stand der bakteriologischen Choleradiagnose. Z. Hygiene Infektionskrankheiten 14:319–338

Koleno C (2019) World War II paratrooper with a Purple Heart and Bronze Star talks about the war. Penn State, Agricultural Sciences (without citation data). https://news.psu.edu/story/592144/2019/11/14/world-war-ii-paratrooper-purple-heart-and-bronze-star-talks-about-war?utm_source=newswire&utm_medium=email&utm_term=598300_HTML&utm_content=11-14-2019-22-57&utm_campaign=Penn%20State%20Today

Krieger LCC (1914) Observations on the use of Ridgway's new color-book. The color of the spores of *Volvaria speciosa* Fr. Mycologia 6(1):29–31

Kumar S, Kumari R (2021) Traditional uses, phytochemistry and biological activities of *Agaricus blazei* Murrill: a comprehensive review. Am J Pharm Pharmacother 6(3):1–5

Kuntze O (1891) Revisio Generum Plantarum Vascularium Omnium atque Cellularium Multarum Secundum Leges Nomenclaturae Internationales: cum enumeratione plantarum exoticarum in itinere mundi collectarum. Pars I & Pars II. Arthur Felix, Leipzig

Lend-lease Act (1941). https://www.archives.gov/milestone-documents/lend-lease-act

Library of Congress (2023a). https://www.loc.gov/classroom-materials/united-states-history-primary-source-timeline/progressive-era-to-new-era-1900-1929/womens-suffrage-in-progressive-era/#:~:text=In%201869%2C%20Anthony%20and%20Elizabeth,gain%20the%20right%20to%20vote

Library of Congress (2023b). https://www.loc.gov/classroom-materials/united-states-history-primary-source-timeline/progressive-era-to-new-era-1900-1929/cities-during-progressive-era/

Lloyd CG (1906a) Letter no. 10: 1–8 "postmarked" Paris

Lloyd CG (1906b) A new rule. Mycol Notes 23:292

Lloyd CG (1907) Concerning the polyporoids. Mycol Notes 27:341–343

Lloyd CG (1908) The result of Kuntzeism. Mycol Notes 31:410

Lloyd CG (1910) Professor W. A. Kellerman. Mycol Notes 36:477–492. (with portrait)

Lloyd CG (1912) Synopsis of the stipitate polyporoids. Bull Lloyd Library Bot Pharm Materia Med Mycol Ser 6:95–208

Lloyd CG (1915a) Synopsis of the section *Apus* of the genus *Polyporus*. Mycol Writ 4:291–392

Lloyd CH (1915b) William Alphonso Murrill. Mycol Writ 4:200. [with portrait].

Lloyd Library (2023). https://lloydlibrary.org/wp-content/themes/Lloyd_wptheme/finding-aids/Curtis%20Gates%20Lloyd%20Papers%201859-1926.pdf

London Blitz (2024). https://en.wikipedia.org/wiki/The_Blitz

Lost Museum Archive (2023). http://lostmuseum.cuny.edu/archive/barnums-american-museum

Lowden RM (1970) William Ashbrook Kellerman's Botanical Expeditions to Guatemala (1905–1908). Published by the Author, Columbus. 31 pp

Lucien Marcus Underwood Wiki (2022). https://en.wikipedia.org/wiki/Lucien_Marcus_Underwood

MacDougal DT (1908) The physiological aspect of a species. Am Nat 42(496):249–252

Maerz A, Paul MA (1930) A dictionary of color. McGraw Hill Co., New York. 207 pp

Maerz A, Paul MA (1950) A dictionary of color, 2nd edn. McGraw Hill Co., New York. 208 pp

Mains EB (1932) Calvin Henry Kauffman. Mycologia 24(3):265–267

McLaughlin DT (1983) Walter Henry Snell. 1889–1980. Mycologia 75(6):941–948

McIlvaine C, Macadam M (1900) One thousand American fungi. lxii + 729 pp. Bowen-Merrill Co., New York

McMorris TC, Anchel M (1965) Fungal metabolites. The structure of the novel sesquiterpenoids illudin-S and -M. J Am Chem Soc 87(7):1594–1600
Medical News Today (2024). https://www.medicalnewstoday.com/articles/216798
Merkel HW (1906) A deadly fungus on the American chestnut. Ann Rep N Y Zool Soc 10:97–103
Merrill ED (1938) Nathaniel Lord Britton. Biographical Memoir. National Academy of Sciences vol 19, Memoir 5: 147–159 [with bibliography by J. H. Barnhart, pp. 160–202]
Methven AS (1987) The Genus *Clavariadelphus* in North America Excluding Mexico. Dissertation ined., Ph.D. Univ. Tennessee, Knoxville, TN. 350 pp
Meyer BS (1983) Botany at the Ohio State University: the first 100 years. Bull Ohio Biol Surv n.s. 6(2):i–vi, 1–177
Mickulas PP (2003) Giving, getting, and growing: philanthropy, science and the New York Botanical Garden, 1888–1929. Dissertation, Ph.D. Rutgers Univ., ProQuest edition. 487 pp
Mickulas P (2007) Britton's botanical empire: the New York Botanical Garden and American botany. 1888–1929. Mem N Y Bot Gard 94:1–316
Mizuna T (2002) Medicinal properties and clinical effects of culinary-medicinal mushroom *Agaricus blazei* Murrill (Agaricomycetideae) (Review) Int. J Med Mushrooms 2(4):14–30
Moore G (2000) An overview of the Torrey Botanical Society's contributions to field work, with discussions on the Society's founding and its founding members. In: Lamont EE (ed) Historial issue: 150 years of the Torrey Botanical Society, Mem. Torrey Bot. Soc., vol 29, pp 70–112
Mueller GM, Wu Q (1997) Mycological contributions of Rolf Singer: field itinerary, index to new taxa, and list of publications. Fieldiana Bot n.s. 38:1–124
Mugwumps Wiki (2023). https://en.wikipedia.org/wiki/Mugwumps
Murrill WA (1897) *Asplenium ebenoides* in Virginia. Fern Bull 5:1–3
Murrill WA (1900) The development of the archegonium and fertilization in the hemlock spruce (*Tsuga canadensis* Carr.). Ann Bot 14:583–607
Murrill WA (1902a) Shade trees. Cornell Univ Agric Exp Stn Bull 205:69–118
Murrill WA (1902b) The Polyporaceae of North America – I. The genus *Ganoderma*. Bull Torrey Bot Club 29:599–608. [Also: Contrib. NYBG no. 29, reprint of above]
Murrill WA (1903a) The Polyporaceae of North America – I. The genus *Ganoderma*. Contrib N Y Bot Gard 29. [Reprint of Murrill, 1902b]
Murrill WA (1903b) The Polyporaceae of North America – II. The genus *Pyropolyporus*. Bull Torrey Bot Club 30(2):109–120. [Also: Contrib. NYBG no. 32, reprint of above]
Murrill WA (1903c) The Polyporaceae of North America – III. The genus *Fomes*. Bull Torrey Bot Club 30(4):225–232. [Also: Contrib. NYBG no. 35, reprint of above]
Murrill WA (1903d) The Polyporaceae of North America – IV. The genus *Elvingia*. Bull Torrey Bot Club 40(5):296–301. [Also: Contrib. NYBG no. 38, reprint of above]
Murrill WA (1903e) The Polyporaceae of North America – V. The genera *Cryptoporus*, *Piptoporus*, *Scutiger* and *Porodisculus*. Bull Torrey Bot Club 30(8):423–434. [Also: Contrib. NYBG no. 41, reprint of above]
Murrill WA (1903f) A historical review of the genera of the Polyporaceae. J Mycol 9:87–102
Murrill WA (1903g) A new family of the Basidiomycetes – Xylophagaceae. Toreya 2(1):7
Murrill WA (1904a) The Polyoraceae of North America – VI. The genus *Polyporus*. Bull Torrey Bot Club 31(1):29–44
Murrill WA (1904b) The Polyporaceae of North America – VII. The genera *Hexagona*, *Grifola*, *Romellia*, *Coltricia* and *Coltriciella*. Bull Torrey Bot Club 31(6):325–348
Murrill WA (1904c) The Polyporaceae of North America – VIII. *Hapalopilus*, *Pycnoporus*, and new monotypic genera. Bull Torrey Bot Club 31(8):415–428
Murrill WA (1904d) The Polyporaceae of North America – IX. *Inonotus*, *Sesia* and monotypic genera. Bull Torrey Bot Club 31:593–610
Murrill WA (1904e) A new species of *Polyporus* from Tennessee. Torreya 4(190):150–151
Murrill WA (1904f) A key to the perennial Polyporaceae of temperate North America. Torreya 4:165–167

Murrill WA (1905a) The Polyporaceae of North America – X. *Agaricus*, *Lenzites*, *Cerrena*, and *Favolus*. Bull Torrey Bot Club 32(2):83–103

Murrill WA (1905b) The Polyporaceae of North America – XI. A synopsis of the brown pileate species. Bull Torrey Bot Club 32(7):353–371

Murrill WA (1905c) The Polyporaceae of North America – XII. A synopsis of the white and bright-colored pileate species. Bull Torrey Bot Club 32(9):469–493

Murrill WA (1905d) The Polyporaceae of North America – XIII. The described species of *Bjerkandera*, *Trametes*, and *Coriolus*. Bull Torrey Bot Club 32(12):633–656

Murrill WA (1905e) A key to the stipitate Polyporaceae of temperate North America – I. Torreya 5(2):28–30

Murrill WA (1905f) A key to the stipitate Polyporaceae of temperate North America – II. Torreya 5(3):43–44

Murrill WA (1905g) A key to the brown sessile Polyporeae of temperate North America. Torreya 5(11):194–195

Murrill WA (1905h) Collecting fungi in Maine. J N Y Bot Gard 6(72):199–202

Murrill WA (1905i) A key to the Agariceae [sic] of temperate North America. Torreya 5(12):213–214

Murrill WA (1906a) A serious chestnut disease. J N Y Bot Gard 7(8):143–153

Murrill WA (1906b) Further remarks on a serious chestnut disease. J N Y Bot Gard 7:203–211

Murrill WA (1906c) A new chestnut disease. Torreya 6(9):186–189

Murrill WA (1906d) The pileate Polyporaceae of central Maine. Torreya 6:34–37

Murrill WA (1906e) A summer in Europe; some foreign botanists and botanical institutions. J N Y Bot Gard 7(82):221–237

Murrill WA (1906f) How Bresadola became a mycologist. Torreya 6:233–234

Murrill WA (1907) Agaricales. Polyporaceae (pars). N Am Flora 9(1):1–72

Murrill WA (1908a) Agaricales. Polyporaceae (conclusio). N Am Flora 9(2):73–132

Murrill WA (1908b) The collections of fungi. J N Y Bot Gard 9(97):1–8

Murrill WA (1908c) Collecting and studying boleti. Torreya 8:50–55

Murrill WA (1908d) A key to the white and bright-colored sessile Polyporeae of temperate North America. I. Torreya 8:14–16

Murrill WA (1908e) A key to the white and bright-colored sessile Polyporeae of temperate North America. II. Torreya 8:28–29

Murrill WA (1908f) A key to the white and bright-colored sessile Polyporeae of temperate North America. III. Torreya 8:130–132

Murrill WA (1908g) Notes on the life and work of Charles C. Frost. Torreya 8:197–200

Murrill WA (1908h) The boleti of the Frost herbarium. Bull Torrey Bot Club 35:517–526

Murrill WA (1908i) The spread of the chestnut disease. J N Y Bot Gard 9:29–30

Murrill WA (1908j) Leaf blight of the plane tree. J N Y Bot Gard 9(102):105–107

Murrill WA (1908k) Collecting fungi at Biltmore. J N Y Bot Gard 9:135–141

Murrill WA (1908l) Edible mushrooms in Bronx Park. J N Y Bot Gard 9(108):205–213

Murrill WA (1909a) Illustrations of fungi – I. Mycologia 1(1):1–3

Murrill WA (1909b) The Boletaceae of North America – I. Mycologia 1(1):4–18

Murrill WA (1909c) Collecting fungi in Jamaica. J N Y Bot Gard 10:21–39

Murrill WA (1909d) Polyporaceae from Japan. Mycologia 1(4):164–170

Murrill WA (1909e) The Boletaceae of North America. I. Mycologia 1(1):4–18

Murrill WA (1909f) The Boletaceae of North America. II. Mycologia 1(4):140–158

Murrill WA (1910a) Report of the Assistant Director [for 1909]. Bull N Y Bot Gard 24(17):169–178

Murrill WA (1910b) Boletaceae. N Am Flora 9(3):133–166

Murrill WA (1910c) Agaricaceae [pars]. N Am Flora 9(3):163–171

Murrill WA (1910d) Collecting fungi in southern Mexico. J N Y Bot Gard 11(123):57–77

Murrill WA (1911a) The Agaricaceae of tropical North America – I. Mycologia 3(1):23–36

Murrill WA (1911b) The Agaricaceaae of tropical North America – II. Mycologia 3(2):79–91

Murrill WA (1911c) The Agaricaceae of tropical North America – III. Mycologia 3(4):189–199

Murrill WA (1911d) The Agaricaceae of tropical North America – IV. Mycologia 3(6):271–282

Murrill WA (1911e) Studying tropical American fungi in European herbaria. J N Y Bot Gard 12:43–54
Murrill WA (1911f) Mushroom poisoning. J N Y Bot Gard 12(141):204–207
Murrill WA (1911g) Report of the Assistant Director. Bull N Y Bot Gard 7:294–300
Murrill WA (1912a) The Agaricaceae of the Pacific coast – I. Mycologia 4(4):205–217
Murrill WA (1912b) The Agaricaceae of the Pacific coast – II. Mycologia 4(5):231–262
Murrill WA (1912c) The Agaricaceae of the Pacific coast – III. Mycologia 4(6):291–308
Murrill WA (1912d) Collecting fungi on the Pacific coast. J N Y Bot Gard 13(145):1–14
Murrill WA (1912e) Collecting fungi on the Pacific coast. J N Y Bot Gard 13:41–44
Murrill WA (1912f) The McClatchie herbarium. J N Y Bot Gard 13(153):147–149
Murrill WA (1912g) Polyporaceae and Boletaceae of the Pacific coast. Mycologia 4(2):91–100
Murrill WA (1912h) The Polyporaceae of Mexico. Bull N Y Bot Gard 8(28):137–153
Murrill WA (1912i) The chestnut canker convention. J N Y Bot Gard 13(147):41–44
Murrill WA (1912j) Why the chestnut canker cannot be controlled by cutting-out method. Pennsylvania Chestnut Blight Conf.: 194–195
Murrill WA (1912k) Collecting fungi in the Adirondacks. J N Y Bot Gard 13:174–178
Murrill WA (1912l) The Agaricaceae of tropical North America. V. Mycologia 4(2):72–83
Murrill WA (1912m) New combinations for tropical agarics. Mycologia 4(6):331–332
Murrill WA (1913a) The Agaricaceae of the Pacific coast – IV. New species of *Clitocybe* and *Melanoleuca*. Mycologia 5(4):206–223
Murrill WA (1913b) The Agaricaceae of tropical North America: VI. Mycologia 5(1):18–36
Murrill WA (1914a) The Underwood library and collection of fungi. J N Y Bot Gard 15:223–224
Murrill WA (1914b) (Agaricales) Agaricaceae [pars]. N Am Flora 10(1):1–76
Murrill WA (1914c) Northern Polypores. Publ. Priv, New York. iv + 64 pp
Murrill WA (1914d) American Boletes. Publ. Priv, New York. iv + 40 pp
Murrill WA (1915a) Western Polypores. Publ. Priv, New York. iv + 40 pp
Murrill WA (1915b) Preliminary list of Upper St. Regis fungi. Mycologia 7(6):297–306
Murrill WA (1915c) Southern Polypores, Publ. Priv, New York. iv + 41 pp
Murrill WA (1915d) Tropical Polypores, Publ. Priv, New York. iv + 117 pp
Murrill WA (1916a) Edible and poisonous mushrooms. A descriptive handbook to accompany the Author's Colored Chart of edible and poisonous mushrooms. Publ. Priv, New York. 76 pp. Hathi Trust: https://babel.hathitrust.org/cgi/pt?id=wu.89101347110&seq=5
Murrill WA (1916b) Illustrations of fungi. J N Y Bot Gard 17(193):7–13
Murrill WA (1916c) A very dangerous mushroom. Mycologia 8(3):186–187
Murrill WA (1916d) Index to illustrations of fungi: I–XXVII. Mycologia 8(1):47–51
Murrill WA (1916e) *Pleurotus*, *Omphalia*, *Mycena*, and *Collybia* published in North American Flora. Mycologia 8(4):218–221
Murrill WA (1916f) Fungi collected at Arkville, New York. Mycologia 8(6):293–399
Murrill WA (1916g) Exploration of Apple Orchard Mountain, Virginia. J N Y Bot Gard 17(204):218–221
Murrill WA (1916h) Report of the Assistant Director. Bull N Y Bot Gard 9(35):310–318
Murrill WA (1916i) *Hygrocybe*, *Camerophyllus*, and *Hygrophorus*. N Am Flora 9:376–396
Murrill WA (1917a) Some fungi collected in Virginia. Mycologia 9(1):34–36
Murrill WA (1917b) The taxonomy of the Agaricaceae. Am J Bot 4(6):315–326
Murrill WA (1918a) Billy, the Boy Naturalist; the True Story of a Naturalist's Boyhood in Virginia just after the Civil War. Publ. Priv, New York. 252 pp. Hathi Trust: https://babel.hathitrust.org/cgi/pt?id=nyp.33433007668738&view=1up&seq=9
Murrill WA (1918b) Three Young Crusoes, Their Life and Adventures on an Island in the West Indies. Publ. Priv, New York. xiv + 224 pp. Hathi Trust: https://babel.hathitrust.org/cgi/pt?id=coo.31924031320066&seq=1
Murrill WA (1918c) The Agaricaceae of tropical North America: VII. Mycologia 10(1):15–33
Murrill WA (1918d) The Agaricaceae of tropical North America: VIII. Mycologia 10(2):62–85

Murrill WA (1918e) Murrill's and Saccardo's names of Polypores compared. Publ. Priv, New York. 31 pp
Murrill WA (1919a) The Naturalist in a Boarding School. Publ. Priv, New York. 275 pp. Hathi Trust: https://babel.hathitrust.org/cgi/pt?id=nyp.33433074788021&view=1up&seq=7
Murrill WA (1919b) The Natural History of Staunton, Virginia. Publ. Priv, New York. 216 pp. Hathi Trust: https://babel.hathitrust.org/cgi/pt?id=hvd.32044106196017&view=1up&seq=1
Murrill WA (1919c) The exhibition of plant forms used in design. J N Y Bot Gard 20(232):75–86
Murrill WA (1920a) The fungi of Blacksburg, Virginia. Mycologia 12(6):322–328
Murrill WA (1920b) An excursion to Mountain Lake, Virginia. Torreya 20(6):116–119
Murrill WA (1920c) Report of the supervisor of public instruction. Bull N Y Bot Gard 10(39):240–246
Murrill WA (1920d) Collecting fungi at Yama Farms. Mycologia 12(1):42–43
Murrill WA (1920e) Light-colored resupinate polypores: I. Mycologia 12(2):77–92
Murrill WA (1920f) Light-colored resupinate polypores: II. Mycologia 12(6):295–308
Murrill WA (1920g) Corrections and additions to the polypores of temperate North America. Mycologia 12(1):6–24
Murrill WA (1920h) Illustrations of fungi: XXXII. Mycologia 12(2):59–61
Murrill WA (1921a) Volume 10 of North American Flora. Mycologia 13(2):118
Murrill WA (1921b) Light-colored resupinate polypores: III. Mycologia 13(2):83–100
Murrill WA (1921c) Two species of *Fuscoporia*. Mycologia 13(2):119
Murrill WA (1921d) The genus *Tinctoporia*. Mycologia 13(2):122–123
Murrill WA (1922a) A fragrant polypore. Mycologia 14(1):46–47
Murrill WA (1922b) The Toronto meeting. Mycologia 14(1):50–52
Murrill WA (1922c) Dark-spored agarics: I. *Drosophila*, *Hypholoma*, and *Pilacre*. Mycologia 14(2):61–70
Murrill WA (1922d) Dark-spored agarics: II. *Gomphidius* and *Stropharia*. Mycologia 14(3):121–142
Murrill WA (1922e) Dark-spored agarics: III. *Agaricus*. Mycologia 14(4):200–221
Murrill WA (1922f) Dark-spored agarics: IV. *Deconica*, *Atylospora*, and *Psathyrella*. Mycologia 14(5):258–278
Murrill WA (1922g) Index to Illustrations of fungi. XXIII–XXXIII. Mycologia 14(6):332–334
Murrill WA (1923a) Virginia fungi. Mycologia 15(5):243–244
Murrill WA (1923b) Springtime in Florida. J N Y Bot Gard 24(283):125–135
Murrill WA (1923c) Florida fungi – I. Mycologia 15(6):278–279
Murrill WA (1923d) Botanical features of Mountain Lake, Virginia. J N Y Bot Gard 24(287):249–256
Murrill WA (1924a) Woodstock fungi. Mycologia 16(1):44–46
Murrill WA (1924b) Report of the supervisor of public instruction. Bull N Y Bot Gard 12(44):199–202
Murrill WA (1924c) Fungi at Woodstock, New York. Torreya 24:47–48
Murrill WA (1924d) Lake Placid fungi. Mycologia 16(2):96–98
Murrill WA (1924e) Dr. Carlos Spegazzini. Mycologia 16(4):200–201
Murrill WA (1924f) Sketches of travel in South America. The Botanical Garden in Rio. Torreya 24(5):91–93
Murrill WA (1925a) Fungi at Lynchburg, Virginia. Mycologia 17(4):183–184
Murrill WA (1925b) Botanizing in Virginia. Mycologia 17(1):44
Murrill WA (1925c) The trees of St. Augustine. J N Y Bot Gard 26(302):36–38
Murrill WA (1925d) Ornamental plants of Gainesville, Florida. J N Y Bot Gard 26(304):80–85
Murrill WA (1925e) Notes on abnormal fruiting in the cinnamon fern. Am Fern J 15(2):66
Murrill WA (1925f) Collecting around St. Augustine, Florida. Mycologia 17(3):127–129
Murrill WA (1926) From Rio to Petropolis. Torreya 26(4):61–62
Murrill WA (1927a) *Spongipellis fissilis*. Mycologia 19(2): 90–91
Murrill WA (1927b) The edibility of *Leotia*. Mycologia 19(2):92
Murrill WA (1927c) The yellow-gilled *Russula*. Mycologia 19(4):151–152

Murrill WA (1927d) Bladderwort and spatter dock. Am Bot 33:51–54
Murrill WA (1929a) Neglected mushrooms. Sci Am 141(1):22–23
Murrill WA (1929b) The oldest living tree. Forest and Stream 1929(4):285. https://archive.org/details/sim_forest-and-stream-a-journal-of-outdoor-life_1929-04_99_4/page/284/mode/2up
Murrill WA (1929c) The horn snake. Forest and Stream 1929(10):777. https://archive.org/details/sim_forest-and-stream-a-journal-of-outdoor-life_1929-10_99_10/page/n3/mode/2up
Murrill WA (1929d) The dignity of matter. The Open Court 1929(1):124–127. https://opensiuc.lib.siu.edu/cgi/viewcontent.cgi?article=4122&context=ocj
Murrill WA (1930a) Bresadola. Mycologia 22(2):49–50
Murrill WA (1930b) The life on other worlds. The Open Court 44(7):446–448
Murrill WA (1930c) The imperious sycamore. Sci. Amer. 143(2):128–129
Murrill WA (1930d) *Pinckneya*. Am Bot 36:1–3
Murrill WA (1937) Flora of Alachua County, Florida (preliminary checklist). Mimeo. Contrib. Univ. Florida + Agric. Exp. Stat., Gainesville. 36 leaves
Murrill WA (1938a) New Florida agarics. Mycologia 30(4):359–371
Murrill WA (1938b) *Agaricus* at Gainesville, Florida. J Elisha Mitchell Sci Soc 54:139–141
Murrill WA (1938c) (Dec.). New Florida polypores. Bull Torrey Bot Club 65:647–661
Murrill WA (1938d) Mushrooms of Alachua County (a preliminary list). Mimeo. Contrib. Univ. Florida + Agric. Exp. Stat. Gainesville. 25 leaves
Murrill WA (1938e) Fungi of Alachua County [cover title]. Basidiomycetes of Alachua County (polypores, puffballs, etc.) [First page title]. Mimeo. Contrib. Univ. Florida + Agric. Exp. Stat. Gainesville. 16 leaves
Murrill WA (1939a) Additions to Florida fungi – I. Bull Torrey Bot Club 66(1):29–37
Murrill WA (1939b) Oligocene Island fungi. Bull Torrey Bot Club 66(3):151–160
Murrill WA (1939c) Some Florida gill-fungi. J Elisha Mitchell Sci Soc 55:361–372
Murrill WA (1940a) Additions to Florida fungi – II. Bull Torrey Bot Club 67(1):57–66
Murrill WA (1940b) Additions to Florida fungi – III. Bull. Torey Bot. Club 67(2):145–154
Murrill WA (1940c) Additions to Florida fungi – IV. Bull Torrey Bot Club 67(3):227–235
Murrill WA (1940d) Additions to Florida fungi – V. Bull Torrey Bot Club 67(4):275–281
Murrill WA (1940e) The violets of Alachua Co., Florida. J Elisha Mitchell Sci Soc 56:367–370
Murrill WA (1940f) Alachua County, Florida, soils and violets. Ecology 21(4):512–513
Murrill WA (1940g) Florida hawthorns (*Crataegus*). Mimeo. Contrib. Herb. Univ. Fla. Agric. Exp. Stat. 16 leaves
Murrill WA (1940h) *Craterellus unicolor* Berk. & Rav. in Florida. Mycologia 32:415–416
Murrill WA (1941) The devil and the deep. Mycologia 33(6):666
Murrill WA (1942a) Spiders of Alachua County, Florida. Florida Entomol. 25(1):7–9
Murrill WA (1942b) Ecologic notes on Florida hawthorns. Ecology 23:121–123
Murrill WA (1942c) Florida boletes (Boletaceae). Mimeo. Contrib. Univ. Florida + Agric. Exp. Stat., Gainesville. 6 leaves
Murrill WA (1942d) Florida collybias (*Collybia*, *Gymnopus*). Mimeo. Contrib. Herb. Univ. Fla. Agric. Exp. Stat. 6 leaves
Murrill WA (1942e) Florida russulas (*Russula*). Mimeo. Contrib. Herb. Univ. Fla. Agric. Exp. Stat. 15 leaves
Murrill WA (1942f) Florida dark-spored agarics. Mimeo. Contrib. Herb. Univ. Fla. Agric. Exp. Stat. 12 leaves
Murrill WA (1942g) Florida polypores (Polyporaceae). Mimeo. Contrib. Herb. Univ. Fla. Agric. Exp. Stat. 45 leaves
Murrill WA (1942h) Florida amanitas (*Amanita*, *Venenarius*). Mimeo. Contrib. Herb. Univ. Fla. Agric. Exp. Stat. 7 leaves
Murrill WA (1942i) Florida lactarias (*Lactarius* Pers.) Mimeo. Contrib. Herb. Univ. Fla. Agric. Exp. Stat. 9 leaves
Murrill WA (1942j) Florida lepiotas. Mimeo. Contrib. Herb. Univ. Fla. Agric. Exp. Stat. 7 leaves.

Murrill WA (1942k) Florida tricholomas. Mimeo. Contrib. Herb. Univ. Fla. Agric. Exp. Stat. 6 leaves
Murrill WA (1942l) Florida *Marasmius* (*Marasmius*, *Marasmiellus*). Mimeo. Contrib. Herb. Univ. Fla. Agric. Exp. Stat. 6 leaves
Murrill WA (1942m) Lower Hymenomycetes of Florida. Mimeo. Contrib. Herb. Univ. Fla. Agric. Exp. Stat. 6 leaves
Murrill WA (1942n) Florida resupinate polypores. Mycologia 34:595–596
Murrill WA (1942o) New fungi from Florida. Lloydia 5:136–157
Murrill WA (1942p) Spiders of Alachua County. Florida Entomol. 25(1):7–9
Murrill WA (1943a) Some southern novelties. Mycologia 35(4):422–433
Murrill WA (1943b) Additions to Florida fungi. Mycologia 35(5):529–537
Murrill WA (1943c) *Ithyphalluis murrillianus* West. Mycologia 35(6):655–656
Murrill WA (1944a) More new fungi from Florida. Lloydia 6:207–228
Murrill WA (1944b) New Florida fungi. Proc. Florida Acad. Sci. 6–7:107–127
Murrill WA (1944c) A few corrections. Mycologia 36(1):121–122
Murrill WA (1944d) *Odontia archeri* (Berk.) Wakefield. Mycologia 36(3):307
Murrill WA (1944e) A new *Russula* spore. Mycologia 36(3):308–309
Murrill WA (1945a) Autobiography. Publ. Priv, Gainesville. 184 pp. Hathi Trust: https://babel.hathitrust.org/cgi/pt?id=coo.31924003517681&view=1up&seq=6
Murrill WA (1945b) Historic foundations of Botany in Florida (and America). Publ. Priv, Gainesville. 51 pp
Murrill WA (1945c) *Hydnum floriforme* in Florida. Mycologia 37(1):156
Murrill WA (1945d) A new polypore from Utah. Mycologia 37(1):156–157
Murrill WA (1945e) Cylindric spores from *Amanita verna*. Mycologia 37(1):159–160
Murrill WA (1945f) Apparent allies of *Amanita verna*. Mycologia 37(2):270–271
Murrill WA (1945g) Common names may be highly important. Mycologia 37(5):636
Murrill WA (1945h) *Fistulina* in Florida. Mycologia 37(6):793
Murrill WA (1945i) *Abortiporus subabortivus* Murr. is valid. Mycologia 37(6):793–794
Murrill WA (1945j) *Boletus tabacinus* Peck. Mycologia 37(6):794
Murrill WA (1945k) Trees. Publ. Priv, Gainesville. 65 pp. Hathi Trust: https://babel.hathitrust.org/cgi/pt?id=mdp.39015068531352&seq=3
Murrill WA (1945l) A guide to Florida plants. Publ. Priv, Gainesville. 89 pp. [sale price = $2]
Murrill WA (1945m) Reptiles good and bad. Publ. Priv, Gainesville. 32 pp
Murrill WA (1945n) A guide to Florida animals. Publ. Priv., Rose Printing Co., Tallahassee. 98 pp. Hathi trust: https://babel.hathitrust.org/cgi/pt?id=wu.89097672158&seq=10
Murrill WA (1945o) New Florida fungi. Quart. J. Florida Acad. Sci. 8(1–4):175–198
Murrill WA (1945p) [1946] More Florida fungi. Lloydia 8: 263–290
Murrill WA (1946a) Flowers. Publ. Priv, Gainesville. 120 pp + illust. [Hathi Trust: https://babel.hathitrust.org/cgi/pt?id=mdp.39015042611494&view=1up&seq=5
Murrill WA (1946b) Notes on Florida fungi. Mycologia 38(1):113–114
Murrill WA (1946c) *Russula inconstans* Murrill. Mycologia 38(2):226
Murrill WA (1946d) *Inocybe hebelomoides* Murrill. Mycologia 38:226
Murrill WA (1946e) A new western polypore. Mycologia 38(3):348–349
Murrill WA (1946f) Familiar trees. Publ. Priv, Gainesville. 174 pp + illust. Hathi Trust: https://babel.hathitrust.org/cgi/pt?id=mdp.39015037957605&view=1up&seq=5
Murrill WA (1946g) Rocks. Publ. Priv, Gainesville. 62 pp
Murrill WA (1946h) Stars. Publ. Priv, Gainesville. 41 pp. [listed in "Pore Fungi" as $2]
Murrill WA (1946i) Pore fungi. Publ. Priv, Gainesville. 110 pp. [Sale price = $2]. Hathi Trust: https://babel.hathitrust.org/cgi/pt?id=mdp.39015006930112&view=1up&seq=6INSERT
Murrill WA (1946j) New and interesting Florida fungi. Lloydia 9:315–330
Murrill WA (1947a) New combinations. Mycologia 39(1):132
Murrill WA (1947b) Red fly-agaric in Florida. Mycologia 39(2):251
Murrill WA (1947c) *Fomes fraxineus* in Florida. Mycologia 39(2):251

Murrill WA (1947d) Ferns. Publ. Priv., Pepper Printing Co., Gainesville. 96 pp. Hathi Trust: https://babel.hathitrust.org/cgi/pt?id=mdp.39015052514810&seq=9
Murrill WA (1947e) Florida polypores. Lloydia 10:242–282
Murrill WA (1947f) Introduced woody plants of the Gainesville region (Preliminary list). Mimeo. Contrib. Herb. Univ. Fla. Agric. Exp. Stat. 12 leaves
Murrill WA (1948a) *Russula tricolor* Murr. Mycologia 40(5):635
Murrill WA (1948b) Florida boletes. Lloydia 11:21–35
Murrill WA (1948c) [1947] Florida polypores. Lloydia 10(1–4):242–282
Murrill WA (1948d) Species of Florida *Lactarius* Fr. Lloydia 11:86–88
Murrill WA (1948e) Florida amanitas. Lloydia 11:99–110
Murrill WA (1949a) Three new fleshy fungi. Mycologia 41:490–491
Murrill WA (1949b) Florida lepiotas. Lloydia 12:56–61
Murrill WA (1949c) Florida tricholomas. Lloydia 12:62–69
Murrill WA (1949d) [1950] Florida clitocybes (*Clitocybe*, *Monadelphus*, *Laccaria*). Lloydia 12:229–232
Murrill WA (1949e) Terrestrial basidiomycete fungi of Florida high-pine woods. Ecology 30:377–382
Murrill WA (1950) Success stories for girls. Pepper Printing Co., Publ. Priv, Gainesville. 68 pp. ["conduct of life;" Girl Scouts of America]
Murrill WA (1951a) New agarics from Florida. Mycologia 43(2):235–239
Murrill WA (1951b) The poroid genus *Dicanthodes* Singer. Mycologia 43(3):376
Murrill WA (1951c) Florida hydnums. Lloydia 14:115–121
Murrill WA (1951d) Florida Gasteromycetes. Lloydia 14(2):187–193
Murrill WA (1951e) Florida rosy-spored agarics. Lloydia 14(2):194–200
Murrill WA (1952) Notes on Florida agarics. Mycologia 44(1):112–114
Murrill WA (1953a) A new species of *Amanita*. Mycologia 45(5):794
Murrill WA (1953b) A valuable drug in a mushroom. Botanical studies on *Clitocybe subilludens* Murrill. Quart. J. Florida Acad. Sci. 16(4):233–234
Murrill WA (1954) *Agaricus blockii* sp. nov. Mycologia 46(1):112
Murrill WA (1955) A new species of *Amanita* from Florida. Mycologia 47(3):427
Murrill (2023) NYBG Library finding guide: https://www.nybg.org/library/finding_guide/archv/murrill_irb.html
Murrill WA, Wolfe CS, Lynch SJ (1942) New Florida hawthorns. Castanea 7:19–30
NatCordCo (2023). http://what-when-how.com/the-merican-economy/national-cordage-company/
National WW II Museum (2024). https://www.nationalww2museum.org/war/articles/pacific-strategy-1941-1944
Nicolson DH (1991) A history of botanical nomenclature. Ann Missouri Bot Gard 78(1):33–56
Nineteenth Amendment Wiki (2023). https://en.wikipedia.org/wiki/Nineteenth_Amendment_to_the_United_States_Constitution
NIST (2024). https://www.nist.gov/blogs/taking-measure/big-inch-fueling-americas-wwii-war-effort
NJCU (2022). https://njcu.libguides.com/lorillard
Normal College (2023). https://library.hunter.cuny.edu/sites/default/files/documents/archives/finding_aids/Normal_College_Collection_Finding_Aid.pdf
NYSE History (2023). https://www.nyse.com/history-of-nyse
Ohio History Central (2023). http://ohiohistorycentral.org/w/Panic_of_1893#:"text=The%20Panic%20of%201893%20was,erupted%20on%20the%20stock%20market
Olivet College (2023). https://www.olivetcollege.edu/about-olivet-college/history/
Osborn HF (1911) The American Museum of Natural History, Its origin, its history, the growth of its departments to December 31, 1909, 2nd edn. Irving Press, New York. 148 pp
Overholts LO (1911) The known Polyporaceae of Ohio. Ohio Nat 11:353–373. Hathi Trust: https://babel.hathitrust.org/cgi/pt?id=uc1.b3080350&seq=9
Overholts LO (1914) The Polyporaceae of Ohio. Ann Missouri Bot Gard 1(1):81–155

P&R Railroad (2023). http://www.american-rails.com/rdg.html
Panic of 1893 Wiki (2023). http://en.wikipedia.org/wiki/Panic_of_1893
Patouillard N (1884) Les Hyménomycètes d'Europe. Anatomie Général et Classification des Champignons Supérieurs. P. Klincksieck, Paris. 166 pp + 4 pls
Patouillard N (1900) Essai Taxonomique sur les Familles et les Genres des Hyménomycètes. Lons-le-Saunier: Impr. et lithographie L. Declume. 184 pp. [Reprint, Asher & Co. 1963, with introductory note by M. A. Donk]
Peck CH (1899) Plants of North Elba. Bull N Y State Mus 6(28):67–266. [with map]
Petersen RH (1975a) Specific and infraspecific names for fungi used in 1821. Part I. Introduction, A and B. Mycotaxon 1:149–188
Petersen RH (1975b) Specific and infraspecific names for fungi used in 1821. Part II. C, and supplement to Part I. Mycotaxon 2:151–165
Petersen RH (1976a) Specific and infraspecific names for fungi used in 1821. Part III. D–G. Mycotaxon 3:239–260
Petersen RH (1976b) Specific and infraspecific names for fungi used in 1821. IV. H–M. Mycotaxon 4:185–210
Petersen RH (1977) Specific and infraspecific names for fungi used in 1821. V. N–Z. Mycotaxon 6:78–126
Petersen RH (1978) Lexemuel Ray Hesler. February 20, 1888–November 20, 1977. Mycologia 70:757–765
Petersen RH (1980) The mycological association of M. J. Berkeley and M. A. Curtis. Biblioth Mycol 72:1–120
Petersen RH (2001) New world botany: Columbia to Darwin. Koeltz Sci. Books. 618 pp
Petersen RH (2016) Making sense of color. Fungi 8:5–25
Petersen RH (2019) North American mycology comes of age: the 19th century participants and their roles. Mycotaxon 134:1–110
Petersen RH (2020a) William Hamilton Gibson: a case of serendipity. Fungi 13(1):1–25
Petersen RH (2020b) A mycologist on mushroom books: 1901. Fungi 13(3):9–17
Petersen RH (2021) Four master teachers who fostered American turn-of-the-(20th)-century mycology and plant pathology. Mycotaxon 136:1–58
Petersen RH (2023) Curtis Gates Lloyd, American mycological narcissist. Mycotaxon 1327:399–434
Petersen RH (2024) Rumph to Donk: pre-Indonesian mycology and a biography. Mycotaxon 137(4, Suppl. 1):99–230
Petersen RH (2025) Joined at the hip: the birthing of the Boyce Thompson Botanical Institute and the Tropical Plant Research Foundation. Huntia. (in press)
Petersen RH, Hughes KW (1998) Mating systems in *Omphalotus* (Paxillaceae, Agaricales). Plant Syst Evol 211:217–229
Petersen RH, Hughes KW (2021) *Collybiopsis* and its type species, *Co. ramealis*. Mycotaxon 136:263–349
Petersen RH, Knudsen H (2015) The mycological legacy of Elias Magnus Fries. IMA Fungus 6(1):99–114
Petersen RH, Knudsen H (2022) Jakob Emanuel Lange: the man and his mushrooms. MycoKeys 89:1–86
Pfister DH (1981) Atkinson, Farlow, and the later starting point debate. Occas Pap Farlow Herb Crypt Bot 16:129–136
Pfister DH, Boise JR, Eifler MA (1990) A bibliography of taxonomic mycological literature 1753–1821. Mycologia Mem 17:1–161
Pierce TJ, Little TJ (2020) The rule, not the exception: slavery and manumission in northeast Tennessee and southwest Virginia. J East Tennessee Hist 92:77–112
Pierre Lorillard Wiki (2022). https://en.wikipedia.org/wiki/Pierre_Lorillard_IV
Pollard CL (1901) The Rochester Code. Bot Gaz 31(4):285–286
Prohibiton Wiki (2023). https://en.wikipedia.org/wiki/Prohibition_in_the_United_States

Pumpelly R (1893) John Strong Newberry. Proc Am Acad Arts Sci 28:394–398
Randolph-Macon History (2023). https://www.rmc.edu/about/history
Ridgway R (1886) Nomenclature of colors for naturalists, and compendium of useful knowledge for ornithologists. Little, Brown & Co, Boston. 129 pp + 17 pls, several colored
Ridgway R (1912) Color standards and color nomenclature. Publ. Priv, Washington, DC. 43 pp + 53 cld. pls
Ridgway Facsimile (1886). https://catalog.lindahall.org/discovery/delivery/01LINDAHALL_INST:LHL/1284877090005961
Ridgway Facsimile (1912). https://catalog.lindahall.org/discovery/delivery/01LINDAHALL_INST:LHL/1284377770005961
Riis J (1890) How the other half lives: studies among the Tenements of New York. Charles Schribner's Sons, New York. 304 pp
Riis Photos (2023). https://mymodernmet.com/jacob-riis-how-the-other-half-lives/#:~:text=In%201890%2C%20Riis%20compiled%20his,realities%20of%20their%20city's%20slums
Robinson BL (1895a) The nomenclature question. A further discussion of the Madison rules. Bot Gaz 20:370–371
Robinson BL (1895b) On the "List of Pteridophyta and Spermatophyta of northeastern America". Prepared by the Nomenclature Committee of the Botanical Club. Bot Gaz 20(3):97–103
Robinson BL (1895c) The nomenclature question. On the application of "once a synonym always a synonym" to binomials. Bot Gaz 20:261–263
Robinson BL (1897) The official nomenclature of the Royal Botanical Garden and Museum of Berlin. Bot Gaz 24:107–110
Robinson BL (1898) Some reasons why the Rochester Nomenclature cannot be regarded as a consistent or stabile system. Bot Gaz 25(6):437–445
Rogers DP (1981) A brief history of mycology in North America. Mycol Soc Am. 86 pp
Rogerson CT (1973) Fred Jay Seaver: 1877–1970. Mycologia 65(4):721–724
Rogerson CT, Samuels GJ (1996) Mycology at The New York Botanical Garden, 1895–1995. Brittonia 48(3):389–398
Romer CD (1993) The nation in depression. J Econ Persp 7(2):19–39
Rose D (2002) William Alphonso Murrill. Mushroom Mag 20(4):1–4. [with portrait and partial bibliography]. https://www.mushroomthejournal.com/greatlakesdata/Authors/Murrill19.html
Roza S, Maniutiu DN, Marta GT, Stela D, Suzana BBF (2016) Research on the biology of the *Agaricus blazei* Murrill mushroom mycelium. J Hort For Biotechnol 20(1):123–126
Rudolph ED (1996) History of the botanical teaching laboratory in the United States. Am J Bot 83(5):661–671
Rusby HH (1906) A historical sketch of the development of botany in New York City. Torreya 6(6):101–111
Rusby HH (1907) The work of Professor Lucien Marcus Underwood. J N Y Bot Gard 8(96):263–269. [with portrait]
Rusby HH (1921) Guide to the economic museum of the New York Botanical Garden. Bull N Y Bot Gard 11(41):1–318
Rydberg PA (1927) Botanizing in the higher Alleghany Mountains. II, West Virginia and Tennessee. J N Y Bot Gard 27(1):33–40
Sachs Wiki (2023). https://en.wikipedia.org/wiki/Julius_von_Sachs
Samuels GJ, Fogel R (2003) Clark Thomas Rogerson, 1918–2001. Mycologia 95(4):773–779
Sarah Sundin (2022). https://www.sarahsundin.com/make-it-do-gasoline-rationing-in-world-war-ii-2/
Schneider BA (1946) [Review] Murrill WA 1945. A guide to Florida animals. Q Rev Biol 21(4):387
Schubert R, Knauer S, Seibt S, Biersack B (2011) Anticancer active illudins: recent developments of a potent alkylating compound class. Curr Med Chem 18(6):790–807
Scudder's Museum Wiki (2023). http://en.wikipedfia.org/wiki/Scudder%27s_Amjerican_Museum
Seaver FJ (1909a) The Hypocreales of North America – I. Mycologia 1(2):41–76
Seaver FJ (1909b) The Hypocreales of North America – II. Mycolgia 1(5):177–207

Seaver FJ (1909c) Notes on North American Hypocreales – I. Mycologia 1(1):19–22

Seaver FJ (1910a) The Hypocreales of North America – III. Mycologia 2(2):48–90

Seaver FJ (1910b) Hypocreales, Fimetariales, Nectriaceae, Hypocreaceae. N Am Flora 3(1):3–64

Seaver FJ (1911) The Hypocreales of North America – IV. Tribe IV. Cordycipiteae. Mycologia 3(5):207–230

Seaver FJ (1926) Mycophagy. Mycologia 18(2):94–95

Seaver FJ (1931) Mycologia. General Index, Volumes 1–24 (1909–1932)

Seth Low Wiki (2023). https://en.wikipedia.org/wiki/Seth_Low

Shear CL (1904) The American Mycological Society. J Mycol 10(1):46–47

Shear CL (1906) American Mycological Society, New Orleans meeting, January 1, 1906. J Mycol 12(2):85–86

Singer R (1942a) Type studies on agarics. Lloydia 5:97–135

Singer R (1942b) Type studies on Basidiomycetes. I. Mycologia 34:64–93

Singer R (1943) Das System der Agaricales III. Ann Mycol 41:1–189

Singer R (1945a) New Boletaceae from Florida (A preliminary communication). Mycologia 37(6):797–799

Singer R (1945b) The Boletineae of Florida with notes on extralimital species. I. The Strobilomycetceae. Farlowia 2(1):97–141

Singer R (1945c) The Boletineae of Florida with notes on extralimital species. II. The Boletaceae (Gyroporoideae). Farlowia 2(2):223–303

Singer R (1946a) Type studies on agarics. II. Lloydia 9:114–131

Singer R (1946b) The Boletineae of Florida with notes on extralimital speies. IV. The lamellate families (Gomphidiaceae, Paxillaceae and Jugasporaceae). Farlowia 2:527–567

Singer R (1947) The Boletineae of Florida with notes on extralimital species. III. The Boletoideae of Florida. Am Midl Nat 37(1):1–135

Singer R (1951) The "Agaricales" (mushrooms) in modern taxonomy. Lilloa 22:1–768

Singer M (1984) Mycologists, and other Taxa. Braunschweig, J. Cramer. 115 pp

Singer R (1986) The Agaricales in modern taxonomy, 4th edn. Koeltz Sci. Books, Konigstein. 981 pp

Singer R, Digilio AP (1952) Review: a dictionary of Color by M.A. Mearz and M.A. Paul. Mycologia 44(2):267–269

Singer R, Smith AH (1946) Proposals concerning the nomenclature of the gill fungi including a list of proposed lectotypes and genera conservanda. Mycologia 38(3):240–298

Sinon S (2022). https://www.nybg.org/blogs/plant-talk/2015/08/history/the-lorillard-jar/

Sledd Wiki (2024). https://en.wikipedia.org/wiki/Andrew_Sledd

Sloan D (1980) Science in New York City. Isis 71(1):35–76

Small JK (1933) Manual of the Southeastern Flora. Publ. Priv, New York. 1554 pp

Small JK (1938) Ferns of the Southeast. Science Press, Lancaster. 517 pp

Smith EF (1892) Suggested by Kuntze's "Revisio Generum Plantarum". Bot Gaz 17(2):62

Smith AH, Hesler LR (1938) Notes on agarics from Tennessee and North Carolina. J Elisha Mitchell Sci Soc 54:261–269

Smith AH, Hesler LR (1942) Studies in North American species of Hygrophorii. Lloydia 5:1–94

Smith-Vikos T (n.d.) Murrill by Smith Vikos (On-line)

Snell WH (1941) The genera of the boletes. Mycologia 33(4):415–423

Spanish American War (2023). https://history.state.gov/milestones/1866-1898/spanish-american-war#:~:text=The%20Spanish%2DAmerican%20War%20of,States%20as%20a%20Pacific%20power

Sparrow FK (1947) (Review) Pore fungi by William Alphonso Murrill. Q Rev Biol 22(1):69

Stamets PE (2000) Techniques for the cultivation of the medical mushroom Royal Sun Agaricus – *Agaricus blazei* Murr. (Agaricomycetideae). Int J Med Mush 2(2):151–160

Stamets P (2005) Mycelium running. Ten Speed Press, Berkeley. 339 pp

Steeples D, Whitten DO (1998) Democracy in desperation: the depression of 1893. Greenwood Press. 280 pp

Stijve TD, Luchini G-F, Goesler W (2001) Simultaneous uptake of rare earth elements, aluminum, iron, and calcium by various macromycetes. Australas Mycol 20:92–98

Stijve TD, Pittet A, Andrey D, Amazonas MALDA, Gessler W (2003) Potential toxic constituents of *Agaricus brasiliensis* (*A. blazei* ss. Heninem.) as compared to other cultivated and wild-growing edible mushrooms. Deutsche Lebensmitteil-Rundschau 99:475–481

Stuckey RL (1978) Index to biographical sketches and obituaries in publications of the Ohio Academy of Science, 1900–1970. Ohio J Sci 70(4):246–255

Sundberg MD (2014) Botanical education in the 1880s and 90s. Part 3. The Botanical Society builds the discipline, 1895–1960. Plant Sci Bull 60(1):28–61

Swett SC (1960) The test of a reformer: a study of Seth Low, New York City Mayor, 1902–1903. N Y Hist Soc Q 44(1):5–41

Tammany Hall Wiki (2023). https://en.wikipedia.org/wiki/Tammany_Hall

Taylor JF (1910) Dr. Kellerman in Guatemala. Ohio State Univ Q 1(2):4–11

Temperate House (2022). https://www.kew.org/read-and-watch/secrets-kew-temperate-house#:~:text=History%20of%20the%20house&text=The%20Temperate%20House%20took%20several,to%20house%20frost%2Dtender%20plants

The Current (Magazine) Wiki (2023). https://en.wikipedia.org/wiki/The_Current_(magazine)

The Eye (2023). http://www.columbiaspectator.com/the-eye/2019/04/15/nearly-a-century-ago-columbias-jewish-applicants-were-sent-to-brooklyn/

Thiers BM, Halling RE (2003) Harry D. Thiers, 1919–2000. Mycologia 95(6):1271–1275

Thom C (1956) George Francis Atkinson, 1854–1918. Nat Acad Sci Biograph Mem 1956:17–44

Thompson Wiki (2023). https://en.wikipedia.org/wiki/William_Gilman_Thompson

Tin Can Tourist Camp (1935). https://www.floridamemory.com/learn/exhibits/photo_exhibits/tin-cans/tincans2.php

Topper J (2011) College Presidents, Public Image, and the Popular Press: a comparative study of Francis L. Patton of Princeton and Seth Low of Columbia, 1888–1902. Iconic Leaders in Higher Education (series), TaylorFrancis.com

Tracy SM, Earle FS (1901) Some new fungi. Bull Torrey Bot Club 28:184–188

Turow S (1977) One L: an inside account of life in the first year at Harvard Law School. Putnam Publ. 288 pp

Tweed Wiki (2023). http://em.wikipedia.org/wiki/William_M_Tweed

Underwood LM (1878) List of ferns growing in the vicinity of Syracuse, Onondaga Co., NY. Case's Bot Index 1(1):80

Underwood LM (1881) Our native ferns and how to study them, Bloomington. 116 pp

Underwood LM (1884) Descriptive catalogue of the North American Hepaticae, north of Mexico. Bull Illinois State Lab Nat Hist 2:1–133

Underwood LM (1888) Some undescribed Hepaticae from California. Bot Gaz 13:112–114. pls. 3–6

Underwood LM (1891) Diseases of the orange in Florida. J Mycol 7:27–36

Underwood LM (1892a) The nomenclature question at Genoa. Bull Torrey Club 19:325–330

Underwood LM (1892b) The international congress at Genoa. Bot Gaz 17:341–347

Underwood LM (1892c) The botanical congress. Bot Gaz 17:425–426

Underwood LM (1893) Index Hepaticarum. Part 1. Mem Torrey Bot Club 4(1):1–91

Underwood LM (1894) The evolution of the Hepaticae. Bot Gaz 19:347–361. Proc Am Assoc Adv Sci 43:259–274

Underwood LM (1896a) On the distribution of North American Helvellales. Minn Bot Stud 9:483–500

Underwood LM (1896b) Mycology in the Southern states. Gard For 9:263–264

Underwood LM (1896c) *Coleosporium Campanulae* (Pers.) Wint. Bull Torrey Bot Club 23:423

Underwood LM (1897a) Notes on the American Hydnaceae – I. Bull Torrey Bot Club 24(4):205–206

Underwood LM (1897b) Some new fungi, chiefly from Alabama. Bull Torrey Bot Club 24:81–86

Underwood LM (1898a) *Selaginella rupestris* and its allies. Bull Torrey Bot Club 25(3):125–133

Underwood LM (1898b) American ferns – I. The ternate species of *Botrychium*. Bull Torrey Bot Club 25(10):521–541

Underwood LM (1898c) Two recently named genera of Basidiomycetes. Bull Torrey Bot Club 25(12):630–631

Underwood LM (1899a) A review of the genera of ferns proposed prior to 1832. Mem Torrey Bot Club 6(4):247–283
Underwood LM (1899b) Moulds, mildews, and mushrooms: a guide to the systematic study of the Fungi and Mycetozoa and their literature. H. Holt, New York. 236 pp
Underwood LM (1900) Why *Dryopteris* and not *Lastraea*? Fern Bull 8:52–55
Underwood LM (1901a) A changed conception of species. Fern Bull 9:49–53
Underwood LM (1901b) The names of our ferns. Bot Gaz 31(5):365–366
Underwood LM (1901c) Mycophagy and its literature. Torreya 1(4):43–46
Underwood LM (1902) Report of the scientific directors. Bull N Y Bot Gard 2:310–312
Underwood LM (1903) The Department of Botany and its relation to the New York Botanical Garden. Contrib Dept Bot Columbia Univ 9:201–225
Underwood LM, Cook OF (1889) A century of illustrative fungi, with generic synopses of the Basidiomycetes and Myxomycetes, Syracuse. 21 pp
Underwood LM, Earle FS (1896a) The distribution of the species of *Gymnosporangium* in the South. Bot Gaz 22:256–258
Underwood LM, Earle FS (1896b) Notes on the pine-inhabiting species of *Peridermium*. Bull Torrey Bot Club 23:400–405
Underwood LM, Earle FS (1897) A preliminary list of Alabama fungi. Alabama Exp Stn Bull 80:111–284. + i–xvii Index
UVA Station (2023). MLBS.Virginia.edu
von Sachs J (1870) Lehrbuch der Botanik, 2nd edn. Verlag von Wilhelm Engelmann
VPI History (2023). https://www.google.com/search?q=Virginia+Polytechnic+Institution%2C+history&rlz=1C1KMZB_enUS578US578&oq=Virginia+Polytechnic+Institution%2C+history&aqs=chrome..69i57j0i22i30j0i390i650l3.11502j0j7&sourceid=chrome&ie=UTF-8
Waksman SA (1947) What is an antibiotic or an antibiotic substance? Mycologia 39:565–569
Ward LF et al (1893) The nomenclature question. Bull Torrey Bot Club 22(7):308–329. [other "Committee" authors; Geo. Vasey, F. I. Knowlton, B.T. Galloway, Erwin F. Smith, Geo. B. Sudworth, Frederick V. Coville]
Warner JN (1962) Sugar cane: an indigenous Papuan cultigen. Ethnology 1(4):405–411
Watson S (1892) On nomenclature. Bot Gaz 17(6):169–170
Weber GF (1961) William Alphonso Murrill. Mycologia 53(6):543–557. [incl. partial bibliography by C. T. Rogerson]
Weber GF (1966) Erdman West (1894–1946). Mycologia 58(2):180–183. [with portrait and bibliography]
West E (1939a) Notes on Florida fungi. I. Mycologia 31(3):423–532
West E (1939b) Preliminary list of myxomycetes from Alachua County. Proc Fla Acad Sci 4:212–217
West E (1941) Notes on Florida fungi. II. Mycologia 33(1):38–49
West E (1945) Notes on Florida fungi. III. Mycologia 37(1):65–79
White VS (1901) The Tylostomataceae of North America. Bull Torrey Bot Club 28:421–444
White CA (1902) Biographical memoir of John Strong Newberry, 1822–1892. Proc Natl Acad Sci 20:1–24. [with portrait and bibliography]
Wikipedia WW II (2024). https://en.wikipedia.org/wiki/Timeline_of_World_War_II_(1944)
Wikiword (2024). https://www.wikiwand.com/en/Victor_Fayod
Willard JT (1909) William Ashbrook Kellerman. Trans Kansas Acad Sci 22:46–47
William Colford Schermerhorn Wiki (2023). https://en.wikipedia.org/wiki/William_Colford_Schermerhorn
Williams RS (1908) Report on botanical exploration in Panama. J N Y Bot Gard 9(105):146–158
World War I Wiki (2023). https://en.wikipedia.org/wiki/World_War_I
Yama Collection (2023). https://nyheritage.org/collections/yama-farms-collection (see especially photo album under "Presidents Conference")
Yama Farms (2023). https://yamafarmsinn.net/Yama_Farms_Inn__A_Home_in_the_Mountains/Welcome.html
Zanoni TA (1980) Otto Kuntze, botanist. I. Biography, bibliography and travels. Brittonia 32(4):551–571

Index

A
A.A. Low & Brothers, 257
Addisonia, 152, 249
Adirondack Mountains, NY, 88, 112, 120, 121, 125, 242
Africa, 100, 156, 198, 229
The Agaricales in modern taxonomy, 201
Agassiz, L., 11, 12, 32, 84
Agriculture Experiment Station at the University of Florida (FLAS), 170
Alabama, 18, 19, 55, 94, 241, 251
Alabama Agriculture and Mechanic College, 238
Alabama Agriculture Experiment Station, 251
Alabama Polytechnic College (Auburn), 12, 40, 47, 49, 239, 240
Alachua County, FL, 170, 181, 185, 189, 193, 220, 272, 273
Alachua General Hospital, 176, 229
Albany, NY, 33, 45, 59, 76, 87, 88, 92, 104, 112, 113, 140, 145, 151
Alsace, France, 119
American Association for the Advancement of Science (AAAS), 51, 71, 73–75, 95, 111, 125, 150, 265, 268–270
American Bryological and Lichenological Society, 243
American Code of Botanical Nomenclature, 74, 76, 120, 124, 127, 190, 221, 226
American Gardener's Magazine (later Magazine of Horticulture), 249
American Midland Naturalist, 224
American Museum, NY, 27, 33, 35
American Museum of Natural History, 27, 31, 33, 44, 158, 246, 249, 265
American Mycological Society, 51, 115
American Naturalist, the, 82
American Nature Study Movement, 147
American Phytopathological Society, 51
Amherst College, MA, 248
Annals of Botany, 23
Annals of the Missouri Botanical Garden, 150–151
Anthony, S.B., 145
Anti-Saloon League, 139
Appalachian Mountain Club, 61
Appalachian Mountains, 3, 56, 159, 204, 268
Apple Orchard Mountain, VA, 3, 12, 125, 136, 150, 159
Arthur, J.C., 43, 74–76, 82, 83, 240, 252, 253, 268
"Ashcan artists", 263
Asheville, NC, 89, 90
Ashland, VA, 9, 171
Atkinson, G.F., 15, 17–23, 51, 52, 76, 94, 103, 108, 116, 121–124, 151, 152, 204, 214, 219, 235, 237–241, 245
Australia, 206
Austria-Hungary, 119
Autobiography, 4, 7, 9, 15, 20, 23, 45, 47, 61, 63, 70, 98, 113, 116, 117, 125, 140, 150, 158, 167, 172, 174, 178, 180, 189, 208, 212, 213, 217, 218, 221, 230

R. H. Petersen, *William Alphonso Murrill: The Naturalist*, Memoirs of The New York Botanical Garden 124, https://doi.org/10.1007/978-3-031-93343-1

B

Bahamas, 94, 244
Bailey, Liberty Hyde, 20–22, 147, 239, 240, 243
Baker, C.F., 251
Baker, G., 27
Barnard College, 36, 44, 246, 259
Barnard, F.A.P., 259
Barnes, C.R., 43, 44, 71, 268
Barnhart, J.H., 50, 189, 244, 267, 271
Barnum, P.T., 28, 33
Bastogne, Belgium, 209
Bedford, VA, 3, 7, 12, 56, 125, 150, 168
Beebe, W., 158
Belgium, 103, 119, 188, 206, 222
Bellows, H.W., 30
Berkeley, M.J. (Rev.), 58, 69, 103
Berlin, Germany, 47, 52, 66, 68, 72, 100, 198, 206, 209, 210
Berte, H.C., 99
Bessey, C.E., 43, 73, 82, 83, 240, 243
Bickmore, A.S., 32, 33
Bigelow, H.E. (GI), 209
Billy, the Boy Naturalist, 135–137
Biltmore, NC, 76, 88–92, 94, 125, 204, 213
Biltmore School of Forestry, 89, 90, 158
Biological Survey Of Alachua County, FL, 181
Blacksburg, VA, 3, 6, 7, 14, 16, 45, 53, 56, 61, 96, 97, 125, 144, 149, 150, 153, 155, 180
Bloomingdale Asylum, NY, 261
Blue Ridge Mountains, NC, 56, 168
Bondarzev, Apollinary Semjonovich, 201
Botanical Bulletin (later Botanical Gazette), 43, 44, 72, 82, 85, 241, 254, 268
Botanical Club of AAAS, 71, 73, 75
Botanical Gazette, 43, 44, 72, 82, 85, 241, 254, 268
Botanical Society of America, 48, 51, 54, 71, 82, 111, 125, 270
Bowling Green Female Seminary, VA, 9
Bowling Green, VA, 9, 10, 15, 23
Boy Scouts of America, 213
Braun, W. von, 206
Brazil, 156, 201, 222
Bresadola, A.G., 53, 54, 68, 116, 151, 182, 198
Brevard, NC, 89, 90
British Association for the Advancement of Science, 74
British Guiana (presentday Guyana), 156, 158
Britton, E.G.K., 15, 140, 214, 241–245, 248
Britton, N.L., 20, 35, 36, 40, 41, 44, 50, 52, 55–58, 63, 70–72, 74, 77, 81–83, 86, 92, 93, 97–99, 104, 111, 112, 117, 118, 126–129, 132, 137–139, 141, 145, 146, 149, 152, 154, 159, 160, 168, 217, 241, 242, 244–248, 251, 261, 262, 264–266, 270, 271
Bronx Park, NY, 37, 39, 56, 58
Bronx Zoo, NY, 37, 58
Brooklyn Bridge, 258, 1883
Brooklyn Institute of Arts and Sciences, 120
Brooklyn, NY, 153, 257, 258, 260, 262, 263
Brooklyn Polytechnic Institute, 257, 268
Brown, A.C., 30, 35–40, 60, 152, 248–249
Buenos Aires, Argentina, 156, 157
Bulgaria, 205
Bulletin of the New York Botanical Garden, 81, 103, 138, 141, 152, 243, 246
Bulletin of the Torrey Botanical Club, 46, 50, 73, 76, 95, 138, 189, 191, 193, 241, 242, 245
Bulletin of the Yama Farms Mycological Club, 145
Burbank, L., 84, 108
Burlingham, G.S., 56, 92, 120, 128, 234
Burlington, Vt, 87
Burma (presentday Myanmar), 206
Burnham, S.H., 140
Burrill, T.J., 85, 249, 250
Burroughs, J., 144, 145
Burt, E.A., 114, 128, 150, 151, 190, 214
Butler, N.M., 40

C

California, 67, 84, 94, 107, 108, 200, 238, 264, 268
Cambridge, MA, 32, 268
Canada, 44, 113, 150, 151, 156, 244, 248
Candolle, A.de, 71, 75
Caribbean sea, 94
Carnegie, A., 36, 63, 263
Carnegie Hall, 262, 1891
Carnegie Institution, 84
Cascadilla Building, Cornell University, 22
Catskill Mountains, NY, 125, 144
Cayuga Lake, NY, 94
Cazenovia Seminary, NY, 266
Central Park, New York City, 32, 35, 37, 38
A Century of Illustrative Fungi, 268
Challenger Expedition, 103, 1875
Champs Élysées, Paris, 205

Chapel Hill, NC, 122, 127, 151
Chappaqua, NY, 97
Chestnut blight, 58–62, 66, 76, 86, 91, 112
Chicago, IL, 10, 39, 40, 43, 44, 82, 85, 107, 145, 201, 249, 254, 256, 268
Chicago World's Fair, 10, 11, 1892
Choate, J., 30
Churchill, W., 195
Cinchona, J., 94, 111
Clements, E.F., 51, 82, 83, 190
Cleveland Medical College, 264
Cleveland, S.G., 258
Clute, W.N., 14
Coker, W.C., 122–124, 127, 128, 150, 151, 188, 190, 235
Cold Spring Harbor Laboratory, MA, 84
Colima, mx, 98
Colorado, 146, 264
Columbia College, 33, 35, 40, 70–73, 245, 257, 259, 265
Columbia College Bulletin, 261
Columbia College of Physicians and Surgeons, 260
Columbian Exposition, Chicago, 39, 256, 1893
"Columbian Gallery of Painting and City Museum", 27
Columbia School of Mines, 31, 35, 245, 264
Columbia University, 12, 23, 36, 43, 44, 52, 54, 88, 92, 104, 105, 107, 123, 153, 246, 248, 257, 259, 262, 268, 270
Comstock, J.H., 21
Confederacy, the, 7, 270
Connecticut, 49, 77, 144, 146, 271
Conservatoire et Jardin Botaniques de Geneve, 246
Cooke, M.C., 69
Cook, O.F., 74, 267, 268
Copenhagen, Denmark, 69, 100
"Corcovado", Brazil, 156, 158
Cornell Herbarium, 22, 45, 116
Cornell University, 15, 16, 18, 27, 94, 123, 129, 252
Cornell University Botany Department, 20, 239, 241
Cotton, A., 101
Cotton Centennial Exposition, New Orleans, 249, 1884
Coulter, J.M., 43, 44, 72, 82, 84, 85, 240, 241, 254, 268
Coville, F.V., 15, 74
Cowles, H.C., 82, 84
Crimea, 205
Crisis of 1907, 76–77, 271
Crooke, J.J., 244, 245
Crystal Palace, the, 28, 29, 35
Cuba, 15, 20, 50, 52, 55, 57, 58, 92, 94, 99, 103, 126, 154, 241, 244, 250, 251, 269, 271
Cuba Fruit Exchange, 251–252
Cuernavaca, mx, 98
Curtis, B.L., 258
Curtis, C.C., 268, 269
Curtis, M.A., 121, 233, 268

D
Dahlem Botanical Garden, Berlin, 100
Daly, Charles (Judge), 30, 39, 56, 93
Damrosch, F., 10
Dana, C.A., 35
Darius Ogden Mills Fund, 152
Dartmouth University, NH, 32
Darwin, C., 84, 147, 167, 214, 244
Delaware Water Gap, PA, 128, 129
De Leon Springs, FL, 167, 169
Deoxyribose nucleic acid (DNA), 234, 235
Department of Public Parks, NY, 35
De Pauw University, Greencastle, Indiana, 269
Dewitt Clinton High School, NY, 23
"Docentry", 141, 146–147, 150, 152, 153, 155, 161
Dodge, W.E., 30
Dudley, W.R., 18, 238, 240, 241
Duggar, B.M., 18, 20, 22, 239
Dunkirk, France, 195
Durand, E.J., 18, 22
Duss, A., 100
Dutch East Indies (presentday Indonesia), 206
Dwight, T.W., 249, 260, 261

E
Earle, F.S., 47, 49–52, 55–58, 94, 95, 103, 234, 249–252, 270, 271
Earle, P., 249, 250
Earle, S.B.S., 250
Earle, W.P., 250
Eaton, A., 42
Eaton, M., 112
Edison, T., 144
Elgin Botanical Garden, NY, 35, 42
Ellis, J. B., 52, 246, 251, 253, 254
Emancipation Proclamation, 7
Engler, H.G.A., 68, 95
Estación Agronómica at Santiago de las Vegas, Cuba, 251
Est. Morelos, mx, 98

European Herbaria, 100, 1910
Evergreen Cemetery, FL, 231
Everhart, B.M., 52, 246, 253

F
Falls Church, VA, 16, 17, 47, 48, 53, 56, 76, 88, 111, 120, 121, 125, 144, 149
Familiar trees, 212, 213, 221, 227
"Famous Four", 144, 145
Fargo, ND, 86
Farlow Herbarium, Harvard University, 201, 222, 225
Farlowia, 225
Farlow, W.G., 43, 76, 94, 103, 152, 201, 222, 225, 240
Fayod, V., 202, 233
Fern Bulletin, 14, 237
Ferns, 12, 14, 15, 20, 22, 23, 48, 50, 52, 53, 72–74, 153, 154, 218, 220, 226, 227, 234, 240, 242, 267, 268, 271
Field Museum, Chicago, 107, 145, 201
"Financial crisis of 1893", 39
Fink, B., 113, 114, 196
Firestone, H., 144
Fitzpatrick, H.M., 51, 129
Florida Academy of Science Journal, 207
Florida Farm Bureau, 175
Florida Social Welfare Board, 229
Florida Times-Union, 213
Flowers, 10, 140, 141, 167, 178, 190, 214, 218, 221, 227, 230, 231, 266
Ford, H., 144
Fordham University, 36, 39
Formosa (presentday Taiwan), 206
Francis Lynde Stetson Fund, 152
Fries, E.M., 49, 65, 66, 69, 76, 100, 202
Frost, C.C., 87
Frye, T.C., 107

G
Gainesville, FL, 9, 154, 167–170, 172–182, 185–188, 190, 191, 195, 196, 199, 200, 202, 203, 207, 208, 213, 215, 220–222, 225, 229, 231, 232, 235, 236
Galerie d'Evolution, Paris, 100, 101
Galerie du Zoologie, Paris, 100, 101
Ganong, W.F., 44
Garden Club of Gainesville, 179
The Genera of Fungi, 190
Generic Synopses of the Basidiomycetes and Myxomycetes, 268
Geological Society of America, 265
Geological Society of London, 265
Geological Survey of New Jersey, 245
Gilbertson, R.L. (GI), 209
"The Gilded Age", 39
Girl Scouts, 23, 69–70, 146, 147, 153–155, 161, 167, 171, 172, 180, 182, 207, 218, 221
Gleason, H.A., 15, 16, 56, 117, 138–142, 155, 160, 246, 247
Godkin, E.L., 30
Godwin, P., 30
Göttingen, Germany, 252
Governor's Island, NY, 31
Gratacap, Louis Pope, 265
Gray, A., 35, 43, 74, 84, 239, 240, 242, 245, 268
Great Depression (The), 173, 174, 183, 197, 200, 202, 209, 214
Great Smoky Mountains National Park, 204
Greeley, Horace, 30
Green, A.H., 30, 32, 33
Green Cove Springs, FL, 167, 169
Greene, E.L., 71–73
Gregory, E.L., 44
Grout, A.J., 56, 243
Guadalajara, mx, 98
Guam, 206
Guatemala City, Guatemala, 255
A Guide to Florida Plants, 215, 216
Guide to the Flowers of Florida, 214
Gulf of Mexico, 197, 251

H
Habsburg family, 129
Hagelstein, R., 170
Haiti, 94
Halling, R.E., 194, 211
Hamburg, Germany, 100
Hariot, P., 66, 100
Harper, R.A., 104, 107, 111, 123
Hartsville, SC, 122
Harvard University, 94
Harz Mountains, Germany, 252
Hedding College, Abington, Illinois, 267
Heim, R., 201
Heinemann, P., 222, 223
Hennings, P.C., 68, 100
Henry Shaw School of Botany, St. Llouis, 114
Hepaticae Americanae, 267
Hesler, L.R., 194, 203, 204, 233–235
Highlands, NC, 235

Hiroshima, Japan, 211
Historic Foundations of Botany, 213
History of North American Birds, 115
Hitler, A., 188, 195, 201, 205, 206, 209
Hohenzollern family, 129
Holland Society of New York, 153, 154
Hollick, C.A., 35, 56, 245, 264, 265, 271
"Honeysuckle Lodge", 171
Horatio Alger stories, 227
Hosack, D., 35, 41, 42
House, H.D., 88, 91, 92, 140, 145, 158
Hovey, C.M., 249
Howe, M.A., 50, 56, 120, 141, 142, 241, 243–246, 268–271
Hull Botanical Laboratory, 82
Hunter College, 153

I
Idaho, 146
Illinois State Laboratory of Natural History, 267
Illinois Wesleyan University, Bloomington, 267
Illudins, 228
Illustrated Flora of the Northern United States and Canada, 248
Indiana Academy of Sciences, 268
Instituto de Botánico of the Fundación Miguel Lillo, Tucumán, Argentina, 222
International Botanical Congress, Brussels, 103, 1910
International Botanical Congress, Genoa, Italy, 71, 269, 1892
International Botanical Congress, Paris, 71, 1867
International Botanical Congress, Stockholm, 221, 1950
International Botanical Congress, Vienna, 95, 1905
International Code of Botanical Nomenclature, 119, 120
International Garden Club, 150
Iowa State University, 86, 112
Italy, 30, 54, 66, 68, 71, 100, 119, 156, 157, 188, 200, 201, 205, 209, 269
Ithaca, NY, 18, 241
Ivy League, 32

J
Jacksonville, FL, 167
Jaczewski, A.L. [Arthrurovich], 68
Jamaica, West Indies, 50, 52, 94, 97, 99, 111, 126, 136, 154, 204, 220, 244
James River, VA, 3, 5, 56
Jardin des Plantes, Paris, 66, 100, 101
Jessup, M.K., 30
Johns Hopkins University, 82
Johnson, D.S., 82
Journal of Mycology, 11, 50, 65, 87, 92, 253–255
Journal of the Elisha Mitchell Scientific Society, 127, 128, 188, 193, 239
Journal of the New York Botanical Garden (JNYBG), 38, 48, 54, 62, 64, 77, 81, 88, 92, 94, 96, 98, 104, 108, 111, 112, 114, 120, 122, 138, 142, 143, 150, 158–160, 168
Just, T., 224

K
Kanosa Camp, NY, 120, 121
Kansas, 253, 254
Kauffman, C.H., 123, 124, 128, 145, 148, 152, 155, 203, 219, 224
Kellerman, D., 252
Kellerman, K.F., 252
Kellerman, S.D., 252, 253
Kellerman, W.A., 11, 50, 252–256
Kelly, H.A., 145, 152
Kentucky, 113, 147, 252, 253, 264
Kephart, H., 4
Kern, F.D., 170
Kingsbury, J.A., 155
Kingston, Jamaica, 94
Kneebone, L.R. (GI), 211
Knight, E.G., 35, 241–242, 245
Koch, H.H.R., 59
Komarov Institute, Leningrad, Russia, 201
Kuntze, C.O., 72, 73, 75, 76, 269

L
Lake Placid, NY, 112–113, 153–155
Lange, J.E., 100, 107
Langlois, A.B. (Abbé), 165
La Plata, Argentina, 156, 157
Lazenby, W.R., 252, 253
Leiden, The Netherlands, 69
Lend-Lease Act, 194, 195, 1941
The life Of Agassiz, 11
Lilloa, 222
Linden, NJ, 150, 198
Linder, D.H., 201

Linnaeus, C., 41, 43, 72, 73, 76, 103, 269
Lloyd Brothers Pharmaceutical Co., Cincinnati, 113
Lloyd, C.G., 19, 21, 52, 65–68, 96–97, 103, 113–115, 119, 121, 124, 126, 127, 140, 144, 148, 155, 156, 165, 166, 196, 201, 214–216, 219, 224, 233, 238, 253, 256
Lloyd Library of Natural History, Cincinnati, Ohio, 224
London, 28, 35, 47, 69, 116, 119, 205, 206, 265
London Botanical Congress, 119, 1915
Lorillard family, 39, 256
Lorillard Mansion, 143
Lorillard, P.A., 256
Lorillard, P. IV, 256–257
Lotsy, J.P., 14, 69
Low, S., 35, 36, 40, 246, 257–263, 271
"Lumper", 127, 204, 233
Luttrell, E.L., 15–17
Luttrell, N.H., 120, 121
Luxembourg, 119
Lvov (aka Lviv), Ukraine, 205
Lyceum of Natural History, NY, 31–32, 265
Lydig, D., 36, 37, 39, 93
Lydig Fund, 37, 93, 153
Lydig, M., 39, 93
Lynchburg, VA, 3, 5, 7, 12, 56, 144, 159, 160, 165–167, 169–173, 180, 185, 187, 193, 202, 221, 230

M

MacArthur, D. (General), 206, 211
MacCracken, H.M. (Rev.), 36, 37
MacDougal, D.T., 50, 63, 82, 83, 86
Magnesia Springs, FL, 186, 187, 199
Magnus, P.W., 68
Maine, 20, 61–63
Mallet, J.W., 11
Manhattan, 23, 30, 31, 38, 42, 45, 49, 51–53, 56, 61, 76, 122, 153, 210, 218, 243–245, 257, 258, 262, 263, 271
Manhattan College, 36
"The Manhattan Project", 210
Mason-Dixon line, 237
Massee, G.E., 69, 101, 103
McClatchie, A.J., 108, 112
McIlvaine, C., 21, 99, 121
McLaughlin, H. ("Boss"), 196, 258
Mediterranean fruit fly, 175
Merkel, H.W., 58, 59
Metropolitan Museum of Art, 27, 141, 142
Mexico, 97–99, 108, 109, 126, 154, 193
Mexico City, mx, 98
Miami University, OH, 113, 114, 196
Michigan, 94, 235, 237, 243, 268
Middlebury College, VT, 114, 151
Millspaugh, C.F., 107, 145
Mississippi Agricultural Experiment Station, 250
Mississippi Valley Horticultural Society, 250
Missouri Botanical Garden, St. Louis, 71, 108, 128, 145, 150, 151
Monroe Doctrine, 20
Montagne, C., 58, 66, 100
Montclair, NJ, 154
Moore, G.T., 35, 145
Morgan, A.P., 252
Morgan, J.P., 35–36, 76, 261
Morris, R.T., 145
Mortzarongo, mx, 98
"Mother oak", 6, 24, 136, 149–150, 160, 231
Moulds, mildews and mushrooms, 271
Mountain Lake, VA, 96, 125, 144, 153, 155
Mount Katahdin, ME, 61
Mount St. Ursula High School, Bronx, NY, 153
Mount Vernon, VA, 38, 149
Mt. Hood, OR, 108
"Mugwump" faction, Republican Party, 258
Murrill, A., 3, 4
Murrill, A.E., 3, 12
Murrill, E.L., 16, 17, 20, 23, 27, 44, 45, 47, 48, 56, 69, 76, 88, 94, 96, 97, 99, 107, 108, 111, 112, 117, 121, 125, 136–139, 146
Murrill, H.A., 3, 12
Murrill, M.D., 3, 10, 219
Murrill, P.S., 3
Murrill, S.L., 3
Murrill, V.D., 3, 12
Murrill, W.A., 3, 4, 6, 7, 9–23, 27, 40, 43–71, 73, 75–77, 81, 83, 86–94, 96–101, 103–142, 144–161, 165–183, 185–208, 211–237, 239–241, 245, 247, 254, 271–274
Museum National d'Histoire Naturelle, Paris, 195
Museum of Comparative Zoology (Harvard Univ.), 32
Mussolini, B., 188, 198, 201
Mycologia, 51, 86, 92–94, 97–101, 104–106, 108, 112, 118, 121–129, 145, 149–154, 156–158, 161, 169, 173, 182, 183, 188, 189, 194, 196, 201, 202, 205, 211, 212, 221, 222, 226
Mycological Notes, 127, 224, 238, 253

Mycological Society of America, 51–52, 93, 115
MycoPortal, 15–17, 21, 45, 47, 49, 50, 53, 56, 69, 76, 88, 94, 97, 98, 104, 107, 108, 112, 116, 118, 120, 138, 149, 158, 165, 168, 169, 171, 173, 178, 186, 187, 191, 195, 198, 202, 203, 207, 220, 226

N
Nagasaki, Japan, 211
National Academy of Science, 196, 264
The Natural History of Staunton, Virginia, 139, 172
"The Naturalist", 7, 8, 10–12, 14–17, 19, 21–24, 27, 32, 35, 41, 55, 63, 64, 70, 81, 82, 88, 94, 97, 107, 108, 113, 115, 117, 118, 128, 135–137, 139–141, 144, 147–150, 154–157, 159, 160, 172, 173, 175, 178, 181–183, 188–190, 194, 203, 207, 215–217, 224, 230, 231, 236, 239, 247, 249, 256
The Naturalist in a Boarding School, 12, 16, 139
Nature Club, Gainesville, FL, 182, 207
Natürlichen Pflanzenfamilien, Die, 68
Neuman's Lake, FL, 199
Neutrality Act (American), 188
Nevada, 140
Newark, NJ, 153
Newberry, J.S., 34, 35, 243, 245, 246, 264–266
"The New Botany", 10, 20, 41–44, 240, 252, 268
New Dorp, Staten Island, NY, 244
New England, 32, 113
New Lyceum of Natural History, NY, 31–32
New York Academy of Medicine, 265
New York Academy of Sciences, 32, 247
New York Botanical Garden (NYBG), 27, 47, 81, 111, 137, 168, 189, 212, 256
New York Botanical Garden Museum Building, 40, 48, 120, 143, 153, 266
New York City, 6, 12, 23, 24, 30, 32, 33, 35, 36, 45, 59, 73, 104, 113, 127, 135, 137, 138, 146, 153, 158, 220, 241, 245, 254, 260, 263, 265, 266, 269
New York City Park Commission., 30
New York Microscopical Society, 265
New York Pharmaceutical Association, 35
New York State Agricultural Experiment Station, Geneva, 240
New York State College of Agriculture, 116, 240
New York State Museum at Albany, 88
New York Stock Exchange, 53
New York Sun, 35
New York Symphony Orchestra, 10
New York Times, 77, 261
New York Tribune, 35
Normal College, 36, 241, 243
North American Flora, 67, 68, 81, 87, 97, 103, 116, 118–120, 124, 126, 127, 129, 140, 148, 150, 151, 189, 199, 222, 233, 241, 270, 271
North Carolina, 8, 76, 88–91, 94, 96, 122, 123, 125, 127, 128, 158, 235, 238–240
North Temperate zone, 96
Notre Dame University, IN, 224
Nova Scotia, Canada, 94, 242

O
O'Brien, M., 67
The Observer, 15, 243
Ocean Springs, MS, 249, 250
Oerted, A., 100
Ohio, 11, 21, 39, 52, 94, 113, 114, 196, 224, 240, 252, 254, 255, 264
Ohio Mycological Bulletin, 254
Ohio Mycological Club, 254
Ohio Naturalist, 113, 256
Ohio Naturalists Club, 256
Ohio Pyle (now Ohiopyle), PA, 60
Ohio State Geological Survey, 265, 1869
Ohio State University, 11, 254
Oklahoma, 198
"Oligocene Island", FL, 192, 193
Olivet College, mi, 237
Olmsted, Frederic Law, 30, 37
Open Court, 173
"Operation Barbarossa", 195
Opium Wars, 257
Oregon, 107, 108, 264
The Origin of Species, 244
Orizaba, mx, 98, 108
Ottoman Empire, 119, 128, 129
Ottoman family, 128
Our Southern Highlanders, 4
Overholts, L.O., 113, 114, 128, 129, 132, 148, 170, 196

P
Pacific Northwestern U.S., 107, 151
Pacific Ocean, 98, 112
Pakawatan Inn, NY, 125
Paris, France, 28, 47, 52, 58, 66, 68, 71, 74, 100, 101, 116, 119, 129, 195, 198, 201, 205, 269
Patouillard, N.T., 66, 68, 100, 202, 233

Patterson, F.W., 58, 61, 128
Patton, G.S. (General), 209
Peaks of Otter, VA, 3, 12, 13, 56, 125, 150, 168
Pearl Harbor, Oahu, Hawai'i, 195, 198, 200
Peck, C.H., 21, 45, 46, 51, 53, 58, 59, 76, 87, 88, 91–93, 99, 104, 107, 108, 112, 113, 145, 151, 157, 194, 204, 233, 235
Pennington, L.H., 120, 128, 199
Pennobscot River, ME, 61
Pennsylvania, 39, 60, 101, 112, 123, 128
Pennsylvania State University, 170, 222
Persoon, C.H., 69, 103, 202, 233, 252
Petropolis, Brazil, 157, 158
Philadelphia Centennial Exposition, 249, 265, 1876
Philippines, 20, 206, 211
Pinchot, G., 89, 147
"Pink Beds", NC, 89, 92, 125, 158
Pintard, J., 27
Piscataquis River, ME, 61
Pleistocene period, 214
Polycystic kidney disease, 230
Polyporaceae of North America, 57
Polytechnik, Zürich, Switzerland, 252
Pore fungi, 218, 219, 221
Prain, D., 116
Prentiss, A., 18–21, 239–241, 252
"Progressive Movement", 258
Puerto Rico, 20, 52, 55, 244

R

Raisinville, mi, 237
Randolph College for women, Lynchburg, VA, 170
Randolph-Macon College, VA, 8, 9, 171
Revisio Generum Plantarum, 72
Rhodes, R. [Gen., Confederate States of America]., 166
Richmond, VA, 8, 9
Ricker, H.W., 61, 62
Rick, J. (Abbé), 158
Ridgway, R., 115, 229
Riis, J., 53, 262
Rio de Janeiro, Brazil, 156, 157
Rio La Plata, 156
Robinson, B.L., 73, 74
Rochester Code of Botanical Nomenclature, 71–74, 269
Rochester, NY, 71, 269
Rocks, 10, 147, 153, 154, 171, 214, 218, 221, 227, 234, 266
Rogerson, C.T. (GI), 150, 211
Rolf's Hall, University of Florida, 186, 191
Romania, 205
Romanov family, 129
Romell, L., 49, 68, 100
Rommel, E., 198
Roosevelt, E., 200
Roosevelt, F.D., 188, 194, 195, 197, 209, 210
Roosevelt, T., 33, 147
Rose, D., 69, 137, 231
Rostrup, O., 100
Rowlee, W.W., 18
Royal Botanical Gardens, Edinburgh, 195
Royal Botanic Gardens, Kew, 35, 101, 102, 160, 195
Royal Horticultural Society, UK, 69
Rusby, H.H., 35, 41, 44, 77, 150, 245
Russia, 119, 128, 199, 201, 205, 206, 211
Rutgers College, NJ, 153
Rydberg, P.A., 167

S

Saccardo, P.A., 157, 240
Sachs, J., 43
"Sacred Harp" singing, 181
San Francisco, CA, 108
San Francisco earthquake, 67, 1906
Santa Cruz Mountains, CA, 107, 108
Sao Paulo, Brazil, 157
"Satchell" Page, 228
Savage, E., 27
Savich, V., 201
Schenck, C.A., 88, 89, 92
Schermehorn, W.C., 36
Schweinitz, L.D. de, 89, 233
Scientific Directors, New York Botanical Garden, 37, 67, 107, 123, 271
Scudder's Museum, NY, 27
Seattle, WA, 107
Seaver, F.J., 86–88, 93, 112, 120, 160, 173, 182, 183, 189, 196, 211, 219, 230
Sebastopol (aka Sevastopol), 205
Sebec Lake, ME, 61
Sebec River, ME, 61
Serbia, 119
Setchell, W.A., 108
Seth Low Junior College, Brooklyn, 263
Shear, C.L., 51
Shenandoah Mountains, VA, 13, 204
"Shiitake", 103, 104
Shull, G.H., 82, 84
Sigma Xi, 23, 208

Singer, R., 132, 149, 196, 197, 200–202, 207, 211, 221, 222, 233, 235
"Singing conventions", 180
Siskiyou Range, CA, 108
Sledd, A.W., 9
Sloane, W.M., 261
Small, G., 70, 247
Small, J.K., 70, 154, 176, 215, 247
Smith, A.H., 204
Smith, E., 72
Smith, J.D., 71
Smithsonian Institution, 115, 216, 264, 268
Snell, W.H., 196
"The Snuff Mill", 49, 256
Society for Plant Morphology and Physiology, 51
South Dakota, 237
Southworth, E.A., 259
Spanish American War, 20, 241
"Spanish flu", 241
Species Concept Symposium, 82, 95, 1908
Species, definition of, 83
Species Plantarum, 72, 76, 269
Spegazzini, C., 156, 157, 172
"Splitter", 127, 204, 233, 234
Stalin, J., 195
Stanford University, 108, 111, 238, 263
Stanton, E.C., 145
Stars, 5, 153, 154, 218, 221, 227, 234, 272
State Agricultural School, Lexington, KY, 252
Staten Island Institute of Arts and Sciences, NY, 247–248
State Normal School, Oshkosh, WI, 252
State Racing Commission, FL, 175
St. Augustine, FL, 167
Staunton, VA, 10, 11, 14–16, 70, 71, 139, 172
St. John's River, FL, 36, 39, 167
St. Louis, MO, 51, 108, 114, 128
Stockbridge, MA, 104
Stockholm, Sweden, 68, 100, 198, 221
Stout, A.B., 111
St. Petersburg Bay, FL, 193
St. Petersburg, Russia, 68, 193, 199
Stratemeyer, E., 137
Student Union Building, University of Florida, 191, 208
Success Stories for Girls, 173, 227
Sullivant Moss Society, 243
Supervisor of Public Instruction, 137, 138, 149, 150, 155, 159, 160, 170, 217
Suwannee River, FL, 193
Sylloge Fungorum, 95, 241
Synopsis Mehodica Fungorum, 76
Syracuse University, NY, 128, 266, 268
Systema Mycologicum, 76, 103

T

Tacoma, WA, 151
Tammany Hall, NY, 27, 30, 146, 258, 262
Tammany Society, NY, 27
The Teaching Botanist, 44
Temperate House, Kew Gardens, 35
Tennessee, 17, 56, 113, 151, 167, 194, 209, 235, 268
Texas, 198
Text-book of Elementary Botany including a Spring Flora, 254
Thaxter, R., 51
Thiers, H.D. (GI), 211
Thompson, W.G., 37, 104
Three Young Crusoes, 16, 17, 136, 138, 165, 172, 217
Tin Can Tourist Camp, Gaiesville, FL, 175–176
Tom's River, NJ, 120
Toronto, Canada, 150
Torreya, 50, 62, 65, 86, 87, 125, 137, 158, 188, 271
Torrey Botanical Club, 31, 33–37, 41, 42, 44–47, 50, 70, 71, 73, 74, 76, 77, 81, 95, 120, 126, 137, 138, 153, 172, 188, 189, 191, 193, 241–243, 245–249, 265, 266, 271
Torrey, J., 33, 34, 40–42, 245
Torrey Professorship of Botany, 104, 107
Tranzschel, V.A., 201
Trees, 11, 22, 23, 50, 52, 58, 59, 61, 66, 76, 92, 116, 136, 150, 154, 178, 185, 190, 207, 212, 213, 215, 221, 227, 239, 254, 272
Trelease, W., 71, 145, 243
Triangle Shirtwaist Fire, 146, 1911
Trient [Trento], Italy, 53, 66, 68, 116, 117, 149
"Tropical School of Botany", 255
Truman, H.S., 210
Tweed, W.M., 30, 31
"Type method", 47, 54, 74, 75, 87, 99, 234, 269

U

Unaka Springs, TN, 56, 151
Underwood, L.M., 266–267, 271
Underwood, M.A.S., 267

Union of Soviet Socialist Republics (USSR), 195, 199, 201
Union Sanitary Commission, Civil War, 264
United States Department of Agriculture (USDA), 51, 52, 55, 58, 61, 74, 128, 198, 243, 250–252, 254, 268
United States District Court for the Southern District of New York, 248
United States Forest Service, 89
United States Geological Survey, 265
University of Chicago, 43, 82, 85, 268
University of Florida, 9, 165, 170, 198, 222
University of Georgia, 158
University of Georgia Mycological Herbarium, 198
University of Illinois, 85, 249
University of Michigan Herbarium, 123, 128, 198
University of North Carolina, 122, 128, 238
University of South Carolina, 122, 238
University of Tennessee, 194
University of Vermont, 87
University of Vienna, Austria, 201
University of Virginia, 11, 97
University of Washington, 107
University of Wisconsin, 71, 123
University Settlement Society, 262
Upper St. Regis Lake, NY, 120, 125
Uppsala, Sweden, 49, 100, 113
Uremia, 230
USDA Bureau of Plant Industry Herbarium, 198
USDA Experiment Station, Cuba, 55, 94, 250
USS Maine, 20
Utah, 146

V
Vanderbilt, C., 35, 36
Vanderbilt, G.W., 89–92
Vassar College, NY, 92
Vassileva, L., 201
Veracruz, mx, 98
"Victory Garden", 199
"Victory Saving Bonds", 200
Virginia, 3–16, 20, 23, 45, 47, 48, 53, 56, 59, 61, 62, 69, 71, 73, 88, 96, 97, 99, 117, 118, 120, 125, 128, 136, 139, 144, 148–151, 153, 155, 159–161, 165, 167–169, 171, 172, 175–177, 183, 185, 204, 213, 220, 221, 230, 242, 268, 271
Virginia Agricultural and Mechanical College (VAMC), 7, 8
Volkeart, E.C., 99, 104–106, 112
Vries, H. de, 69, 83
Vulcan Popocatepetl, mx, 98

W
Walden, W.I., 27, 171
Ward, L.F., 74
Warm Springs, GA, 209
Warsaw, Poland, 205
Washington, dc, 10, 12, 17, 50, 51, 55, 58–61, 71, 74, 75, 88, 95, 107, 111, 117, 125, 128, 149, 151, 216, 250
The Water Birds of North America, 115
Watson, S., 72
The Way the Other Half Lives, 53
"Weathersfield Elm", CT, 144
Weber, G.F., 3, 8, 11, 16, 21, 23, 69, 117, 121, 160, 167, 169, 170, 172, 175–178, 185–187, 191, 193, 194, 199, 207, 208, 228–231
Weichert, R., 231
Wesleyan Female Institute, VA, 10
West, E., 169, 170, 173, 175, 185–187, 189, 191, 193, 205, 225
West Virginia, 96, 242
White, V.S., 46
Wiegand, K.M., 19, 22
Wild Flower Preservation Society of America, 243
Willamette Valley, OR, 108
Williams, R.S., 92, 111, 118, 146, 231
Wilson, P., 63, 98
Wilson, W., 128
Wisconsin, 71, 73, 104, 123, 242, 252, 270
Woman's Christian Temperance Union, 139
Woodstock, NY, 145, 153–155, 266
World War I, 128, 139, 144, 146, 175, 176, 188
World War II, 68, 181, 194, 214
Wright, C., 58
Wyoming, 146

Y
Yama Farms, 144–145, 149, 151, 154, 155, 161, 220
Yama Farms Inn, 144, 155
Yama Farms Mycological Club, 145, 155, 161
Yasuda, A., 112
Young Men's Christian Association (YMCA), 12
Yucatan, mx, 97, 255

Z
Zacapa, G., 255
Zeller, S.B., 107

The manufacturer's authorised representative in the EU is Springer Nature Customer Service Centre GmbH, Europaplatz 3, 69115 Heidelberg, Germany. If you have any concerns regarding our products, please contact ProductSafety@springernature.com

Printed and bound by CPI Group (UK) Ltd, Croydon, CR0 4YY
01/10/2026
02225968-0009